Nature-Based Solutions for Restoration of Ecosystems and Sustainable Urban Development

Nature-Based Solutions for Restoration of Ecosystems and Sustainable Urban Development

Special Issue Editor

Thomas Panagopoulos

MDPI • Basel • Beijing • Wuhan • Barcelona • Belgrade • Manchester • Tokyo • Cluj • Tianjin

Special Issue Editor
Thomas Panagopoulos
University of Algarve
Portugal

Editorial Office
MDPI
St. Alban-Anlage 66
4052 Basel, Switzerland

This is a reprint of articles from the Special Issue published online in the open access journal *Sustainability* (ISSN 2071-1050) (available at: https://www.mdpi.com/journal/sustainability/special_issues/Nature-Based_Solutions).

For citation purposes, cite each article independently as indicated on the article page online and as indicated below:

LastName, A.A.; LastName, B.B.; LastName, C.C. Article Title. *Journal Name* **Year**, *Article Number*, Page Range.

ISBN 978-3-03936-242-4 (Pbk)
ISBN 978-3-03936-243-1 (PDF)

Cover image courtesy of Thomas Panagopoulos.

Contents

About the Special Issue Editor

Thomas Panagopoulos (Dr.) is a professor of landscape architecture with specialization in landscape restoration. He received his MSc in renewable natural resources in 1992 and Ph.D. in forestry and natural environment in 1995. He is currently a member of the Research Centre for Tourism, Sustainability and Well-Being, and a member of the coordinating body of the Ph.D. program in innovation and land management. He was the department head of landscape architecture and the master's degree director at the University of Algarve, Portugal. He is a reviewer and member of the editorial board of several reputed international journals on sustainability and environmental management. He has acted as a principal investigator, co-principal investigator, and investigator for projects with total approved funding of over 8 million euros. This is the result of his research strategy, which crosses numerous disciplinary boundaries to create a holistic transdisciplinary approach to science, and his multicultural background in fostering research at an international level. He has vast experience working in many European and private projects. Currently, he is coordinating research for BIODES, improving life in a changing urban environment through biophilic design; RESTORE, rethinking sustainability towards a regenerative economy; TrailGazerBid, enhancing natural and cultural assets to stimulate economic development; Euroguadiana, European laboratory of transborder governance. He has also helped many cities develop their sustainability plan. From 2011 to 2017, he was on the executive board of UNISCAPE (the Network of Universities for the Implementation of the European Landscape Convention).

 *sustainability*

Review

Stakeholders' Engagement on Nature-Based Solutions: A Systematic Literature Review

Vera Ferreira [1,*]**, Ana Paula Barreira** [2]**, Luís Loures** [3]**, Dulce Antunes** [4] **and Thomas Panagopoulos** [1,*]

[1] Research Center for Tourism, Sustainability and Well-being, University of Algarve, Campus de Gambelas, 8000 Faro, Portugal

[2] Center for Advanced Studies in Management and Economics, Faculty of Economics, University of Algarve, Campus de Gambelas, 8000 Faro, Portugal; aprodrig@ualg.pt

[3] Research Center for Endogenous Resource Valorization, Polytechnic Institute of Portalegre (IPP), 7300-110 Portalegre, Portugal; lcloures@ipportalegre.pt

[4] Mediterranean Institute for Agriculture, Environment and Development, FCT, University of Algarve, Campus of Gambelas, 8000 Faro, Portugal; mantunes@ualg.pt

[*] Correspondence: vlferreira@ualg.pt (V.F.); tpanago@ualg.pt (T.P.); Tel.: +00351-289800900 (T.P.); Fax: +00351-289818419 (T.P.)

Received: 30 November 2019; Accepted: 9 January 2020; Published: 15 January 2020

Abstract: Cities are facing a broad range of social and environmental challenges due to the current pressure of global urbanization. Nature-based solutions aim to utilize green infrastructure to improve people's health and wellbeing. The design of urban environments must embrace the individual ideals of citizens and stakeholders which can only be achieved if effective methods of communication, involvement, and feedback are ensured. Such a procedure creates trust during its implementation, helping to take ownership and stewardship of processes and sites. This systematic literature review explores the current state of the art regarding citizen and stakeholder participation in nature-based solutions (NBS). The search on the SCOPUS database identified 142 papers in total that met the inclusion criteria. The participation analysis was separated in two areas: (a) analysis of perceptions, preferences, and perspectives of citizens and stakeholders, and (b) analysis of the participation process, including challenges and opportunities, motivations, methods and frameworks, and collaborative governance. The results revealed that stakeholder and citizen participation or collaboration in nature-based solutions is increasingly recognized as promising; however, research in several related domains is still lacking.

Keywords: nature-based solutions; green infrastructure; stakeholder participation; collaborative governance; urban sustainability; citizen perceptions

---

1. Introduction

Due to the current pressure of global urbanization, quality of life and sustainability of European cities have gained political impetus in the last decade. Cities are facing a broad range of challenges, such as climate change, human health issues, social inequity and poverty, degradation, loss of natural capital and the provision of ecosystem services, and an enhanced readiness to deal with disasters (e.g., floods) [1,2].

The concept of biophilia advanced the idea that contact with nature plays a fundamental role in human physical and mental wellbeing [3]. Additionally, there was an emerging need of using natural components and their multiple functions, to increase sustainable development into the cities, dealing with recognized issues. Key challenges for sustainable cities are to significantly increase their resources and efficiency in addressing issues relating to transportation, climate change, and water and air quality.

Such actions should exert profound economic, social, and environmental impacts, resulting in a better quality of life (including health and social cohesion), jobs, and growth.

The importance of nature and its functions in cities have been studied for many years, using different metaphors, such as urban forests (UF), ecosystem services (ES), urban green spaces (UGS), biophilic urbanism (BU), green infrastructure (GI) and, more recently, nature-based solutions (NBS) [3–6]. While ES are often valued in terms of immediate benefits to human well-being and economy, and UF, UGS, BU, and GI focus on the provision of these ES through biodiversity protection, NBS simultaneously addresses diverse societal challenges in the long-term, allowing benefits to people and the environment itself [7]. Nature-based solutions have largely evolved from previous ecosystem-based concepts and/or principles (e.g., ecosystem services, green infrastructures, ecosystem-based management, and natural capital), but it also pays attention to the social and economic benefits of resource-efficient and universal solutions that combine technical, business, finance, governance, regulatory, and social innovation [8]. Parker and Baro [9] reviewed the literature dealing with GI and identified that the concept is diffuse and imprecise, with a focus on environmental, ecological, and social planning and policy, neglecting its economic, health, and wellbeing effects, as well as its performance. Given that NBS is broad in definition and scope, it can be considered as an umbrella for the previously- mentioned concepts [6]. Sarabi et al. [10] reviewed the literature incorporating the concept and concluded that it remains ambiguous and fragmented—perhaps due, in no small part, to the fact that articles addressing the concept only started appearing quite recently (the first references dating back only to 2015). Relying on Sarabi et al. [10], we feel the definition of NBS proposed by the EC [1] assumes an understanding of the concept similar to the approach followed in the current work: "actions inspired by, supported by or copied from nature and which aim to help societies address a variety of environmental, social and economic challenges in sustainable ways".

The stakeholder involvement in urban green infrastructure (GI), including urban forest (UF) or urban green space (UGS), is not new and has greatly advanced in recent decades [11–13]. However, looking at the more recent NBS concept the human context is critical and is gaining increasing attention from scientists and practitioners as potential solutions to enhance socio-ecological systems resilience. Recent work suggests more attention should be given to the incorporation of local and indigenous knowledge into formulating and applying solutions [8,14]. Though, the involvement of locals is still rarely adopted; mainly resulting from the general perception that multi-stakeholder initiatives slow down urban planning and policy development processes due to a lack of consensus and different sectoral interests [8].

Theoretical Framework and Rationale for the Review

Some authors have presented literature reviews on the topic of GI [9,15–17] or NBS [10]; however, those works deviated from the focus provided in the current review. Parker and Simpson [15] present a systematic quantitative review on how public green infrastructure contributes to city livability, informing urban planners, policymakers, and researchers about the psychological, physiological, general well-being, and wider societal benefits that humans receive as a result of experiencing GI. Parker and Boro [9] reviewed the publications dealing with GI but they also looked mainly at the benefits generated by GI. The work of Verkataramanan et al. [16] is centered on the role of GI for stormwater and flood management and its impacts on human health and social wellbeing. Zuniga-Teran and Gerlak [17] reviewed publications on urban green spaces to assess how those infrastructures promote social justice. Sarabi et al. [10] aimed to develop a conceptual theory for NBS based on the current state of the art; as such, their analysis excluded publications dealing with the technical and physical dimensions of NBS. All these previous works did not specifically examine the participation of citizens and stakeholders in the processes of NBS.

Figure 1 represents a conceptual understanding of NBS, their benefits, planning process, and implementation. The various links within and between ecological and social systems are accounted for in the process. We consider that NBS needs to be embedded in the existing policies, supported by urban planning and adopted in joint dialogues between policy, society, and science. Urban planning can contribute with relevant spatial information for decision support concerning the choice for the location of projects and how they relate with societal challenges, the identification of alternative solutions, and the measurement of their respective impacts [18]. Nature-based solutions can enhance natural capital and promote biodiversity while delivering other co-benefits [7]. NBS operationalizes the concept of ecosystem services that are the contributions (co-benefits) that ecosystems, in combination with other inputs, make to human well-being.

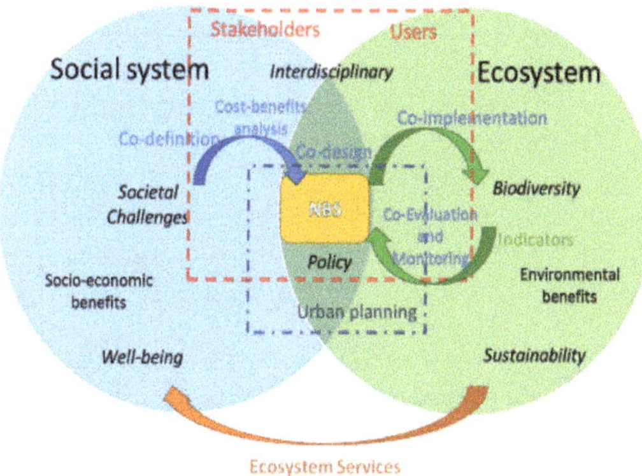

Figure 1. Conceptual understanding of nature-based solutions. Source: Own elaboration.

The focus of this systematic literature review is to bring up the role played by the participation of stakeholders and citizens on the identification of problems and solutions for urban green areas. Previous works have revised citizens' participation in the context of GI, e.g., [19,20]; but they have adopted parcel approaches by considering only either users or citizens actively involved in initiatives of GI.

The application of NBS serves as the primary focus of this literature review. However, as noted, it is a recent concept that, of necessity, includes the human component—indeed, stakeholder engagement is essential to the success of such initiatives. This literature review, therefore, includes the other concepts mentioned above, aiming thereby to provide a complete picture of public participation in NBS processes. The objective is to understand the progress of citizen participation and stakeholders over the years. We investigate the perceptions, preferences, and perspectives of different actors, the participation drivers and motivations, the participation methods and frameworks, the collaborative governance, and the participation challenges and opportunities. This analysis is performed aiming to answer the following research questions:

- (RQ1) How are the perceptions, preferences, and perspectives of the citizens and stakeholders taken into consideration in the literature addressing NBS?
- (RQ2) Which motivations trigger the citizens' and stakeholders' engagement?
- (RQ3) What are the main benefits and costs sought by citizens and stakeholders resulting from the participation processes of NBS?
- (RQ4) What are the major difficulties and opportunities raised by the engagement of citizens/stakeholders?
- (RQ5) Which approaches are predominant in collaborative governance to involve citizens and stakeholders in participatory processes of NBS?

2. Materials and Methods

A systematic literature review was conducted to provide an overview of research to date related to the citizen and stakeholder participation and engagement on NBS. This systematic literature review is based on the Preferred Reporting Items for Systematic Reviews (PRISMA) guidelines (http://prisma-statement.org/).

2.1. Identification

A search for publications addressing NBS-related concepts, as well as stakeholder and citizen participation, was conducted in June of 2019 using the Scopus search engine. Scopus was selected due to its broader coverage compared to other academic search engines. The following combination of keywords was performed using Boolean operators:

TITLE-ABS-KEY ("nature-based solution" OR "green infrastructure" OR "biophilic infrastructure" OR "urban green space" OR "biophilic design" OR "urban forest" OR "urban biodiversity") AND TITLE-ABS-KEY ("stakeholder" OR "public" OR "citizen" OR "resident" OR "community" OR "expert") AND TITLE-ABS-KEY ("engagement" OR "participation" OR "perceptions" OR "perspectives" OR "involvement" OR "collaboration" OR "preferences") AND TITLE-ABS-KEY ("urban").

The initial search yielded a total of 814 records; however, 147 records were excluded by source and document type. We limited our research to publications on journals, excluding conference proceedings, book series, books, and trade publications. Only journal articles were included in the systematic literature review because we want to focus on high-quality empirical studies ensured by the peer-review process undertaken by academic journals. Literature review articles dealing with the topic of urban green solutions, but not addressing the issue of citizens' and stakeholders' participation, were not included in this systematic review; however, they were considered in the discussion to highlight the main contribution of the current work. Additionally, we only included articles written in the English language, thus excluding 30 records in various other languages. The excluded publications were organized by criterion of exclusion.

2.2. Screening and Eligibility Criteria

Explicit inclusion and exclusion criteria were defined prior to screening of abstracts and full texts; Figure 2 details the inclusion and exclusion process according to the PRISMA flow diagram. To be included, articles needed to indicate that the public or stakeholders were engaged in some stage of NBS process. First, we screened the abstracts; 287 records were excluded because they provided no evidence of such participation. Posteriorly, we attempt to access the full text of those articles and only 294 were available. After reading the full-text article, a total of 142 papers were included. The reasons for exclusion were:

- Conceptual articles without evidence of empirical work;
- Not relevant with respect to participatory processes (i.e., without analysis of opportunities and challenges, methods, approaches, motivations, perceptions, and preferences);
- Studies outside the urban context.

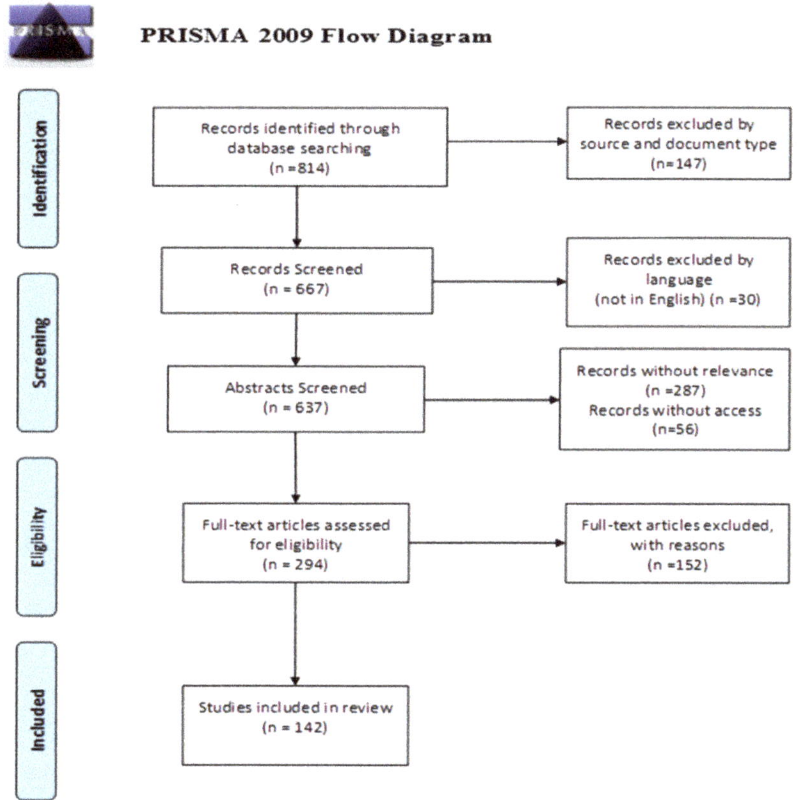

Figure 2. Preferred Reporting Items for Systematic Reviews (PRISMA) expression of the systematic literature review.

3. Results

3.1. General Characteristics of the Body of Research

3.1.1. Temporal Progression of the Research on the Issue

The research methodology did not impose a restriction publication year. However, before 2000 there are few papers, without online access, that mentioned public participation or perceptions on urban forests. Only in 2000 and 2001, as displayed in Figure 3, appeared the first papers (with online access) about the relevance of public perceptions and preferences on urban forest [11,21]. A paper published in 2005 relates a collaborative urban forest-planning initiative in Helsinki, Finland, that was begun in 1995 [12]. The term "urban green space" related to resident perceptions appeared in 2006 [22,23]. The concept of "green infrastructure" related to public participation in urban planning emerges in 2009 [13]. "Nature-based solution" appears for the first time in 2016 [14].

Figure 3 shows an exponential increase of publications in this field after 2015, with 78.2% of the articles being published after this year. This finding suggests that citizens and stakeholder participation in NBS constitutes a growing research area.

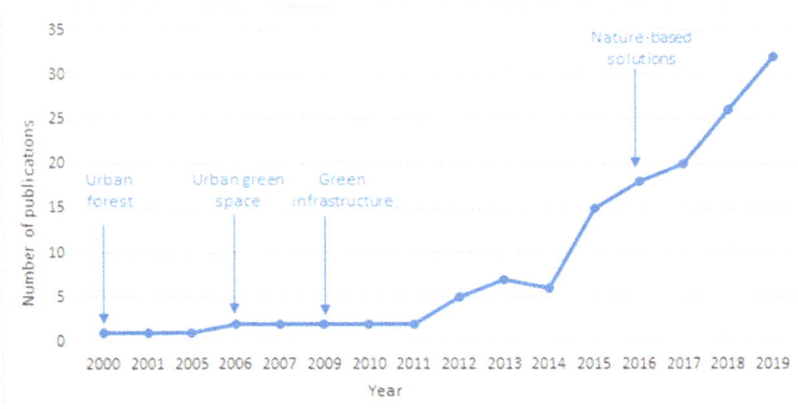

Figure 3. Number of selected published articles, per year, on citizen and stakeholder participation in NBS.

3.1.2. Geographical Distribution of Research on the Issue

The distribution of publications by country is presented in Figure 4 (for countries with more than two study cases).

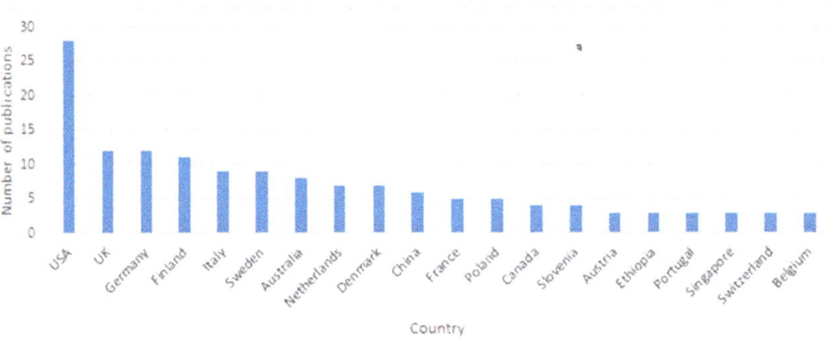

Figure 4. The distribution by country of the selected publications.

The United States plays a leading role in the research on citizen and stakeholder participation and engagement in NBS, with a total of 28 (19.7%) publications, followed by the United Kingdom and Germany (both with 8.5%). A total of 47 countries were accounted for in the selected articles, encompassing applications mainly in Europe (113 study cases corresponding to 76 publications and 28 countries) but also in North America (33 studies/publications and 3 countries) and Asia (22 studies corresponding to 21 publications and 10 countries). Of the 76 European publications, it is interesting to stress that only 12 have applications in southern countries (Portugal, Italy, and Greece) despite the fact that they face a higher risk in the near future of hazards relating to climate change [24]. Curiously, we find only one publication addressing this issue in the countries of South America [25].

The first published studies [11,21] were from the United Kingdom and Nigeria and, in 2019, there were 32 published papers, comprehending 24 countries and five continents.

3.1.3. Coverage of the Issue by Journals

Figure 5 shows the distribution of the selected research articles per journal. Regarding this distribution, Urban Forestry and Urban Greening is the leading source of published articles (30.3% of publications), followed by Landscape and Urban Planning (12.0% of publications) and Sustainability (7.0%). A total of 45 different journals were identified.

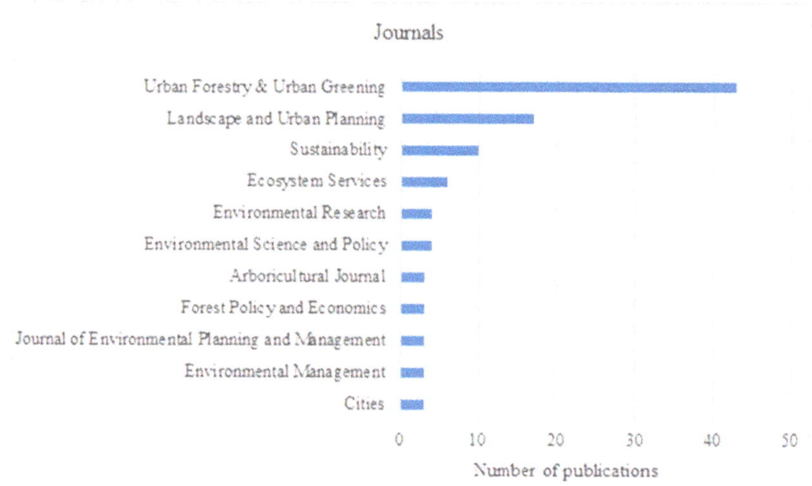

Figure 5. The main journals publishing on the issue of citizen and stakeholder participation on NBS.

3.1.4. General Focus of the Articles

The published articles were further explored to better address the research questions. They were separated into two general areas according to their main objective: (a) analysis of perceptions, preferences, and perspectives of the stakeholders, and (b) analysis of the participation process. While some articles paid attention to the perceptions, preferences, and perspectives of stakeholders as an important component of participation process, other studies focused on the analysis of elements of the participatory process. Within these two areas, articles can be grouped into sub-areas according to specific objectives, as shown in Table 1. This table also shows the total number of publications per area (n). Almost half the studies (45.8%) reported stakeholders' and citizens' perceptions, preferences, and perspectives regarding NBS—in particular, the level of satisfaction (through the identification of benefits and costs) and the preferences (through the identification of preferred attributes). The remaining body of publications (54.2%) analyzes the participatory process, with more studies on methods and frameworks and fewer on the main drivers and participant motivations.

In addition to the identification of focus areas, a data visualization technique known as "Word Cloud" was produced for each of the areas. The most common and obvious words in these two areas ("urban", "green", "management", "planning", and "environmental") were not included in this analysis to make clear the differences between them. The word clouds are presented in Figure 6. Looking at the publications that focus on the perceptions, preferences, and perspectives of the stakeholders, some of the most frequent words were "park," "trees," "landscape," "ecosystem" and "services". Words that emerged when examining the participatory processes were "social", "community", "participation", "governance", and "policy". These results highlight that the literature focused on the first area deals with the cognitive image of the ideal NBS developed by citizens and stakeholders, whereas the literature addressing the second area is more concerned with public participation in practice.

Table 1. The areas and subareas of reviewed papers that were grouped according to the central aim of the study. The number of publications (n) is shown.

Area	Sub-Area	Aim
Analysis of perceptions, preferences and perspectives on NBS (n = 65)	Benefits and costs (n = 35)	Focus on citizen and/or stakeholder perceptions of benefits and costs of NBS
	Attributes (n = 16)	Focus on citizen and/or stakeholder preferences of specific attributes and design of NBS
	NBS challenges (n = 14)	Focus on citizen and/or stakeholder viewpoints of challenges on NBS implementation
Analysis of participation processes on NBS (n = 77)	Drivers and motivations (n = 14)	Analyzes drivers and motivation for participation
	Methods, tools, and frameworks (n = 26)	Analyzes methods, tools, or frameworks for participation
	Collaborative governance and interactions (n = 17)	Analyzes participation in terms of governance and existing interactions
	Challenges and opportunities (n = 20)	Analyzes challenges and opportunities presented in the participatory process

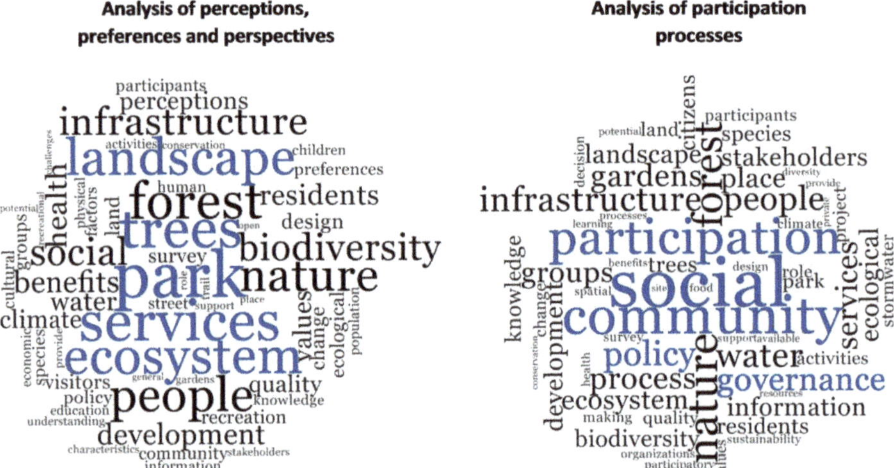

Figure 6. Word cloud arranged by area from the selected 142 papers.

3.2. Citizens' and Stakeholders' Perceptions, Preferences, and Perspectives

NBS effective governance and accuracy establishment increasingly require public input, and urban planners and policymakers are progressively aware of the need to take the perceptions and experiences into consideration. Understanding stakeholders' perceptions of, and preferences for green spaces and engaging them in the planning process can potentially bring benefits to residents and urban planners [23,26–28]. In addition to this understanding, considering the perceptions and preferences of citizens is seen as a first step in promoting and facilitating effective citizen participation and governance.

3.2.1. Perceived Benefits and Costs

Perceptions of the Benefits

The perceived benefits of NBS are related to the ecosystem services (ES) provided. Some researchers have used an ES framework to create a more common language for the valuation of various benefits that NBS can provide [29,30]. The identification of perceived ES can provide guidelines and practical advice on urban design and management actions; mapping them is becoming a key tool for guiding decision-making [29]. The mapped perceptions enable the localization of the most highly valued ecosystems in a landscape and allow for identification of critical focal areas for ES management. Visual methods (photomontages) have also been shown useful to investigate residents' perceptions [31].

Table 2 presents the main benefits perceived by citizens and stakeholders according to the selected articles.

Table 2. The mainly perceived benefits of NBS by citizens and stakeholders as presented in the selected publications.

Perceived Benefits (n = 34)		Authors
Social Benefits (n = 27)	Aesthetics, scenic views and proximity to nature (n = 13)	Coles and Bussey 2000 [11]; Huang 2014 [32]; Barau 2015 [33]; Buchel and Frantzeskaki 2015 [29]; Conedera et al. 2015 [34]; Qiu and Nielsen 2015 [35]; Rupprecht et al. 2015 [36]; Larson et al. 2016 [37]; Ives et al. 2017 [38]; Korpilo et al. 2018 [39]; Panagopoulos et al. 2018 [40]; Campbell-Arvai 2019 [28]; Guenat et al. 2019 [41].
	Quality of life (n = 4)	Sanesi and Chiarello 2006 [22]; Conedera et al. 2015 [36]; Panagopoulos et al. 2018 [40]; Gwedla and Shackleton 2019 [42].
	Physical and mental well-being (n = 12)	Coles and Bussey 2000 [11]; Peckham et al. 2013 [43]; Buchel and Frantzeskaki 2015 [29]; Yen et al. 2016 [44]; Faivre et al. 2017 [45]; Duan et al. 2018 [46]; Keith et al. 2018 [27]; Nath et al. 2018 [47]; Panagopoulos et al. 2018 [40]; Zwierzchowska et al. 2018 [48]; Campbell-Arvai 2019 [28]; Gashu et al. 2020 [49].
	Sociocultural (n = 6)	Huang 2014 [32]; Barau 2015 [33]; Buchel and Frantzeskaki 2015 [29]; Conedera et al. 2015 [34]; Zwierzchowska et al. 2018 [48]; Gashu et al. 2020 [49].
Environmental Benefits (n = 21)	Recreational and exercise (n = 10)	Popoola and Ajewole 2001 [21]; Barnhill and Smardon 2012 [26]; Buchel and Frantzeskaki 2015 [29]; Giannakis et al. 2016 [50]; Larson et al. 2016 [37]; Yen et al. 2016 [44]; Ives et al. 2017 [38]; Keith et al. 2018 [27]; Meyer and Schulz 2018 [51]; Nath et al. 2018 [47].
	Biodiversity and wildlife (n = 9)	Peckham et al. 2013 [43]; Shwartz et al. 2014 [52]; Rupprecht et al. 2015 [36]; Giannakis et al. 2016 [50]; Meyer and Schulz 2017 [51]; Korpilo et al. 2018 [39]; Campbell-Arvai 2019 [28]; Wang et al. 2019 [53].
	Shade (n = 5)	Conway and Yip 2016 [54]; Paul and Nagendra 2017 [55]; Fernandes et al. 2019 [56]; Guenat et al. 2019 [41]; Gwedla and Shackleton 2019 [42].
	Better air quality and climate regulation (n = 13)	Sanesi and Chiarello [22]; Peckham et al. 2013 [43]; Buchel and Frantzeskaki 2015 [29]; Rupprecht et al. 2015 [36]; Conway and Yip 2016 [54]; Giannakis et al. 2016 [50]; Yen et al. 2016 [44]; Faivre et al. 2017 [45]; Paul and Nagendra 2017 [55]; Duan et al. 2018 [46]; Fernandes et al. 2019 [56]; Guenat et al. 2019 [41]; Miller and Montalto 2019 [30].
	Water runoff mitigation (n = 3)	Barnhill and Smardon 2012 [26]; Paul and Nagendra 2017 [55]; Miller and Montalto 2019 [30].
Economic Benefits (n = 8)	Food provision (n = 3)	Barau 2015 [33]; Guenat et al. 2019 [41]; Gwedla and Shackleton 2019 [42].
	Wood provision (n = 2)	Popoola and Ajewole 2001 [21]; Meyer and Schulz 2017 [51].
	Increase in property value (n = 3)	Jim and Chen 2006 [23]; Yen et al. 2016 [44]; Panagopoulos et al. 2018 [40].

Perception of the Risks That Can Cause Costs/Disservices

Compared to ecosystem services, fewer studies have considered ecosystem disservices. An understanding of perceived risks among NBS, by citizens and stakeholders, is fundamental for their effective planning, implementation, and management [54]. Table 3 presents the main threats referred to in the screened research papers.

Table 3. The mainly perceived risks of NBS by citizens and stakeholders as presented in the selected publications.

Perceived Risks (n = 9)	Authors
Danger (e.g., crime and vandalism) (n = 6)	Sanesi and Chiarello 2006 [22]; Ostoić et al. 2017 [57]; Keith et al. 2018 [27]; Campbell-Arvai 2019 [28]; Fernandes et al. 2019 [56]; Gwedla and Shackleton [42].
Dirtiness (e.g., leaves in autumn or bird excrement) (n = 4)	Conway and Yip 2016 [54]; Ostoić et al. 2017 [57]; Fernandes et al. 2019 [56]; Gwedla and Shackleton 2019 [42].
Attraction of unwanted animals/insects (n = 3)	Jim and Chen 2006 [23]; Conway and Yip 2016 [54]; Campbell-Arvai 2019 [28].
Limited Access/Environmental injustice (n = 2)	Ostoić et al. 2017 [57]; Keith et al. 2018 [27].
Damage (e.g., person, property) (n = 2)	Conway and Yip 2016 [54]; Campbell-Arvai 2019 [28].
Allergies (n = 1)	Gwedla and Shackleton 2019 [42].
Economic costs (e.g., construction and maintenance) (n = 2)	Conway and Yip 2016 [54]; Campbell-Arvai 2019 [28].
Invasive species of plants (n = 1)	Campbell-Arvai 2019 [28]
Contamination (e.g., soil through chemicals and dirty water use) (n = 1)	Guenat et al. 2019 [41].

3.2.2. Preferred Attributes for the Design of NBS

Urban policy has failed to provide specific design guidance for NBS. Public perception surveys enabled urban planners to identify preferred green infrastructure attributes and use this information in their urban planning framework [58]. Table 4 shows the most mentioned characteristics and attributes preferred by citizens and stakeholders for NBS design according to the analyzed publications.

Table 4. The most relevant preferences for design of NBS reported by citizens and stakeholders, as presented in the selected publications.

Preferences for Design	Authors
Tree or flower abundance, biodiversity (n = 10)	Koo et al. 2013 [60]; Zhang et al. 2013 [61]; Baur et al. 2016 [62]; Arnberger et al. 2017 [63]; Derkzen et al. 2017 [64]; Pietrzyk-Kuszynska et al. 2017 [65]; Ayala-Azcárraga et al. 2019 [66]; Hwang et al. 2019 [67]; Rahnema et al. [58]; Ramer et al. 2019 [68].
Increase in fauna (n = 5)	Caula 2009 [69]; Koo et al. 2013 [60]; Ayala-Azcárraga et al. 2019 [66]; Hwang et al. 2019 [67]; Ramer et al. 2019 [68].
Water, streams, and fountains (n = 4)	Arnberger et al. 2017 [63]; Karanikola et al. 2017 [70]; Menconi and Grohmann 2018 [71]; Rahnema et al. 2019 [58].
Walkways, stepping stone corridors (n = 5)	Zhang et al. 2013 [61]; Karanikola et al. 2017 [70]; Ayala-Azcárraga et al. 2019 [66]; Hwang et al. 2019 [67]; Shams and Barker 2019 [72].
Security (n = 3)	Zhang et al. 2013 [61]; Baur et al. 2016 [62]; Shams and Barker 2019 [72].
Cleanliness and proper maintenance (n = 3)	Baur et al. 2016 [62]; Pietrzyk-Kaszyńska et al. 2017 [65]; Shams and Barker 2019 [72].
Naturalness and wilderness areas (n = 5)	Zhang et al. 2013 [61]; Baur et al. 2016 [62]; Hwang et al. 2019 [67]; Rahnema et al. 2019 [58]; Shams and Barker 2019 [72].
Accessibility, distance to home or to city center (n = 6)	Zhang et al. 2013 [61]; Arnberger et al. 2017 [63]; Derkzen et al. 2017 [64]; Pietrzyk-Kuszynska et al. 2017 [64]; Ayala-Azcárraga et al. 2019 [66]; Shams and Barker 2019 [72].
Information signs and environmental education (n = 5)	Caula 2009 [69]; Koo et al. 2013 [60]; Karanikola et al. 2017 [70]; Pietrzyk-Kuszynska et al. 2017 [65]; Shams and Barker 2019 [72].
Facilitate social interactions (Seats, tables, picnic or barbecue areas, shelters) (n = 5)	Zhang et al. 2013 [61]; Karanikola et al. 2017 [70]; Menconi and Grohmann 2018 [71]; Ayala-Azcárraga et al. 2019 [66]; Shams and Barker 2019 [72].

Table 4. *Cont.*

Preferences for Design	Authors
Kids playground (n = 3)	Zhang et al. 2013 [61]; Menconi and Grohmananan 2018 [71]; Ayala-Azcárraga et al. 2019 [66].
Sports and recreational facilities (n = 3)	Karanikola et al. 2017 [70]; Menconi and Grohmann 2018 [71]; Shams and Barker 2019 [72].
Connectivity to places of interest (e.g., parks, restaurants, shops, monuments) (n = 2)	Pietrzyk-Kuszynska et al. 2017 [65]; Shams and Barker 2019 [72].
Green space into buildings (n = 2)	Tsantopoulos et al. 2018 [59]; Xue et al. 2019 [3].

Xue et al. [3] found that stakeholders are more likely to prefer biophilic design strategies that focus on immediate human spatial experience in buildings. Additionally, Tsantopoulos et al. [59] investigated public perceptions and attitudes toward GI on buildings and found that most of the citizens were keen to improve aesthetics through green roofs, trellises, or vertical gardens.

3.2.3. Perspectives on the NBS Challenges

The lack of knowledge and awareness literature that explores the citizens' and stakeholders' perspectives and perceptions regarding challenges for NBS. Table 5 shows the main challenges reported by citizens and stakeholders for the implementation and maintenance of NBS based on the selected publications.

Table 5. The main NBS challenges identified by citizens and stakeholders, as reported in the selected publications

NBS challenges	Authors
Lack of knowledge and awareness about the environmental problems and their possible solutions and impacts (n = 8)	Lamichhane and Thapa 2012 [73]; Keeley et al. 2013 [74]; Hoyle et al. 2017 [75]; Furlong et al. 2018 [76]; Khoshkar et al. 2018 [77]; Onori et al. 2018 [78]; Girma 2019 [79]; Molla and Mekonnen 2019 [80].
Lack of evidence of the success and efficacy of the solutions (n = 1)	Kabisch et al. 2016 [14].
Lack of political support/guidance (n = 8)	Lamichhane and Thapa 2012 [73]; Keeley et al. 2013 [74]; Zivojinovic and Wolfslehner 2015 [81]; Furlong et al. 2018 [76]; Khoshkar et al. 2018 [77]; Girma 2019 [79]; Lähde and Marino 2019 [82]; Molla and Mekonnen 2019 [80].
Financial constraints and lack of funding (n = 8)	Lamichhane and Thapa 2012 [73]; Keeley et al. 2013 [74]; Rall et al. 2015 [83]; Zivojinovic and Wolfslehner 2015 [81]; Furlong et al. 2018 [76]; Khoshkar et al. 2018 [77]; Di Marino et al. 2019 [84]; Girma 2019 [79].
Lack of engagement due low social cohesion (n = 7)	Lamichhane and Thapa 2012 [73]; Rall et al. 2015 [83]; Zivojinovic and Wolfslehner 2015 [81]; Kabisch et al. 2016 [14]; Hoyle et al. 2017 [75]; Bissonnette et al. 2018 [85]; Girma 2019 [79].
Lack of skilled personnel/technical and scientific knowledge (n = 3)	Keeley et al. 2013 [74]; Zivojinovic and Wolfslehner 2015 [81]; Girma 2019 [79].
Maintenance and monitoring (n = 4)	Lamichhane and Thapa 2012 [73]; Rall et al. 2015 [83]; Keeley et al. 2017 [74]; Khoshkar et al. 2018 [77].

Lack of knowledge and awareness, as well as political support, financial constraints, and lack of public engagement, are the most mentioned challenges. Lack of knowledge and awareness is responsible for negative views of trees and vegetation; it may result in limiting public support for urban green solutions, particularly on private land [76,80]. Vandalism may also explain low support for NBS [57]. Furthermore, the lack of awareness from municipal, regional and national government agencies about the benefits of such green solutions has affected budget al.ocation for their development [79]. Collaboration of nonprofit organizations and voluntary community groups and partnerships can raise funds for the development of those solutions [86].

According to Kabisch et al. [14], NBS need to be recognized and developed as proactive investments and supported as such in planning procedures; they should also be fostered in joint dialogues among policy, society, and science. To avoid the loss of desired project qualities and minimize the related conflicts of interests between the different stakeholders, municipalities in and around Stockholm have increased citizen participation and collaboration between planners and developers, as a way of overcoming the challenges faced with the implementation of NBS [77]. The results of this test were the minimization of conflicting interests between involved and affected actors, and the decrease of green quality management issues. According to Young [86], this type of initiative benefits from being launched early in an administration term, providing long-term advancement and protection of GI or NBS investments.

3.3. Citizens' and Stakeholders' Participation in the Processes of NBS

3.3.1. Opportunities and Challenges Found in the NBS Practices

Opportunities

As previously mentioned, citizen and stakeholder participation in NBS has been reported as crucial for the success of those solutions. Such participation usually involves people and organized groups that can influence the decision-making processes. In this section, firstly, we analyze the opportunities that emerge from actual community participation in NBS that were identified in the selected publications. Table 6 summarizes them.

Table 6. The opportunities for NBS that emerge from citizen and stakeholder participation, as reported in the selected publications.

Opportunities (n = 16)	Authors
Promote social cohesion (cooperative working, mutual learning, and experience-sharing) (n = 6)	Chou et al. 2017 [87]; Fors et al. 2018 [88]; Harper et al. 2018 [89]; Kosová et al. 2018 [90]; Ugolini et al. 2018 [91]; Rolf et al. 2019 [92].
Add-value to urban natural and social capital (n = 2)	Dennis and James 2016 [93]; Dennis and James 2016 [94].
Increase biodiversity (n = 3)	Mabelis and Maksymiuk 2009 [13]; Dennis and James 2016 [94]; Fischer et al. 2019 [95].
Contextualize functions with ecosystem services (n = 1)	Dennis and James 2016 [93].
Develop initiatives of environmental education (n = 3)	Moskell and Allred 2013 [96]; Chou et al. 2017 [87]; Fischer et al. 2019 [95].
Intensify the public acceptability, confidence, consciousness and sense of belonging (n = 4)	Sipilä and Tyrväinen 2005 [12]; Fors et al. 2018 [88]; Gulsrud et al. 2018 [97]; Rolf et al. 2019 [92].
Influence social learning and innovation (n = 6)	Travaline et al. 2015 [98]; Dennis et al. 2016 [99]; Chou et al. 2017 [87]; Gulsrud et al. 2018 [97]; Kosová et al. 2018 [90]; Ugolini et al. 2018 [91].
Benefit from multifunctionality (n = 2)	Belmeziti et al. 2018 [100]; Rolf et al. 2019 [92].
Connect people with nature (n = 3)	Chou et al. 2017 [87]; Gulsrud et al. 2018 [97]; Fors et al. 2018 [88].
Establish long-term partnerships to attain funding (n = 1)	Ugolini et al. 2018 [91].
Prevent conflicts (n = 2)	Sipilä and Tyrväinen 2005 [12]; Rolf et al. 2019 [92].

Green infrastructure planning needs to rely on collaborative and participatory approaches to enhance ecosystem services at all scales [85]. The public involvement can play a fundamental role in enhancing the productivity of urban green spaces, increasing the value of ecosystem services. Stakeholder participation ensures identification of ES from the beginning and the links between them and greenspace components, helping urban planners and managers to improve the multifunctionality of solutions [93,100]. Work has been carried out which demonstrates that biodiversity increases proportionally to levels of user participation [94].

The different actors bring different forms of knowledge to the process. Brink and Wamsler [101] stressed the importance of collaboration between local government and citizens in order to address climate impacts more effectively. The concept of "biodiverse edible schools," that link food production and consumption with local biodiversity, presented by Fischer et al. [95], adopts a long-term engagement of stakeholders from various domains to improve healthy food and environmental education at the school. Participation is anticipated to produce better policies by encouraging the exchange of information and ideas and by promoting collaborative learning about problems and their potential solutions [90].

Partnerships with local actors, especially through community groups, can encourage trust, while facilitating ecosystem stewardship and social learning as critical factors for socioecological resilience [97].

Challenges

Several barriers hinder public participation; these mostly include challenges related to the cultural domain. The most typical problem is that of poor social mobilization, reflected in the fact that urban residents often perceive GI solution stewardship as the responsibility of government (either local, state, or federal), not their own [96]. Table 7 presents a resume of the main identified challenges that have prevented stakeholders and citizens from being engaged and participating in NBS projects.

Table 7. The challenges faced by citizens and stakeholders that prevent their participation in processes of NBS, identified in the selected publications

Challenges (n = 10)	Authors
Deal with conflicting points of view and interests (n = 3)	Sipilä and Tyrväinen 2005 [12]; Cousins 2017 [102]; Ugolini et al. 2018 [91].
Understand the hierarchies of institutions and bureaucracies (n = 5)	Mattijssen et al. 2017 [103]; Mensah et al. 2017 [104]; Gulsrud et al. 2018 [97]; Liu and Jensen 2018 [105]; Ugolini et al. 2018 [91].
Overtake the lack of political support (n = 2)	Gulsrud et al. 2018 [97]; Liu and Jensen 2018 [105].
Feel the involvement as being time consuming and expensive (n = 3)	Sipilä and Tyrväinen 2005 [12]; Mabelis and Maksymiuk 2009 [13]; Travaline et al. 2015 [98].
Overcome the poor flow of information and social mobilization (n = 2)	Moskell and Allred 2013 [96]; Mensah et al. 2017 [104].
Maintain continuity of the collaboration (n = 1)	Mattijssen et al. 2017 [103].

Sipilä and Tyrväinen [12] found that participatory approaches are more demanding with respect to time and resources than conventional planning. The same authors mentioned other difficulties: the type of people interested in participating (with limited representativeness); the low number of participants; increased conflict between opposing stakeholders; and too-high expectations from participants, resulting in disappointment over compromise.

To avoid the challenges related to divergence of interests among stakeholders, Ugolini et al. [91] indicated the need for a "common language"—i.e., "speaking green"—which may accommodate diverse priorities and concerns. Early dialogue and partnership with citizens and stakeholders provides the opportunity to identify common goals [77].

Usually, the lack of political support is related to low awareness of politicians and key administrators. Bureaucracy is a challenge encountered in collaborations with public administrations [91]; it can also have a hindering or discouraging impact on the activities of citizens [103].

Yamaki [106] suggested that social media would be essential to overcoming some participation challenges related with the exchange of information and knowledge or poor social mobilization.

An important challenge that remains is to achieve continuity in the engagement of citizens; in this regard, Mattijssen et al. [103] highlight the importance of attracting new volunteers over time to maintain a critical mass.

3.3.2. Identified Drivers/Motivations for Public Participation

Looking at the selected articles, the main motivations and drivers for citizen and stakeholder participation on the NBS can be organized into three reasons as summarized in Table 8.

Table 8. The motivations and related drivers invoked by citizens and stakeholders to participate in process of NBS, as identified in the selected publications.

Motivations/Drivers		Authors
Environmental (n = 7)	Environment protection and contribution to sustainability (n = 5)	Asah and Blahna 2012 [107]; Shan 2012 [108]; Zare et al. 2015 [109]; Chelleri et al. 2016 [25]; Beery et al. 2018 [110].
	Characteristics of the physical environment (n = 2)	Fors 2019 [111]; Murphy 2019 [112].
Communal (n = 8)	Protect the community/improve collective health (n = 2)	Zare et al. 2015 [109]; Beery et al. 2018 [110].
	Promote social interactions (n = 2)	Asah and Blahna 2012 [107]; Zare et al. 2015 [109].
	Bring neighbors to participate and be part of the experiences (n = 5)	Green et al. 2012 [113]; Lewis et al. 2018 [114]; Lieberherr and Green 2018 [115]; Lim 2018 [116]; Fors 2019 [111].
Personal (n = 8)	Possibility to learn from and experience environmentally friend solutions (n = 3)	Asah and Blahna 2012 [107]; Chelleri et al. 2016 [25]; Lewis et al. 2018 [114].
	Interest in gardening (n = 2)	Fors 2019 [111]; Petrovic et al. 2019 [117].
	Sense of place and attachment (n = 3)	Murphy 2019 [112]; Petrovic et al. 2019 [117]; Romolini 2019 [118].
	Proximity to disturbance and effects on residential properties (n = 2)	Hunter et al. 2011 [119]; Fors 2019 [111].

Civic consciousness has been increasing; in recent years, more and more people started to display a positive attitude and strong willingness toward participation on NBS [108,110]. Some citizens are motivated by the prospect of improving their physical and/or mental health, and others are only moved by collective purposes (e.g., building friendly relationships) [109]. Usually, the degree of engagement is directly connected to the individual's proximity to the disturbance [119], and some types of environment are more difficult for people to involve in [111].

Some authors have found that most of the citizens are willing to participate in NBS planning, implementation, and management to ensure social, and environmental benefits, e.g., [25,108]; some found further that citizens were willing to contribute financially, e.g., [64,69]. Zare et al. [109] found that people were more willing to participate as a "membership in a public conservation committee" than in practical management activities. According to Green et al. [113], for example, to engage or encourage citizens as stormwater managers, we must go beyond the technical aspects and invest in social factors to motivate behavioral change. It is also acknowledged that, as participants share their experiences, neighbors become more willing to trust in the program of NBS [113,116].

Green infrastructure solutions have contributed to enhancing ecosystem governance and increasing a sense of place, especially in relation to urban gardens and urban agricultural areas (where residents directly participate in their management, maintenance, and monitoring) [117]. Romolini [118] found a positive relationship between public interest in collaborate and place attachment, i.e., individuals' psychological and emotional connection to urban green spaces precedes their involvement in processes of NBS.

3.3.3. Methods, Tools, and Frameworks

A range of methods and tools have been used to engage stakeholders and citizens in urban GI and NBS. Table 9 displays the various methods and tools used in the selected publications of this literature review.

Table 9. Methods/tools used in the selected publications to involve citizens and stakeholders as participants on NBS

Participation Methods/Tools			Authors
Social media (n = 3)			Afzalan and Muller 2014 [120]; Guerrero et al. 2016 [121]; Yamaki 2016 [106].
E-Tools/virtual tool (n = 2)			Shwartz et al. 2013 [122]; Møller et al. 2019 [123].
GIS-based tools (n = 8)		PPGIS	Janse and Konijnendijk 2007 [124]; Hawthorne et al. 2015 [125]; Raymond et al. 2016 [126]; Rall et al. 2019 [127].
		SolVES	Sun et al. 2019 [128].
		VGI	Guerrero et al. 2016 [121]; Møller et al. 2019 [123].
		3D visualization	Neuenschwander et al. 2014 [129].
Focus Group (n = 3)			Nilsson et al. 2007 [130]; Kangas et al. 2014 [131]; Sturiable et al. 2018 [132].
Workshop (n = 4)			Janse and Konijnendijk 2007 [124]; Bellamy et al. 2017 [133]; Assmuth et al. 2017 [134]; van der Jagt et al. 2019 [135].
Questionnaire/Survey/Q methodology (n = 10)			Janse and Konijnendijk 2007 [124]; Kangas et al. 2014 [131]; Hawthorne et al. 2015 [125]; Lindemann and Briege 2016 [136]; Raymond et al. 2016 [126]; Sun and Hall 2016 [137]; Jayasooriya et al. 2019 [138]; Lafortezza and Giannico 2019 [139]; Møller et al. 2019 [123]; Rall et al. 2019 [127].
Interviews (n = 4)			Nordström et al. 2010 [140]; Kangas et al. 2014 [131]; Beumer and Martens 2015 [141]; Sturiable and Scuderi 2018 [132].
Meetings (n = 5)			Nordström et al. 2010 [140]; Afzalan and Muller 2014 [120]; O'Donnell et al. 2018 [142]; Sturiable and Scuderi 2018 [132]; Lafortezza and Giannico 2019 [139].
Visual methods (n = 4)			Qiu et al. 2013 [143]; Lindemann and Briege 2016 [136]; Rink and Arndt 2016 [31]; Sun et al. 2019 [128].
Learning Alliances (n = 2)			O'Donnell et al. 2018 [142]; van der Jagt et al. 2019 [135]
Living Labs (n = 3)			Bellamy et al. 2017 [133]; Lafortezza and Giannico 2019 [139]; van der Jagt et al. 2019 [135]

Questionnaires and surveys have been the most common tool used in the participatory processes (Table 9). The reason may lie in the fact that these tools are reproducible, comparable, and easy to implement for the collection of citizens' perceptions, preferences, and viewpoints, being at the same time useful to support NBS planning and the decision-making process. Some authors used Q-methodology that is based on a limited set of perceptions and viewpoints that people have on a certain topic, to provide information (similarities and differences) within a range of perceptions and viewpoints [131]. Visual methods (e.g., photo-realistic visualizations) have been linked to questionnaires and surveys to help in the design process [31,136].

Methods including geographic information systems have been increasing due to the incorporation of socio-spatial information in strategic green space planning. An example is public participation geographic information systems (PPGIS) [127], that enhance citizens or stakeholders to identify locations on a map of various aspects such as perceptions, preferences, or values, and particularly associate them with ecosystem services.

E-tools have received attention because of their potential to connect government and citizens and facilitate interaction between them. Møller et al. [123] studied three map-based e-tools, i.e., users share information on digital maps (so-called Volunteered Geographic Information (VGI)).

Social media platforms have increasingly been used as a fundamental tool to facilitate collaboration and interactions among stakeholders on NBS, enhancing social learning that fosters social capital, resource mobilizations, and consensus building [106].

Other methods have shown the potential of participatory processes, such as workshops, focus groups, and meetings. In a workshop, Assmuth et al. [134] used an innovative and "out-of-the-box" method of opinion elicitation, a role chair session (where experts were asked to step into the shoes of other population groups, based on their prior knowledge to interpret the positions of the groups while sitting in a chair for the respective group addressed).

Some of the selected publications presented a structured methodological framework for citizen or stakeholder participation in NBS, including specific methods and steps to follow. Further frameworks have been developed to map priority areas for green infrastructure [133] and to integrate stakeholders' perceptions and knowledge regarding this important component for urban landscape [132,137,139] and add remotely-sensed data, such as high-resolution satellite images and Laser Imaging Detection and Ranging (LiDAR) [139]. In this context, Beumer and Martens [141] speak about "BIMBY (Biodiversity in my (back) yard)," an indicator framework for assessing biodiversity in domestic gardens in the way that it explicitly combines ecological factors, cultural elements and citizen's preferences. Other authors used platforms for the mobilization and co-production of knowledge at different scales, known as Learning Alliances (LA) [135,142] and Living Labs (LL) [133,135,139], and combine a whole set of participation tools, such as workshops or focus groups, and use them structurally through the process, concerning the engaged group, time, scale, aim, and the expected outcome.

3.3.4. Collaborative Governance and Actors' Interactions Through the Decision-Making Process

In collaborative urban planning, one of the key questions is how the different actors interact and how their input can be integrated into planning and decision-making process. Looking at the participatory processes on the different NBS, diverse forms of collaborative governance and interaction between actors have been identified; these are summarized in Table 10.

Table 10. Type of collaborative governance and actors' interactions through the process of NBS, as identified in the selected publications.

Collaborative Governance	Authors
Top-down approach and a central-government decision process (n = 6)	Rosol 2010 [144]; Faehnle et al. 2014 [145]; Skandrani et al. 2015 [146]; Gasperi et al. 2016 [147]; Kronenberg et al. 2016 [148]; Shifflet et al. 2019 [149].
Bottom-up and citizen-led approaches (n = 6)	Rosol 2010 [144]; Cvejić et al. 2015 [150]; Skandrani et al. 2015 [146]; Gasperi et al. 2016 [147]; Jerome 2017 [151]; van der Jagt et al. 2017 [152].
Public-private interactions (n = 5)	Young 2011 [86]; Milanovič and Foški 2015 [153]; Brink and Wamsler 2016 [31]; Simić et al. 2017 [154]; Buijs et al. 2019 [155].
Cross-sectoral partnerships (n = 4)	Ugolini et al. 2015 [156]; Schifman et al. 2017 [157]; van der Jagt 2017 [152]; Frantzeskaki 2019 [158].

Top-down approaches are usually led and supported by national, regional or local governments or other public institutions and green spaces are usually managed in a centralized manner, with low (or very low) levels of civic involvement [147,148]. Gasperi et al. [147] exposed not participatory top-down projects that resulted in unsuccessful initiatives, as they did not respond to community expectations and needs.

Bottom-up and citizen-led approaches are usually driven by citizens or non-governmental actors, enhancing strong local community engagement on green infrastructure planning and management. Although these approaches can strongly depend on the support of the local government [147]. Many examples of community-led systems are urban gardens or urban agriculture [147].

On urban vegetation governance, there have been efforts to move from top-down planning and decision-making towards involving a broad range of non-governmental actors and recognizing local needs and expertise based on citizens' everyday experiences, to overcome barriers and foster innovation [145]. Schifman et al. [157] developed a framework (FrASH) to integrate networks of stakeholders and organizations into GI projects to enhance multifunctionality. This work is centered on the combination of inputs from organizations to reach collaborative decision-making.

Recent research examines and proposes other multi-stakeholder governance processes such as public-private interactions [155] and cross-sectoral partnerships [156]. Public-private partnerships are encouraged as they carry top-down guidelines with the contributions of private sector. Cross-sectoral partnerships require the reconfiguration of the relationships between state, market actors, civil society, and science. Ugolini et al. [156] pointed out that the involvement of experts from academia has been a great benefit, since academics can offer scientific knowledge and innovation, research experience, expertise, and problem solutions.

4. Discussion

The selection of publications allowed us to identify two main areas of research with almost equal relevance: (a) publications analyzing the perceptions, preferences and perspectives of citizens and stakeholders in order to engage them into participation by uncovering their cognitive image of NBS as prior action and (b) publications that analyze the participatory process itself. In the first area, the benefits and costs sought by those involved or affected by the configuration of NBS emerge as the principal subarea of research. In the second area, the research on methods and tools is predominant. The contrast between the two domains becomes visible through analyzing the most common words as presented in the word cloud (see Figure 6): park, spaces, trees, landscape, and nature, for the first area, versus management, planning, community, governance and participation, for the second area.

Among the benefits perceived by citizens and stakeholders from NBS, the majority of the literature addresses the social benefits, closely followed by the environmental benefits and only a few studies look at the economic benefits in order to analyze various themes. This bias that favors the social and environmental benefits is also referred to by Parker and Baro [9]. Concerning the social benefits, studies stood out that deal with aesthetics, scenic views, and proximity to nature, followed by studies addressing physical and mental well-being and recreation/exercise. Studies on the environmental benefits focus on better air quality and climate regulation, as well as on biodiversity and wildlife. Surprisingly, only a small number of publications (n = 4) dealt with the benefits from NBS to quality of life in cities.

Regarding the citizens' and stakeholders' perceived risks of NBS that can generate costs/disservices, we observe that the number of publications is significantly inferior to the one dealing with the perceived benefits. The publications dealing with the risks of citizen and stakeholder participation are distributed diffusely across 10 categories, which signals an absence of a comprehensive understanding of the costs associated with NBS. The tendency of the literature to focus on the benefits, avoiding looking at the risks, has been pointed out by those skeptical about the participatory approaches to NBS. Fors et al. [20] highlighted that there is a rhetoric around public participation, as they failed to find empirical evidence of a convergence between expected and achieved benefits from the involvement of citizens and stakeholders and its effects for the quality of the implemented solutions. It is worth mentioning that, among the identified risks in the selected publications, all categories with more than one publication had at least one article in 2019 (with the exception of environmental justice). The risk related to environmental injustice has only three publications, none of them in 2019, which seems to indicate that this subtopic has been receiving scant attention from the researchers. This observation deserves consideration, as one of the objectives of NBS should be to reduce social injustices. The small number of studies dealing with how NBS impacts on social justice may lie in the fact that it often is difficult to bring those with low income into the participation process. Zuniga-Teran and Gerlak [17] suggested two main reason for this: (a) such families may not be able to afford using their time in tasks of social engagement and (b) the absence or dismantling of certain NGOs that may represent this group.

The preferences for attributes identified in the surveyed literature, as reported by citizens and stakeholders and related to the design of NBS, showed a spread over 14 categories; these highlight the diversity of interests involved in any process of participation. The abundance of trees or flowers and the increase of urban biodiversity is clearly the attribute most preferred by society, as reported in the selected publications. Fors et al. [20] also found that one of the factors that make users of green spaces participate is the desire to have a higher area covered with trees or healthier trees. Interestingly, the citizens' and stakeholders' preferences for design that take into consideration the natural environment and wilderness areas only appear in publications of 2019. This may be a sign of an increasing awareness in the populace of the need to respect ecological balances and plan the use of land according to environmentally-sustainable rules.

Looking at the second area of research identified in this review: the participation process in NBS and, in particular, the opportunities this process brings, a pulverization of opportunities is identified; we organized these into 11 categories. The most referred opportunities derived from the implementation of participation processes were the increase of social cohesion, followed by the influence of such participation for social learning and the identification of innovative solutions. Surprisingly, the opportunity to prevent conflicts was scarcity addressed in the literature (only two publications; see Table 6), being acknowledged as one of the limitations for successfully implement NBS, e.g., [10,19]. Citizens and stakeholders report it as one of the challenges emerging from the participation processes; however, the issue is only analyzed in three articles (see Table 7).

Other challenges that need to be overpass seem to capture more the attention of the researchers, namely, the difficulty of the participants in NBS processes to understand how public entities are organized and how to deal with bureaucracies. In contrast with Sarabi et al. [10] that identified nine publications dealing with the inadequacy of financial resources, this challenge is not reported in the selected publication included in this work when associated to the implementation of NBS (Table 7), appearing only in eight publications when perceptions are assessed (see Table 5). Only a few works looked at the challenge related to time spent on participation processes as a drawback (since citizens with low income may find it difficult to devote time to this type of process, which implies that those more vulnerable in society may find their needs overlooked). This suggests that the implementation of NBS is not contributing yet to bring social inclusiveness. This appears as one of the major gaps in the literature and one of the aspects in which the intention of involving all the layers of the society has been failing.

Contrasting the challenges identified after the implementation of NBS processes (Table 7) with those that were expected (Table 5), it is observed that almost completely different lists of challenges (with the exception of lack of political support) are derived. This divergence in the identified challenges signals that expectations regarding the implementation of NBS are detached from the real challenges faced through their implementation. Fors et al. [20] have pointed out how concepts in theory can be distant from the ones supported by empirical evidence. They called for the "defeat of generalizations" and for the involvement of participants in all the various phases present in processes of NBS.

The literature focused on how citizens' and stakeholders' motivations to participate are limited and almost equally distributed across environmental, communal, and personal drivers (see Table 8).

Several methods and tools for citizen and stakeholder participation were identified; these represent various levels of involvement and influence on the NBS process. The most used tools, in the selected publications, were questionnaires or surveys, a consultation-based method, useful to measure the needs and views of citizens. Tools, such as workshops and focus groups, were found to be less used;

however, they were useful in bringing together decision-makers and interest groups and initiating mutual exchange of views, experiences, expertise, etc. [124]. These tools are usually used in Learning Alliances (LA) and Lab Living (LL) platforms for a more in-depth participation [135]. This type of participation enables an iterative process of knowledge exchange through a decision-making process. Additionally, LA and LL facilitate the process of collaborative learning across different scales, enabling partnerships and bottom-up innovation [135,142].

The bottom-up and citizen-led approach was the most often identified form of involving citizens and stakeholders in the NBS process, despite the fact that several articles also dealt with top-down approaches and centralized decision processes and public-private partnerships. Buijs et al. [19] proposed the concept of "mosaic governance" that goes beyond the dichotomy between top-down or bottom-up approaches, defining it as an approach that considers the specificity of the context and a balance between the expert knowledge and the autonomy of social initiatives.

5. Conclusions

Nature-based solutions have been studied around the world, and some existing reviews have paid attention to these urban green solutions in a general way, relating environmental and social benefits, opportunities, challenges, and barriers [9,10,15–17]. Important review research has focused on participation contributions, considering specific urban green spaces thus a comprehensive looking at the various concepts that encompass NBS approaches is still lacking [15,20]. Facing the ambiguity around terms and concepts (urban forest, biodiversity, urban green space, green infrastructure, biophilic infrastructure, and nature-based solutions) usually used to describe urban green projects, we present here an exercise of comprehensively include all of them. Existing review research has focused on specific terms, underlooking important developments in the area. Trying to assemble all the important publications and associated nomenclatures, our research aimed to find, organize, and highlight the most important findings from participatory experiences involving stakeholders, citizens, residents, community, and public entities. Further, this investigation pays special attention to stakeholders' and citizens' perceptions, preferences and perspectives as an a priori condition for public participation. In this manner, this research also aims to increase the awareness of managers and planners about the importance of such input to improve the quality, accuracy, and efficacy of planning and management.

Stakeholder and citizen participation or collaboration on urban green projects is increasingly recognized as promising means. The present review aimed to provide a synopsis of current research knowledge and gaps and, as well, an overview on research progress and emerging trends. From the analysis of the gathered publications, several gaps were identified that call for future investigations:

1. The lack of research applied to countries of southern Europe (which are in the forefront of climate changes) as well as the almost nonexistent research applied to Africa and South America, which preclude the establishment of a comprehensive theoretical and empirical knowledge of the participation processes during their several steps from the conceptualization to the implementation and management of NBS.
2. Despite the bulk of the literature dealing with perceptions, preferences, and perspectives of citizens and stakeholders engaged in participation processes of NBS and their anticipated benefits, only a few studies pay attention to economic benefits and those raising the quality of life in cities.
3. Few studies looked at the risks perceived by citizens and stakeholders due their involvement in NBS, and in particular, how NBS are perceived as contributing to reduce social injustice.
4. Remaining to be explored is the possibility of using the participatory process in NBS to prevent conflicts between the various interests involved.
5. New studies are needed aiming to interconnect the theoretical conceptions and the practice of participation processes in NBS, in order to adjust the citizens' and stakeholders' expected difficulties and the ones faced in reality—mitigating, in accordance, eventual frustrations of those

involved and promoting the maintenance of collaboration during the life cycle of the implemented NBS as well as in future projects.

6. Future research should evaluate the contribution of participatory processes for the quality of decisions, the building of public trust in the decision-making process, and for the success of implemented social-learning strategies.

Author Contributions: Conceptualization: V.F., L.L., A.P.B., D.A., and T.P.; methodology: V.F., L.L., A.P.B., D.A., and T.P.; formal analysis: V.F.; investigation: V.F.; writing—original draft preparation: V.F. and A.P.B.; writing—review and editing: T.P.; supervision: T.P, and L.L. All authors have read and agreed to the published version of the manuscript.

Funding: This research was funded by the Foundation for Science and Technology (FCT), grant number PTDC/GES-URB/31928/2017.

Acknowledgments: This paper was financed by the FCT- Fundação para a Ciência e Tecnologia through project "Improving life in a changing urban environment through Biophilic Design".

Conflicts of Interest: The authors declare no conflict of interest.

References

1. European Commission. *Towards an EU Research and Innovation Policy Agenda for Nature-Based Solutions & Re-Naturing Cities: Final Report of the Horizon 2020 Expert Group on Nature-Based Solutions and Re-Naturing Cities*; European Commission: Brussels, Belgium, 2015.

2. Lafortezza, R.; Sanesi, G. Nature-based solutions: Settling the issue of sustainable urbanization. *Environ. Res.* **2018**, *172*, 394–398. [CrossRef] [PubMed]

3. Xue, F.; Gou, Z.; Lau, S.S.Y.; Lau, S.K.; Chung, K.H.; Zhang, J. From biophilic design to biophilic urbanism: Stakeholders' perspectives. *J. Clean. Prod.* **2019**, *211*, 1444–1452. [CrossRef]

4. Pauleit, S.; Zölch, T.; Hansen, R.; Randrup, T.B.; van den Bosch, C.K. *Nature-Based Solutions and Climate Change—Four Shades of Green*; Springer International Publishing: Cham, Switzerland, 2017; pp. 29–49.

5. Lafortezza, R.; Chen, J.; van den Bosch, C.K.; Randrup, T.B. Nature-based solutions for resilient landscapes and cities. *Environ. Res.* **2018**, *165*, 431–441. [CrossRef] [PubMed]

6. Escobedo, F.J.; Giannico, V.; Jim, C.Y.; Sanesi, G.; Lafortezza, R. Urban forests, ecosystem services, green infrastructure and nature-based solutions: Nexus or evolving metaphors? *Urban For. Urban Green.* **2019**, *37*, 3–12. [CrossRef]

7. Eggermont, H.; Balian, E.; Azevedo, M.N.; Beumer, V.; Brodin, T.; Claudet, J.; Fady, B.; Grube, M.; Keune, H.; Lamarque, P.; et al. Nature-based solutions: New influence for environmental management and research in Europe. *GAIA* **2015**, *24*, 243–248. [CrossRef]

8. Raymond, C.M.; Frantzeskaki, N.; Kabisch, N.; Berry, P.; Breil, M.; Nita, M.R.; Geneletti, D.; Calfapietra, C. A framework for assessing and implementing the co-benefits of nature-based solutions in urban areas. *Environ. Sci. Policy* **2017**, *77*, 15–24. [CrossRef]

9. Parker, J.; de Baro, M.E.Z. Green infrastructure in the urban environment: A systematic quantitative review. *Sustainability* **2019**, *11*, 3182. [CrossRef]

10. Sarabi, S.E.; Han, Q.; Romme, A.G.; de Vries, B.; Wendling, L. Key Enablers of and Barriers to the Uptake and Implementation of Nature-Based Solutions in Urban Settings: A Review. *Resources* **2019**, *8*, 121. [CrossRef]

11. Coles, R.W.; Bussey, S.C. Urban forest landscapes in the UK—progressing the social agenda. *Landsc. Urban Plan.* **2000**, *52*, 181–188. [CrossRef]

12. Sipilä, M.; Tyrväinen, L. Evaluation of collaborative urban forest planning in Helsinki, Finland. *Urban For. Urban Green.* **2005**, *4*, 1–12. [CrossRef]

13. Mabelis, A.A.; Maksymiuk, G. Public participation in green urban policy: Two strategies compared. *Int. J. Biodivers. Sci. Manag.* **2009**, *5*, 63–75. [CrossRef]

14. Kabisch, N.; Frantzeskaki, N.; Pauleit, S.; Naumann, S.; Davis, M.; Artmann, M.; Haase, D.; Knapp, S.; Korn, H.; Stadler, J.; et al. Nature-based solutions to climate change mitigation and adaptation in urban areas: Perspectives on indicators, knowledge gaps, barriers, and opportunities for action. *Ecol. Soc.* **2016**, *21*, 39. [CrossRef]

15. Parker, J.; Simpson, G.D. Public green infrastructure contributes to city livability: A systematic quantitative review. *Land* **2018**, *7*, 161. [CrossRef]
16. Venkataramanan, V.; Packman, A.I.; Peters, D.R.; Lopez, D.; McCuskey, D.J.; McDonald, R.I.; Miller, W.M.; Young, S.L. A systematic review of the human health and social well-being outcomes of green infrastructure for stormwater and flood management. *J. Environ. Manag.* **2019**, *246*, 868–880. [CrossRef] [PubMed]
17. Zuniga-Teran, A.A.; Gerlak, A.K. A multidisciplinary approach to analyzing questions of justice issues in urban greenspace. *Sustainability* **2019**, *11*, 3055. [CrossRef]
18. Albert, C.; Schröter, B.; Haase, D.; Brillinger, M.; Henze, J.; Herrmann, S.; Gottwald, S.; Guerrero, P.; Nicolas, C.; Matzdorf, B. Addressing societal challenges through nature-based solutions: How can landscape planning and governance research contribute? *Landsc. Urban Plan.* **2019**, *182*, 12–21. [CrossRef]
19. Buijs, A.E.; Mattijssen, T.J.; Van der Jagt, A.P.; Ambrose-Oji, B.; Andersson, E.; Elands, B.H.; Steen Møller, M. Active citizenship for urban green infrastructure: Fostering the diversity and dynamics of citizen contributions through mosaic governance. *Curr. Opin. Environ. Sustain.* **2016**, *22*, 1–6. [CrossRef]
20. Fors, H.; Molin, J.F.; Murphy, M.A.; van den Bosch, C.K. User participation in urban green spaces—For the people or the parks? *Urban For. Urban Green.* **2015**, *14*, 722–734. [CrossRef]
21. Popoola, L.; Ajewole, O. Public perceptions of urban forests in Ibadan, Nigeria: Implications for environmental conservation. *Arboric. J.* **2001**, *25*, 1–22. [CrossRef]
22. Sanesi, G.; Chiarello, F. Residents and urban green spaces: The case of Bari. *Urban For. Urban Green.* **2006**, *4*, 125–134. [CrossRef]
23. Jim, C.Y.; Chen, W.Y. Perception and attitude of residents toward urban green spaces in Guangzhou (China). *Environ. Manag.* **2006**, *38*, 338–349. [CrossRef] [PubMed]
24. Forzieri, G.; Alessandra, B.; e Silva, F.B.; Herrera, M.A.M.; Leblois, A.; Lavalle, C.; Jeroen, C.J.H.; Feyen, L. Escalating impacts of climate extremes on critical infrastructures in Europe. *Glob. Environ. Chang.* **2018**, *48*, 97–107. [CrossRef] [PubMed]
25. Chelleri, L.; Kua, H.W.; Sánchez, J.P.R.; Md Nahiduzzaman, K.; Thondhlana, G. Are people responsive to a more sustainable, decentralized, and user-driven management of urban metabolism? *Sustainability* **2016**, *8*, 275. [CrossRef]
26. Barnhill, K.; Smardon, R. Gaining ground: Green infrastructure attitudes and perceptions from stakeholders in Syracuse, New York. *Environ. Pract.* **2012**, *14*, 6–16. [CrossRef]
27. Keith, S.J.; Larson, L.R.; Shafer, C.S.; Hallo, J.C.; Fernandez, M. Greenway use and preferences in diverse urban communities: Implications for trail design and management. *Landsc. Urban Plan.* **2018**, *172*, 47–59. [CrossRef]
28. Campbell-Arvai, V. Engaging urban nature: Improving our understanding of public perceptions of the role of biodiversity in cities. *Urban Ecosyst.* **2019**, *22*, 409–423. [CrossRef]
29. Buchel, S.; Frantzeskaki, N. Citizens' voice: A case study about perceived ecosystem services by urban park users in Rotterdam, the Netherlands. *Ecosyst. Serv.* **2015**, *12*, 169–177. [CrossRef]
30. Miller, S.M.; Montalto, F.A. Stakeholder perceptions of the ecosystem services provided by Green Infrastructure in New York City. *Ecosyst. Serv.* **2019**, *37*, 100928. [CrossRef]
31. Rink, D.; Arndt, T. Investigating perception of green structure configuration for afforestation in urban brownfield development by visual methods-A case study in Leipzig, Germany. *Urban For. Urban Green.* **2016**, *15*, 65–74. [CrossRef]
32. Huang, S.C.L. Park user preferences for establishing a sustainable forest park in Taipei, Taiwan. *Urban For. Urban Green.* **2014**, *13*, 839–845. [CrossRef]
33. Barau, A.S. Perceptions and contributions of households towards sustainable urban green infrastructure in Malaysia. *Habitat Int.* **2015**, *47*, 285–297. [CrossRef]
34. Conedera, M.; Del Biaggio, A.; Seeland, K.; Moretti, M.; Home, R. Residents' preferences and use of urban and peri-urban green spaces in a Swiss mountainous region of the Southern Alps. *Urban For. Urban Green.* **2015**, *14*, 139–147. [CrossRef]
35. Qiu, L.; Nielsen, A.B. Are perceived sensory dimensions a reliable tool for urban green space assessment and planning? *Landsc. Res.* **2015**, *40*, 834–854. [CrossRef]
36. Rupprecht, C.D.D.; Byrne, J.A.; Ueda, H.; Lo, A.Y. "It's real, not fake like a park": Residents' perception and use of informal urban green-space in Brisbane, Australia and Sapporo, Japan. *Landsc. Urban Plan.* **2015**, *143*, 205–218. [CrossRef]

37. Larson, L.R.; Keith, S.J.; Fernandez, M.; Hallo, J.C.; Shafer, C.S.; Jennings, V. Ecosystem services and urban greenways: What's the public's perspective? *Ecosyst. Serv.* **2016**, *22*, 111–116. [CrossRef]

38. Ives, C.D.; Oke, C.; Hehir, A.; Gordon, A.; Wang, Y.; Bekessy, S.A. Capturing residents' values for urban green space: Mapping, analysis and guidance for practice. *Landsc. Urban Plan.* **2017**, *161*, 32–43. [CrossRef]

39. Korpilo, S.; Virtanen, T.; Saukkonen, T.; Lehvävirta, S. More than A to B: Understanding and managing visitor spatial behaviour in urban forests using public participation GIS. *J. Environ. Manag.* **2018**, *207*, 124–133. [CrossRef]

40. Panagopoulos, T.; Tampakis, S.; Karanikola, P.; Karipidou-Kanari, A.; Kantartzis, A. The usage and perception of pedestrian and cycling streets on residents' well-being in Kalamaria, Greece. *Land* **2018**, *7*, 100. [CrossRef]

41. Guenat, S.; Dougill, A.J.; Kunin, W.E.; Dallimer, M. Untangling the motivations of different stakeholders for urban greenspace conservation in sub-Saharan Africa. *Ecosyst. Serv.* **2019**, *36*, 100904. [CrossRef]

42. Gwedla, N.; Shackleton, C.M. Perceptions and preferences for urban trees across multiple socio-economic contexts in the Eastern Cape, South Africa. *Landsc. Urban Plan.* **2019**, *189*, 225–234. [CrossRef]

43. Peckham, S.C.; Duinker, P.N.; Ordóñez, C. Urban forest values in Canada: Views of citizens in Calgary and Halifax. *Urban For. Urban Green.* **2013**, *12*, 154–162. [CrossRef]

44. Yen, Y.; Wang, Z.; Shi, Y.; Soeung, B. An assessment of the knowledge and demand of young residents regarding the ecological services of urban green spaces in Phnom Penh, Cambodia. *Sustainability* **2016**, *8*, 523. [CrossRef]

45. Faivre, N.; Fritz, M.; Freitas, T.; de Boissezon, B.; Vandewoestijne, S. Nature-Based Solutions in the EU: Innovating with nature to address social, economic and environmental challenges. *Environ. Res.* **2017**, *159*, 509–518. [CrossRef] [PubMed]

46. Duan, J.; Wang, Y.; Fan, C.; Xia, B.; de Groot, R. Perception of urban environmental risks and the effects of urban green infrastructures (UGIs) on human well-being in four public green spaces of Guangzhou, China. *Environ. Manag.* **2018**, *62*, 500–517. [CrossRef] [PubMed]

47. Nath, T.K.; Zhe Han, S.S.; Lechner, A.M. Urban green space and well-being in Kuala Lumpur, Malaysia. *Urban For. Urban Green.* **2018**, *36*, 34–41. [CrossRef]

48. Zwierzchowska, I.; Hof, A.; Iojă, I.C.; Mueller, C.; Poniży, L.; Breuste, J.; Mizgajski, A. Multi-scale assessment of cultural ecosystem services of parks in Central European cities. *Urban For. Urban Green.* **2018**, *30*, 84–97. [CrossRef]

49. Gashu, K.; Gebre-Egziabher, T.; Wubneh, M. Local communities' perceptions and use of urban green infrastructure in two Ethiopian cities: Bahir Dar and Hawassa. *J. Environ. Plan. Manag.* **2020**, *63*, 287–316. [CrossRef]

50. Giannakis, E.; Bruggeman, A.; Poulou, D.; Zoumides, C.; Eliades, M. Linear parks along urban rivers: Perceptions of thermal comfort and climate change adaptation in Cyprus. *Sustainability* **2016**, *8*, 1023. [CrossRef]

51. Meyer, M.A.; Schulz, C. Do ecosystem services provide an added value compared to existing forest planning approaches in Central Europe? *Ecol. Soc.* **2017**, *22*, 6. [CrossRef]

52. Shwartz, A.; Turbé, A.; Simon, L.; Julliard, R. Enhancing urban biodiversity and its influence on city-dwellers: An experiment. *Biol. Conserv.* **2014**, *171*, 82–90. [CrossRef]

53. Wang, Y.; Kotze, D.J.; Vierikko, K.; Niemelä, J. What makes urban greenspace unique—Relationships between citizens' perceptions on unique urban nature, biodiversity and environmental factors. *Urban For. Urban Green.* **2019**, *42*, 1–9. [CrossRef]

54. Conway, T.M.; Yip, V. Assessing residents' reactions to urban forest disservices: A case study of a major storm event. *Landsc. Urban Plan.* **2016**, *153*, 1–10. [CrossRef]

55. Paul, S.; Nagendra, H. Factors influencing perceptions and use of urban nature: Surveys of park visitors in Delhi. *Land* **2017**, *6*, 27. [CrossRef]

56. Fernandes, C.O.; da Silva, I.M.; Teixeira, C.P.; Costa, L. Between tree lovers and tree haters. Drivers of public perception regarding street trees and its implications on the urban green infrastructure planning. *Urban For. Urban Green.* **2019**, *37*, 97–108. [CrossRef]

57. Ostoić, S.K.; van den Bosch, C.C.; Vuletić, D.; Stevanov, M.; Živojinović, I.; Mutabdžija-Bećirović, S.; Lazarević, J.; Stojanova, B.; Blagojević, D.; Stojanovska, M.; et al. Citizens' perception of and satisfaction with urban forests and green space: Results from selected Southeast European cities. *Urban For. Urban Green.* **2017**, *23*, 93–103. [CrossRef]

58. Rahnema, S.; Sedaghathoor, S.; Allahyari, M.S.; Damalas, C.A.; Bilali, H. El Preferences and emotion perceptions of ornamental plant species for green space designing among urban park users in Iran. *Urban For. Urban Green.* **2019**, *39*, 98–108. [CrossRef]

59. Tsantopoulos, G.; Varras, G.; Chiotelli, E.; Fotia, K.; Batou, M. Public perceptions and attitudes toward green infrastructure on buildings: The case of the metropolitan area of Athens, Greece. *Urban For. Urban Green.* **2018**, *34*, 181–195. [CrossRef]

60. Koo, J.C.; Park, M.S.; Youn, Y.C. Preferences of urban dwellers on urban forest recreational services in South Korea. *Urban For. Urban Green.* **2013**, *12*, 200–210. [CrossRef]

61. Zhang, H.; Chen, B.; Sun, Z.; Bao, Z. Landscape perception and recreation needs in urban green space in Fuyang, Hangzhou, China. *Urban For. Urban Green.* **2013**, *12*, 44–52. [CrossRef]

62. Baur, J.W.R.; Tynon, J.F.; Ries, P.; Rosenberger, R.S. Public attitudes about urban forest ecosystem services management: A case study in Oregon cities. *Urban For. Urban Green.* **2016**, *17*, 42–53. [CrossRef]

63. Arnberger, A.; Allex, B.; Eder, R.; Ebenberger, M.; Wanka, A.; Kolland, F.; Wallner, P.; Hutter, H.P. Elderly resident's uses of and preferences for urban green spaces during heat periods. *Urban For. Urban Green.* **2017**, *21*, 102–115. [CrossRef]

64. Derkzen, M.L.; van Teeffelen, A.J.A.; Verburg, P.H. Green infrastructure for urban climate adaptation: How do residents' views on climate impacts and green infrastructure shape adaptation preferences? *Landsc. Urban Plan.* **2017**, *157*, 106–130. [CrossRef]

65. Pietrzyk-Kaszyńska, A.; Czepkiewicz, M.; Kronenberg, J. Eliciting non-monetary values of formal and informal urban green spaces using public participation GIS. *Landsc. Urban Plan.* **2017**, *160*, 85–95. [CrossRef]

66. Ayala-Azcárraga, C.; Diaz, D.; Zambrano, L. Characteristics of urban parks and their relation to user well-being. *Landsc. Urban Plan.* **2019**, *189*, 27–35. [CrossRef]

67. Hwang, Y.H.; Yue, Z.E.J.; Ling, S.K.; Tan, H.H.V. It's ok to be wilder: Preference for natural growth in urban green spaces in a tropical city. *Urban For. Urban Green.* **2019**, *38*, 165–176. [CrossRef]

68. Ramer, H.; Nelson, K.C.; Spivak, M.; Watkins, E.; Wolfin, J.; Pulscher, M.L. Exploring park visitor perceptions of 'flowering bee lawns' in neighborhood parks in Minneapolis, MN, US. *Landsc. Urban Plan.* **2019**, *189*, 117–128. [CrossRef]

69. Caula, S.; Hvenegaard, G.T.; Marty, P. The influence of bird information, attitudes, and demographics on public preferences toward urban green spaces: The case of Montpellier, France. *Urban For. Urban Green.* **2009**, *8*, 117–128. [CrossRef]

70. Karanikola, P.; Panagopoulos, T.; Tampakis, S. Weekend visitors' views and perceptions at an urban national forest park of Cyprus during summertime. *J. Outdoor Recreat. Tour.* **2017**, *17*, 112–121. [CrossRef]

71. Menconi, M.E.; Grohmann, D. Participatory retrofitting of school playgrounds: Collaboration between children and university students to develop a vision. *Think. Sk. Creat.* **2018**, *29*, 71–86. [CrossRef]

72. Shams, I.; Barker, A. Barriers and opportunities of combining social and ecological functions of urban greenspaces—Users' and landscape professionals' perspectives. *Urban For. Urban Green.* **2019**, *39*, 67–78. [CrossRef]

73. Lamichhane, D.; Thapa, H.B. Participatory urban forestry in Nepal: Gaps and ways forward. *Urban For. Urban Green.* **2012**, *11*, 105–111. [CrossRef]

74. Keeley, M.; Koburger, A.; Dolowitz, D.P.; Medearis, D.; Nickel, D.; Shuster, W. Perspectives on the use of green infrastructure for stormwater management in cleveland and milwaukee. *Environ. Manag.* **2013**, *51*, 1093–1108. [CrossRef] [PubMed]

75. Hoyle, H.; Jorgensen, A.; Warren, P.; Dunnett, N.; Evans, K. "Not in their front yard" The opportunities and challenges of introducing perennial urban meadows: A local authority stakeholder perspective. *Urban For. Urban Green.* **2017**, *25*, 139–149. [CrossRef]

76. Furlong, C.; Phelan, K.; Dodson, J. The role of water utilities in urban greening: A case study of Melbourne, Australia. *Util. Policy* **2018**, *53*, 25–31. [CrossRef]

77. Khoshkar, S.; Balfors, B.; Wärnbäck, A. Planning for green qualities in the densification of suburban Stockholm—Opportunities and challenges. *J. Environ. Plan. Manag.* **2018**, *61*, 2613–2635. [CrossRef]

78. Onori, A.; Lavau, S.; Fletcher, T. Implementation as more than installation: A case study of the challenges in implementing green infrastructure projects in two Australian primary schools. *Urban Water J.* **2018**, *15*, 911–917. [CrossRef]

79. Girma, Y.; Terefe, H.; Pauleit, S. Urban green spaces use and management in rapidly urbanizing countries: The case of emerging towns of Oromia special zone surrounding Finfinne, Ethiopia. *Urban For. Urban Green.* **2019**, *43*, 126357. [CrossRef]

80. Molla, M.B.; Mekonnen, A.B. Understanding the local values of trees and forests: A strategy to improve the urban environment in Hawassa City, Southern Ethiopia. *Arboric. J.* **2019**, *41*, 1–13. [CrossRef]

81. Živojinović, I.; Wolfslehner, B. Perceptions of urban forestry stakeholders about climate change adaptation—A Q-method application in Serbia. *Urban For. Urban Green.* **2015**, *14*, 1079–1087. [CrossRef]

82. Lähde, E.; Di Marino, M. Multidisciplinary collaboration and understanding of green infrastructure Results from the cities of Tampere, Vantaa and Jyväskylä (Finland). *Urban For. Urban Green.* **2019**, *40*, 63–72. [CrossRef]

83. Rall, E.L.; Kabisch, N.; Hansen, R. A comparative exploration of uptake and potential application of ecosystem services in urban planning. *Ecosyst. Serv.* **2015**, *16*, 230–242. [CrossRef]

84. Di Marino, M.; Tiitu, M.; Lapintie, K.; Viinikka, A.; Kopperoinen, L. Integrating green infrastructure and ecosystem services in land use planning. Results from two Finnish case studies. *Land Use Policy* **2019**, *82*, 643–656. [CrossRef]

85. Bissonnette, J.F.; Dupras, J.; Messier, C.; Lechowicz, M.; Dagenais, D.; Paquette, A.; Jaeger, J.A.G.; Gonzalez, A. Moving forward in implementing green infrastructures: Stakeholder perceptions of opportunities and obstacles in a major North American metropolitan area. *Cities* **2018**, *81*, 61–70. [CrossRef]

86. Young, R.F. Planting the living city: Best practices in planning green infrastructure—Results from major U.S. cities. *J. Am. Plan. Assoc.* **2011**, *77*, 368–381. [CrossRef]

87. Chou, R.J.; Wu, C.T.; Huang, F.T. Fostering multi-functional urban agriculture: Experiences from the champions in a revitalized farm pond community in Taoyuan, Taiwan. *Sustainability* **2017**, *9*, 2097. [CrossRef]

88. Fors, H.; Jansson, M.; Nielsen, A.B. The impact of resident participation on urban woodland quality-a case study of Sletten, Denmark. *Forests* **2018**, *9*, 670. [CrossRef]

89. Harper, R.W.; Huff, E.S.; Bloniarz, D.V.; DeStefano, S.; Nicolson, C.R. Exploring the characteristics of successful volunteer-led urban forest tree committees in Massachusetts. *Urban For. Urban Green.* **2018**, *34*, 311–317. [CrossRef]

90. Kozová, M.; Dobšinská, Z.; Pauditšová, E.; Tomčíková, I.; Rakytová, I. Network and participatory governance in urban forestry: An assessment of examples from selected Slovakian cities. *For. Policy Econ.* **2018**, *89*, 31–41. [CrossRef]

91. Ugolini, F.; Sanesi, G.; Steidle, A.; Pearlmutter, D. Speaking "Green": A worldwide survey on collaboration among stakeholders in urban park design and management. *Forests* **2018**, *9*, 458. [CrossRef]

92. Rolf, W.; Pauleit, S.; Wiggering, H. A stakeholder approach, door opener for farmland and multifunctionality in urban green infrastructure. *Urban For. Urban Green.* **2019**, *40*, 73–83. [CrossRef]

93. Dennis, M.; James, P. Considerations in the valuation of urban green space: Accounting for user participation. *Ecosyst. Serv.* **2016**, *21*, 120–129. [CrossRef]

94. Dennis, M.; James, P. User participation in urban green commons: Exploring the links between access, voluntarism, biodiversity and well being. *Urban For. Urban Green.* **2016**, *15*, 22–31. [CrossRef]

95. Fischer, L.K.; Brinkmeyer, D.; Karle, S.J.; Cremer, K.; Huttner, E.; Seebauer, M.; Nowikow, U.; Schütze, B.; Voigt, P.; Völker, S.; et al. Biodiverse edible schools: Linking healthy food, school gardens and local urban biodiversity. *Urban For. Urban Green.* **2019**, *40*, 35–43. [CrossRef]

96. Moskell, C.; Allred, S.B. Residents' beliefs about responsibility for the stewardship of park trees and street trees in New York City. *Landsc. Urban Plan.* **2013**, *120*, 85–95. [CrossRef]

97. Gulsrud, N.M.; Hertzog, K.; Shears, I. Innovative urban forestry governance in Melbourne? Investigating "green placemaking" as a nature-based solution. *Environ. Res.* **2018**, *161*, 158–167. [CrossRef]

98. Travaline, K.; Montalto, F.; Hunold, C. Deliberative Policy Analysis and Policy-making in Urban Stormwater Management. *J. Environ. Policy Plan.* **2015**, *17*, 691–708. [CrossRef]

99. Dennis, M.; Armitage, R.P.; James, P. Appraisal of social-ecological innovation as an adaptive response by stakeholders to local conditions: Mapping stakeholder involvement in horticulture orientated green space management. *Urban For. Urban Green.* **2016**, *18*, 86–94. [CrossRef]

100. Belmeziti, A.; Cherqui, F.; Kaufmann, B. Improving the multi-functionality of urban green spaces: Relations between components of green spaces and urban services. *Sustain. Cities Soc.* **2018**, *43*, 1–10. [CrossRef]

101. Brink, E.; Wamsler, C. Collaborative governance for climate change adaptation: Mapping citizen–municipality interactions. *Environ. Policy Gov.* **2018**, *28*, 82–97. [CrossRef]

102. Cousins, J.J. Infrastructure and institutions: Stakeholder perspectives of stormwater governance in Chicago. *Cities* **2017**, *66*, 44–52. [CrossRef]

103. Mattijssen, T.J.M.; van der Jagt, A.P.N.; Buijs, A.E.; Elands, B.H.M.; Erlwein, S.; Lafortezza, R. The long-term prospects of citizens managing urban green space: From place-making to place-keeping? *Urban For. Urban Green.* **2017**, *26*, 78–84. [CrossRef]

104. Mensah, C.A.; Andres, L.; Baidoo, P.; Eshun, J.K.; Antwi, K.B. Community Participation in Urban Planning: The Case of Managing Green Spaces in Kumasi, Ghana. *Urban Forum* **2017**, *28*, 125–141. [CrossRef]

105. Liu, L.; Jensen, M.B. Green infrastructure for sustainable urban water management: Practices of five forerunner cities. *Cities* **2018**, *74*, 126–133. [CrossRef]

106. Yamaki, K. Role of social networks in urban forest management collaboration: A case study in northern Japan. *Urban For. Urban Green.* **2016**, *18*, 212–220. [CrossRef]

107. Asah, S.T.; Blahna, D.J. Motivational functionalism and urban conservation stewardship: Implications for volunteer involvement. *Conserv. Lett.* **2012**, *5*, 470–477. [CrossRef]

108. Shan, X.Z. Attitude and willingness toward participation in decision-making of urban green spaces in China. *Urban For. Urban Green.* **2012**, *11*, 211–217. [CrossRef]

109. Zare, S.; Namiranian, M.; Feghhi, J.; Fami, H.S. Factors encouraging and restricting participation in urban forestry (Case study of Tehran, Iran). *Arboric. J.* **2015**, *37*, 224–237. [CrossRef]

110. Beery, T. Engaging the private homeowner: Linking climate change and green stormwater infrastructure. *Sustainability* **2018**, *10*, 4791. [CrossRef]

111. Fors, H.; Wiström, B.; Nielsen, A.B. Personal and environmental drivers of resident participation in urban public woodland management—A longitudinal study. *Landsc. Urban Plan.* **2019**, *186*, 79–90. [CrossRef]

112. Murphy, A.; Enqvist, J.P.; Tengö, M. Place-making to transform urban social–ecological systems: Insights from the stewardship of urban lakes in Bangalore, India. *Sustain. Sci.* **2019**, *14*, 607–623. [CrossRef]

113. Green, O.O.; Shuster, W.D.; Rhea, L.K.; Garmestani, A.S.; Thurston, H.W. Identification and induction of human, social, and cultural capitals through an experimental approach to stormwater management. *Sustainability* **2012**, *4*, 1669–1682. [CrossRef]

114. Lewis, O.; Home, R.; Kizos, T. Digging for the roots of urban gardening behaviours. *Urban For. Urban Green.* **2018**, *34*, 105–113. [CrossRef]

115. Lieberherr, E.; Green, O.O. Green infrastructure through citizen stormwater management: Policy instruments, participation and engagement. *Sustainability* **2018**, *10*, 2099. [CrossRef]

116. Lim, T.C. An empirical study of spatial-temporal growth patterns of a voluntary residential green infrastructure program. *J. Environ. Plan. Manag.* **2018**, *61*, 1363–1382. [CrossRef]

117. Petrovic, N.; Simpson, T.; Orlove, B.; Dowd-Uribe, B. Environmental and social dimensions of community gardens in East Harlem. *Landsc. Urban Plan.* **2019**, *183*, 36–49. [CrossRef]

118. Romolini, M.; Ryan, R.L.; Simso, E.R.; Strauss, E.G. Visitors' attachment to urban parks in Los Angeles, CA. *Urban For. Urban Green.* **2019**, *41*, 118–126. [CrossRef]

119. Hunter, M.R. Impact of ecological disturbance on awareness of urban nature and sense of environmental stewardship in residential neighborhoods. *Landsc. Urban Plan.* **2011**, *101*, 131–138. [CrossRef]

120. Afzalan, N.; Muller, B. The Role of social media in green infrastructure planning: A case study of neighborhood participation in park siting. *J. Urban Technol.* **2014**, *21*, 67–83. [CrossRef]

121. Guerrero, P.; Møller, M.S.; Olafsson, A.S.; Snizek, B. Revealing cultural ecosystem services through instagram images: The potential of social media volunteered geographic information for urban green infrastructure planning and governance. *Urban Plan.* **2016**, *1*, 1–17. [CrossRef]

122. Shwartz, A.; Cheval, H.; Simon, L.; Julliard, R. Virtual garden computer program for use in exploring the elements of biodiversity people want in cities. *Conserv. Biol.* **2013**, *27*, 876–886. [CrossRef]

123. Møller, M.S.; Olafsson, A.S.; Vierikko, K.; Sehested, K.; Elands, B.; Buijs, A.; van den Bosch, C.K. Participation through place-based e-tools: A valuable resource for urban green infrastructure governance? *Urban For. Urban Green.* **2019**, *40*, 245–253. [CrossRef]

124. Janse, G.; Konijnendijk, C.C. Communication between science, policy and citizens in public participation in urban forestry-Experiences from the Neighbourwoods project. *Urban For. Urban Green.* **2007**, *6*, 23–40. [CrossRef]

125. Hawthorne, T.L.; Elmore, V.; Strong, A.; Bennett-Martin, P.; Finnie, J.; Parkman, J.; Harris, T.; Singh, J.; Edwards, L.; Reed, J. Mapping non-native invasive species and accessibility in an urban forest: A case study of participatory mapping and citizen science in Atlanta, Georgia. *Appl. Geogr.* **2015**, *56*, 187–198. [CrossRef]

126. Raymond, C.M.; Gottwald, S.; Kuoppa, J.; Kyttä, M. Integrating multiple elements of environmental justice into urban blue space planning using public participation geographic information systems. *Landsc. Urban Plan.* **2016**, *153*, 198–208. [CrossRef]

127. Rall, E.; Hansen, R.; Pauleit, S. The added value of public participation GIS (PPGIS) for urban green infrastructure planning. *Urban For. Urban Green.* **2019**, *40*, 264–274. [CrossRef]

128. Sun, F.; Xiang, J.; Tao, Y.; Tong, C.; Che, Y. Mapping the social values for ecosystem services in urban green spaces: Integrating a visitor-employed photography method into SolVES. *Urban For. Urban Green.* **2019**, *38*, 105–113. [CrossRef]

129. Neuenschwander, N.; Wissen Hayek, U.; Grêt-Regamey, A. Integrating an urban green space typology into procedural 3D visualization for collaborative planning. *Comput. Environ. Urban Syst.* **2014**, *48*, 99–110. [CrossRef]

130. Nilsson, K.; Åkerlund, U.; Konijnendijk, C.C.; Alekseev, A.; Caspersen, O.H.; Guldager, S.; Kuznetsov, E.; Mezenko, A.; Selikhovkin, A. Implementing urban greening aid projects—The case of St. Petersburg, Russia. *Urban For. Urban Green.* **2007**, *6*, 93–101. [CrossRef]

131. Kangas, A.; Heikkilä, J.; Malmivaara-Lämsä, M.; Löfström, I. Case Puijo-Evaluation of a participatory urban forest planning process. *For. Policy Econ.* **2014**, *45*, 13–23. [CrossRef]

132. Sturiale, L.; Scuderi, A. The evaluation of green investments in urban areas: A proposal of an eco-social-green model of the city. *Sustainability* **2018**, *10*, 4541. [CrossRef]

133. Bellamy, C.C.; van der Jagt, A.P.N.; Barbour, S.; Smith, M.; Moseley, D. A spatial framework for targeting urban planning for pollinators and people with local stakeholders: A route to healthy, blossoming communities? *Environ. Res.* **2017**, *158*, 255–268. [CrossRef] [PubMed]

134. Assmuth, T.; Hellgren, D.; Kopperoinen, L.; Paloniemi, R.; Peltonen, L. Fair blue urbanism: Demands, obstacles, opportunities and knowledge needs for just recreation beside Helsinki Metropolitan Area waters. *Int. J. Urban Sustain. Dev.* **2017**, *9*, 253–273. [CrossRef]

135. Van der Jagt, A.P.N.; Smith, M.; Ambrose-Oji, B.; Konijnendijk, C.C.; Giannico, V.; Haase, D.; Lafortezza, R.; Nastran, M.; Pintar, M.; Železnikar, Š.; et al. Co-creating urban green infrastructure connecting people and nature: A guiding framework and approach. *J. Environ. Manag.* **2019**, *233*, 757–767. [CrossRef] [PubMed]

136. Lindemann-Matthies, P.; Brieger, H. Does urban gardening increase aesthetic quality of urban areas? A case study from Germany. *Urban For. Urban Green.* **2016**, *17*, 33–41. [CrossRef]

137. Sun, N.; Hall, M. Coupling human preferences with biophysical processes: Modeling the effect of citizen attitudes on potential urban stormwater runoff. *Urban Ecosyst.* **2016**, *19*, 1433–1454. [CrossRef]

138. Jayasooriya, V.M.; Ng, A.W.M.; Muthukumaran, S.; Perera, B.J.C. Multi Criteria Decision Making in Selecting Stormwater Management Green Infrastructure for Industrial Areas Part 1: Stakeholder Preference Elicitation. *Water Resour. Manag.* **2019**, *33*, 627–639. [CrossRef]

139. Lafortezza, R.; Giannico, V. Combining high-resolution images and LiDAR data to model ecosystem services perception in compact urban systems. *Ecol. Indic.* **2019**, *96*, 87–98. [CrossRef]

140. Nordström, E.M.; Eriksson, L.O.; Öhman, K. Integrating multiple criteria decision analysis in participatory forest planning: Experience from a case study in northern Sweden. *For. Policy Econ.* **2010**, *12*, 562–574. [CrossRef]

141. Beumer, C.; Martens, P. Biodiversity in my (back)yard: Towards a framework for citizen engagement in exploring biodiversity and ecosystem services in residential gardens. *Sustain. Sci.* **2015**, *10*, 87–100. [CrossRef]

142. O'Donnell, E.C.; Lamond, J.E.; Thorne, C.R. Learning and Action Alliance framework to facilitate stakeholder collaboration and social learning in urban flood risk management. *Environ. Sci. Policy* **2018**, *80*, 1–8. [CrossRef]

143. Qiu, L.; Lindberg, S.; Nielsen, A.B. Is biodiversity attractive?—On-site perception of recreational and biodiversity values in urban green space. *Landsc. Urban Plan.* **2013**, *119*, 136–146. [CrossRef]

144. Rosol, M. Public Participation in post-fordist urban green space governance: The case of community gardens in Berlin. *Int. J. Urban Reg. Res.* **2010**, *34*, 548–563. [CrossRef] [PubMed]

145. Faehnle, M.; Bäcklund, P.; Tyrväinen, L.; Niemelä, J.; Yli-Pelkonen, V. How can residents' experiences inform planning of urban green infrastructure? Case Finland. *Landsc. Urban Plan.* **2014**, *130*, 171–183. [CrossRef]
146. Skandrani, Z.; Prévot, A.C. Beyond green-planning political orientations: Contrasted public policies and their relevance to nature perceptions in two European capitals. *Environ. Sci. Policy* **2015**, *52*, 140–149. [CrossRef]
147. Gasperi, D.; Pennisi, G.; Rizzati, N.; Magrefi, F.; Bazzocchi, G.; Mezzacapo, U.; Stefani, M.C.; Sanyé-Mengual, E.; Orsini, F.; Gianquinto, G. Towards regenerated and productive vacant areas through urban horticulture: Lessons from Bologna, Italy. *Sustainability* **2016**, *8*, 1347. [CrossRef]
148. Kronenberg, J.; Pietrzyk-Kaszyńska, A.; Zbieg, A.; Żak, B. Wasting collaboration potential: A study in urban green space governance in a post-transition country. *Environ. Sci. Policy* **2015**, *62*, 69–78. [CrossRef]
149. Shifflett, S.D.; Newcomer-Johnson, T.; Yess, T.; Jacobs, S. Interdisciplinary collaboration on green infrastructure for urban watershed management: An Ohio case study. *Water* **2019**, *11*, 738. [CrossRef]
150. Cvejić, R.; Železnikar, Š.; Nastran, M.; Rehberger, V.; Pintar, M. Urban agriculture as a tool for facilitated urban greening of sites in transition: A case study. *Urbani Izziv* **2015**, *26*, S84–S97. [CrossRef]
151. Jerome, G. Defining community-scale green infrastructure. *Landsc. Res.* **2017**, *42*, 223–229. [CrossRef]
152. Van der Jagt, A.P.N.; Szaraz, L.R.; Delshammar, T.; Cvejić, R.; Santos, A.; Goodness, J.; Buijs, A. Cultivating nature-based solutions: The governance of communal urban gardens in the European Union. *Environ. Res.* **2017**, *159*, 264–275. [CrossRef]
153. Milanovič, N.P. Green infrastructure and urban revitalisation in Central Europe: Meeting environmental and spatial challenges in the inner city of Ljubljana, Slovenia. *Urbani izziv* **2015**, *26*, 50–64.
154. Simić, I.; Stupar, A.; Djokić, V. Building the green infrastructure of Belgrade: The importance of community greening. *Sustainability* **2017**, *9*, 1183. [CrossRef]
155. Buijs, A.; Hansen, R.; Van der Jagt, S.; Ambrose-Oji, B.; Elands, B.; Lorance Rall, E.; Mattijssen, T.; Pauleit, S.; Runhaar, H.; Stahl Olafsson, A.; et al. Mosaic governance for urban green infrastructure: Upscaling active citizenship from a local government perspective. *Urban For. Urban Green.* **2019**, *40*, 53–62. [CrossRef]
156. Ugolini, F.; Massetti, L.; Sanesi, G.; Pearlmutter, D. Knowledge transfer between stakeholders in the field of urban forestry and green infrastructure: Results of a European survey. *Land Use Policy* **2015**, *49*, 365–381. [CrossRef]
157. Schifman, L.A.; Herrmann, D.L.; Shuster, W.D.; Ossola, A.; Garmestani, A.; Hopton, M.E. Situating green infrastructure in context: A framework for adaptive socio-hydrology in cities. *Water Resour. Res.* **2017**, *53*, 10139–10154. [CrossRef]
158. Frantzeskaki, N. Seven lessons for planning nature-based solutions in cities. *Environ. Sci. Policy* **2019**, *93*, 101–111. [CrossRef]

Article

The (Re)Insurance Industry's Roles in the Integration of Nature-Based Solutions for Prevention in Disaster Risk Reduction—Insights from a European Survey

Roxane Marchal [1,*,†], Guillaume Piton [2], Elena Lopez-Gunn [3], Pedro Zorrilla-Miras [3], Peter van der Keur [4], Kieran W. J. Dartée [5], Polona Pengal [6], John H. Matthews [7], Jean-Marc Tacnet [2], Nina Graveline [8], Monica A. Altamirano [9], John Joyce [10], Florentina Nanu [11], Ioana Groza [11], Karina Peña [5], Blaz Cokan [6], Sophia Burke [12] and David Moncoulon [1]

[1] Caisse Centrale de Réassurance, Department R&D, Cat & Agriculture Modelling, 157 boulevard Haussmann, 75008 Paris, France; dmoncoulon@ccr.fr

[2] University Grenoble Alpes, IRSTEA, Centre de Grenoble, UR ETNA, 2 rue de la Papeterie, 38402 Saint-Martin-d'Hères, France; guillaume.piton@irstea.fr (G.P.); jean-marc.tacnet@irstea.fr (J.-M.T.)

[3] I-Catalist, C/Borni 20, Las Rozas, 28232 Madrid, Spain; elopezgunn@icatalist.eu (E.L.-G.); pzorrilla-miras@icatalist.eu (P.Z.-M.)

[4] Geological Survey of Denmark and Greenland (GEUS), Department of Hydrology, Øster Voldgade 10, DK-1350 Copenhagen, Denmark; pke@geus.dk

[5] Field Factors, Van der Burghweg 1, 2628 CS Delft, The Netherlands; kieran@fieldfactors.com (K.W.J.D.); kp@fieldfactors.com (K.P.)

[6] Institute for Ichthyological and Ecological Research–Revivo, Šmartno 172, 2383 Šmartno pri Slovenj Gradcu, Slovenia; polona.pengal@ozivimo.si (P.P.); blaz.cokan@ozivimo.si (B.C.)

[7] Alliance for Global Water Adaptation (AGWA), 7640 NW Hood View Circle, Corvallis, OR 97330, USA; johoma@alliance4water.org

[8] University Montpellier, Institut National de la Recherche Agronomique (INRA), UMR Innovation, 2 Place Pierre Viala, 34000 Montpellier, France; nina.graveline@inra.fr

[9] Deltares, Boussinesqweg 1, 2629 HV Delft, The Netherlands; monica.altamirano@deltares.nl

[10] Stockholm International Water Institute (SIWI), Linnégatan 87A, 115 23 Stockholm, Sweden; john.joyce@siwi.org

[11] Business Development Group (BDG), 80 Plantelor Str. Sector 2, 030167 Bucharest, Romania; florentina.nanu@bdgroup.ro (F.N.); ioana.groza@bdgroup.ro (I.G.)

[12] Ambiotek CIC, Leigh-On-Sea, Essex, Southend-on-Sea SS9 1ED, UK; sophia.burke@ambiotek.com

* Correspondence: rmarchal@ccr.fr

† Formerly, Univ. Grenoble Alpes, IRSTEA, Centre de Grenoble, UR ETNA, 2 rue de la Papeterie, 38402 Saint-Martin-d'Hères, France.

Received: 1 October 2019; Accepted: 30 October 2019; Published: 6 November 2019

Abstract: Nature-based solutions (NBS) are increasingly being considered as an option to reduce societies' vulnerability to natural hazards, creating co-benefits while protecting ecosystem services in a context of changing climate patterns with more frequent and extreme weather events. The reinsurance and insurance industries are increasingly cited as sectors that can play a role to help manage risks, by improving disaster risk reduction (DRR) and loss prevention. This paper investigates how the (re)insurance industry could support the transition from a paradigm focused on ex-post responses to ex-ante risk reduction measures including NBS, in line with the Sendai Framework. This paper presents the results of a series of 61 interviews undertaken with the (re)insurance sector and related actors under the EU H2020 Nature Insurance Value Assessment and Demonstration (NAIAD) project. Methods based on a Grounded Theory approach indicate how this sector can play different roles in loss prevention, including ecosystem-based disaster risk reduction (eco-DRR). Results illustrate how the (re)insurance industry, under these roles, is gradually innovating by having a better understanding of hazards and mitigation. The findings of the study contribute to wider discussions such as the possibility of new arrangements like natural insurance schemes and evidence-based assessment of avoided damage costs from green protective measures, in Europe and beyond.

Keywords: natural hazard insurance; climate change adaptation; disaster risk reduction; nature-based solutions; nature assurance scheme; insurance value of ecosystems

1. Introduction

Following the 2015 fundamental year of key agreements (Sendai Framework, Paris-COP21, Agenda2030, Addis Ababa Action Agenda) and in the light of first publications on the projected climate change impacts on the insurance industry [1,2], one of the main current discussions considering climate change has moved towards the nascent increasing interest in nature-based solutions (NBS), and more globally towards preventive management [3,4]. Preventive measures are diverse, from planning (land-uses, building codes), mitigation (property level) and protection, (structural measures, NBS) to preparedness and recovery (soft) measures (early warning, emergency measures and insurance, etc.). NBS are defined by the European Commission to be "actions to help societies address a variety of environmental, social and economic challenges in sustainable ways, simultaneously providing human well-being and biodiversity benefits. They are actions which are inspired by, supported by or copied from nature, and which protect, manage and restore natural or modified ecosystems" [5]. There are a variety of nature-based measures that could provide solutions to growing societal challenges such as climate change, sustainable development and natural disasters [6–10].

Damage from natural disasters are expected to largely increase in Europe by 2050 as a result of climate change and increased vulnerability exposure. In mainland France, Caisse Centrale de Réassurance (CCR) has estimated that the insured property damage will rise by 50%, if no preventive measures are implemented [11]. At the urban level, the city of Copenhagen experienced a catastrophic cloudburst event in July 2011 causing almost 1 billion Euro in damage and an estimated damage in the same order of magnitude is expected for the near future in case no climate adaptation is implemented [12]. The insurance umbrella organization in Denmark (Insurance & Pension) analyzed the cloudburst data for this event in Copenhagen to support economic valuation of urban flooding [13]. There are several policies that link insurance to climate change, risk management and sustainable development objectives [14,15]. Despite the growing number of policies, there has been little research examining how effectively the industry is integrating these topics and what their new operational roles are [16–20]. Since 2016, research has mainly focused on innovative ways to link climate change adaptation (CCA) with disaster risk reduction (DRR) and NBS across different sectors with the reinsurance and insurance industries, as a key global player for risk management [17,21,22].

The core insurance business is based on risk transfer, which means shifting the financial consequences of risks from a household, a company, or a community to an insurer, who receives a premium payment in return for having to reimburse their clients after a disaster occurs. Less known by citizens, a similar risk transfer is possible for insurance firms themselves in order to secure their assets in case of an extreme disaster via reinsurance companies working on a larger, often international scale. The survey focuses only on interviews with non-life (re)insurance companies. In the paper we stressed that "insurance companies" comprise both reinsurance and insurance companies. Premiums are computed based on historical data for a similar risk and hazard. The industry relies on catastrophe loss risk modelling to understand the effects of hazards, vulnerability and damage. When disaster strikes, insurance companies play a crucial role in post-event recovery through compensations. Yet, the insurance industry is moving to earlier phases of the DRR cycle e.g., risk analysis, preparedness and early warning [23], and especially towards assessing the potential of prevention to reduce damage costs in addition to emergency relief after a disaster event. For example, with more accurate information on hazards, with early warning message to policyholders, and recently introduced research on protective measures effectiveness, decision-makers are then capable to respond in a pro-active way to climate change challenges and disaster risk management [24].

The links between disaster insurance and nature-based-solutions have now merged into the concepts of "insurance value of ecosystems" (IVE) and "nature insurance value" [25]. As well as the concept of ecosystem-based disaster risk reduction (Eco-DRR) defined as "decision-making activities [...] that recognize the role of ecosystems in supporting communities to prepare for, cope with and recover from disaster situations" [26,27]. The latter concept was used during the interviews to incorporate all alternative solutions for sustainable risk management.

The paper is framed within the NAIAD H2020 project (Nature Insurance value: Assessment and Demonstration), which aims to deepen the scientific knowledge on the insurance value of ecosystems and on NBS to reduce the human and economic costs of risks associated with water (floods and droughts). Concepts, tools and methods have been developed and tested with local stakeholders to support mainstreaming of those solutions with replicable methods at demonstration sites scales from urban to catchment across Europe. The main aim of the paper is to draw the first-hand knowledge from the sector on how the European insurance sector considers ecosystems as a potentially reliable means to reduce risk, including how the national disasters insurance systems differ from each other.

Hence, this paper is focused on understanding the role of insurance in prevention through NBS under climate change. In other words, considering NBS as a potential tool to secure affordability of insurance contracts from the perspective of the insurance industry itself. Therefore, the current state-of-knowledge about integrating NBS within the insurance industry is investigated. The different natural hazard insurance schemes in Europe have been analyzed [28–30]. A literature review was undertaken on these topics to provide the state-of-the-art and to identify knowledge gaps. This study addresses the research gaps identified, by examining the current insurance knowledge, visions and expertise through 61 semi-directed interviews in ten European countries. The study offers new insights on the role that the insurance industry could play to help address DRR and CCA goals. Survey questions were collectively designed by NAIAD partners based on the literature review focusing on hazards under climate change, prevention and NBS. Using Grounded Theory [31], the findings explore the roles of insurance in earlier phases of DRR. In doing so, this study seeks to contribute with new work and a dialogue with the sector on roles and initiatives by the insurance sector in disaster risk management through synergies with other actors involved in the field. Firstly, how the insurance industry integrates and manages risks under climate change is investigated. Secondly, the current understanding of the concept of NBS to mitigate natural hazards and the insurance value of ecosystems is examined. Finally, the different and new roles that the insurance industry could play before disasters strike are highlighted.

2. Materials and Methods

2.1. Data Acquisition

This paper presents the results of semi-directed interviews conducted over a six-month period (from August to December 2017), substantiated by a NAIAD taskforce. The interviews have been performed after a large literature review related to the linkages between the insurance sector, natural ecosystems and disaster risk reduction. The literature review was initiated based on papers and reports published by previous European projects (i.e., Placard, Enhance, Esmeralda, Eklipse, Operas, Openness, Oppla, SmarteST, CascEff, Know4DRR, Imprex. For the core reference projects.) on the topics. After this first round, the review was completed by a literature surveillance of the most recent related publications (from 2015 to 2019) using the next keywords: "European natural hazard insurance", "insurance and risk reduction", "insurance and ecosystem services", ''value of ecosystem", "insurance value of ecosystem", "nature-based solutions", "natural infrastructure", "green infrastructure", "green measures", "ecosystem-based disaster risk reduction", "nature-based solution market", "ecosystem services and insurance system", "disaster insurance and nature-based solutions". This review identified gaps in the literature in this research area, some specific articles that met the inclusion criteria are considered as core references [3,16–20,32–35]. These documents are not articles from periodicals, as well as documents on the evaluation of insured damages related to natural hazards, reports on the consequences of climate change on property insured damages or reports on the assessments of preventive measures. Some of these documents were written by insurance companies in the frame of partnerships with scientists, or as in-house research and development studies not published in periodicals. Nevertheless, the low number of publications found indicates a knowledge gap in the journals in this area.

Thus, a questionnaire was developed to address the knowledge gaps that had arisen during the literature-review by answering questions. The questionnaire gathered 58 questions divided into 8 sections (Appendix A) to better understand needs, gaps, challenges and opportunities for the European insurance sector. The scope of the NAIAD Insurance survey was to engage discussion with the insurance sector on DRR, NBS and IVE topics but not to gather monetary elements. During the interviews a tailored selection of the 58 questions was asked, selected according to the interviewee's profile. Semi-directed interviews, by nature, permit a long time frames for discussion, during which people could answer the questions included in the questionnaire. For example, the questions on how the national insurance scheme works were asked only if the functioning of the natural hazards insurance scheme had not been clarified by prior literature review. The latest document providing an up-to-date of the European insurance systems was published in October 2017 by the European Commission, during the development of the interviews [36]. As an example, Germany, the Netherlands, Slovenia and Switzerland are not covered in the document and the existing literature was not updated, therefore specific questions about the functioning of these schemes were specifically asked to the interviewees from those countries. In parallel of the interviews, 11 country fact sheets "how well do you know European natural hazard insurance systems?" have been created to visually summarize the schemes (Figure 1). In this context, the interviews also helped to validate the fact sheets with the participants and to clarify further less known insurance schemes.

Some closed questions that call for a yes-or-no answer or multiple-choice answer were asked to provide a first idea on the interviewees' knowledge or interest. It helped to skip questions or on contrary, to take a longer time for discussion. This allowed also for a quantitative analysis based on the percentage of responses related to the closed questions. The post-modern Grounded Theory for quantitative analysis has however not been applied [37].

The interviewers targeted were in order of priority, the insurance sector (31% are primary insurance and 22% are reinsurance companies), academic experts, banks, large project developers and ministries. Other stakeholders like non-governmental organizations, cities and landowners have also been contacted to provide a broader, more balanced and complementary knowledge. The recruitment of participants and interviews was carried out over the period June to December 2017, by the NAIAD taskforce for the survey and by the NAIAD demonstration sites partners (DEMOs). A snowball method was used for the selection of the interviewees, starting from the work and contacts developed by the NAIAD project partners.

The 61 interviewees were selected in ten European countries (Denmark, Germany, Spain, Slovenia, France, Italy, Romania, Netherlands, United-Kingdom and Switzerland) and at the European level to provide a wide overview on the insurance industry and to understand current practices throughout the EU (Table 1). Figure 2 indicates the panel of interviewees according to their sectors and on the right the interviewees in relation to the insurance sector.

In relation to informed consent, NAIAD's ethical and safety requirements were followed to ensure the full freedom of expression and to guarantee that data collected during the interviews was strictly confidential and only accessible to a limited number of members of the NAIAD taskforce. All participants gave their signed informed consent before participation in the interviews. The raw material of these interviews will never be published and the recordings have been deleted after transcriptions. The quotations available in the "Conclusions" part are strictly confidential and have been anonymized.

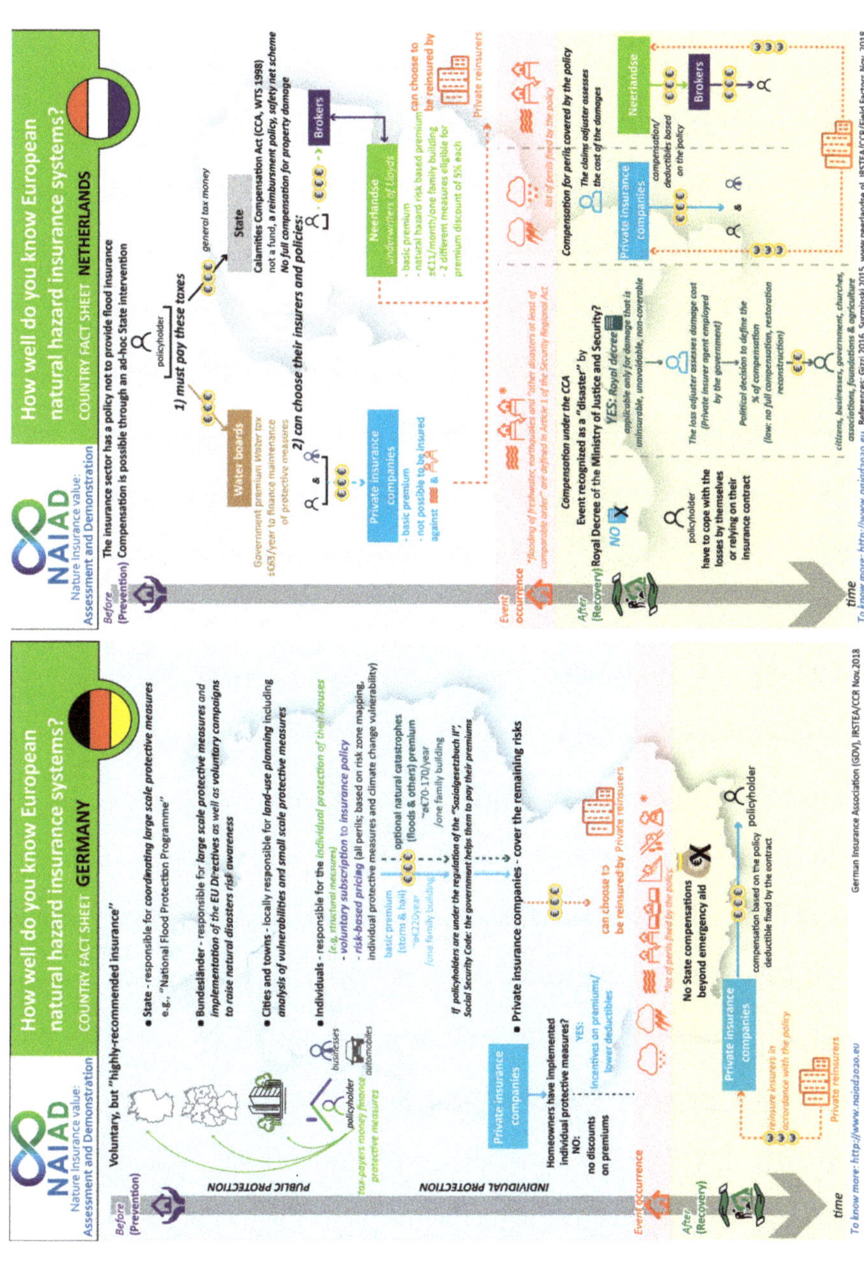

Figure 1. On the left, the country fact sheet for Germany based on the interviews and on the right, the one for the Netherlands based on the literature and up-dated with the interviews. The 11 NAIAD Country Fact Sheet are available in the Supplementary Materials. (source: Authors' own).

Table 1. Number of interviews per country (source: Authors' own).

Country	Number
Denmark	1
Germany	2
Spain	7
Slovenia	13
France	12
Italy	2
Romania	7
Netherlands	6
United-Kingdom	6
Switzerland	3
European level	2

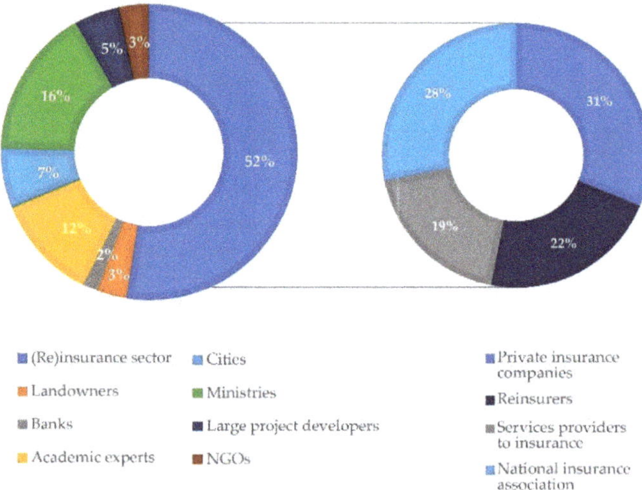

Figure 2. On the left, the panel of interviewees and on the right the interviewees in relation to the insurance sector (source: Authors' own).

2.2. Research Method

In-person interviews were recorded, transcribed verbatim and translated to English to facilitate analysis between the European team of authors. The 61 interviews lasted on average 60 minutes; the shortest interview lasted 30 minutes and the longest 95 minutes. All interviews were analyzed based a Grounded Theory approach. A sample of 25 to 40 or more interviews are required to apply this method. The Grounded Theory methodology is based on an in-depth analysis and summarizes qualitative textual data to broaden explanation of a process which is at its infancy or non-existent [31]. Grounded Theory avoids forcing theory and preconceptions into data, by translating the responses of participants through different steps (coding process and case-based memos writing). The coding process is based on six precise codification steps to be followed for each interview [38], (Table 2).

At the same time, memos are written to gather relevant information and the assumptions identified during the coding procedures. These memos were used during theoretical coding. Then, at the end of the process, codes and memos were mixed and compared to the core categories. Thus, theories and memos were then continuously compared and grounded on the data [39]. Figure 3 summarizes these different steps and their relationship to Grounded Theory.

The Table 3 below presents an example of the coding processes and the Figure 4 presents the process for axial and theoretical codings.

Table 2. Six codification steps of the Grounded Theory (source: Authors' own).

Code Name	What the Code Highlights?	Code Characteristics
"In Vivo"	Action oriented, capture behavior or processes. Can provide imaginary, symbols or metaphors.	Direct language of respondent, use "quotation", the terms used by the participants themselves.
"Process Coding"	Actual or conceptual actions—routines, rituals.	"-ings", what people do (rather than have).
"Initial Coding" or so-called "Open Coding"	Extractions of relevant concepts (labelling), deeper analysis, being open to selective coding. Those codes are grouped into similar concepts. All the codes are gathered within a codebook.	Based on the lines and paragraphs from the data. To code directly from the data and not to force data into preconception.
"Focused Coding" or so-called "Selective Coding"	Highlight major categories and themes from the data (core variable analysis) theory—memos as a basis for the formulation of the final reports.	Frequency and significant codes are needed to develop categories (larger segments of data).
"Axial Coding"	Relationship between categories and codes. Links between one data to another and comparison between the data. Those concepts generated categories which are the basis to write memos (highlight hypotheses about connections between categories, new questions, ideas, relationship between codes). Memos have to be seen as "intellectual workspace for documenting analysis"—all the memos formed memo banks.	To design diagrams of temporal/spatial and cause/effect relationships of the phenomenon (clustering codes into new or more specific codes).
"Theoretical Coding"	To identify conflict, obstacles, problems, issues. To integrate and synthesize the categories to create new theories. Consequently, all theories are identified and organized allowing for comparison between them and data, theoretical coding.	To find core categories. To condense into a few words that seem to explain, what the research is about.

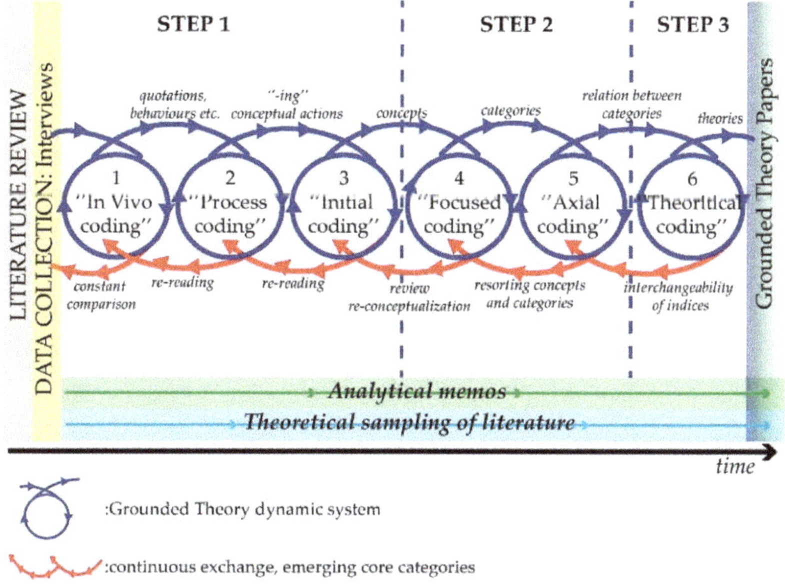

Figure 3. Grounded Theory, methodology explanation from data to results (source: Authors' own).

Table 3. Example of coding processes (source: Authors' own).

Raw Data	Initial Coding	Focused Coding	Theoretical Coding
Q: What is the (re)insurance company conceptual understanding of insurance value of ecosystem? I am little bit surprised, because I do not understand the meaning of insurance value of ecosystems (IVE), and there seems to be a little bit of confusion concerning this term. I would understand the IVE, as the value of risk reduction. Q: What are tangible examples of projects to address existing or potential climate risks your organization has pursued? Our company is addressing future risks by taking into account climate change. It is a part of our DNA, of our responsibility. There are some examples of projects to address risks: there are the risk management aspects, which is the communication to our customers, we try to highlight, anticipate, adapt and mitigate climate change risks. On the investment side, we are investing in green bonds. It requires wider demonstration of nature-based solutions (NBS) effectiveness in order to increase investments in that side. Finally, the importance of practical programs to understand risk exposure and resilience.	New concept, confusing but trying to define it Classification of examples to address risks Awareness on climate change Risk management for communicating Already investing in green bonds Practical programs on hazard and resilience	Confusion Asking for evidence-based findings on NBS Awareness of the on-going business changes	The process of making sense of evidence on prevention and construction of knowledge in partnerships (practical programs) The process of using and developing roles in both ex-post (risk management) and ex-ante (resilience)

Case-based memo

This was quite an eye-opening interview in the sense that after several performed interviews and discussions with the other interviewers, we had a concern with the IVE concept which creates confusion, positive and negative feelings. This interview was key in our consideration of the IVE concept. So, my question really is, shall we change the terminology to be more understandable and better fitted to the insurance sector? I am so glad we had this interview as the interviewee was very practical and open. The other key element, is that during the interview, when we enter in the questions of the section 7 to the end, I learn so much and the element of responses of some questions (such as in the examples above). I definitely learned that insurers' business is changing, at the end of the day the on-going reflection within the industry on the issues of climate change and loss prevention. During the interview, there was an explanation of the "roles" of the industry, right now I really need to take care with these "roles" to understand what they are and if other interviewees referred to that … Maybe the axial coding will highlight this within interviews and by comparison between them. As this would largely have impacts on the business model and on the mainstreaming of NBS. On the other hand, it is interesting that the interviewee provides examples on their daily job to integrate climate change into their catastrophe models and on loss prevention, especially on flood events. It was also interesting to gather points of view on financing aspects, notably on green bonds and elements on required regulations. So, I guess that the company of this insurer is really advanced on that topic and had a trusted experience. I tried to highlight differences between the natural hazard insurance schemes; the interviewee neutrally explains how it works. After the different interviews, for now, there is no differences in the understanding or about loss prevention between countries. It is more related to the state of advancement, dedicated research (I will not forget the fact that some of participants cannot explain due to confidentially reasons) between companies.

The memos are written after each interview and the responses to the questions are coded in a table following the Grounded Theoryframework. The constant comparison between the coding processes are exemplified in the following example of axial and theoretical coding. The conceptual memos, categories, subcategories and theorization are the findings presented in 3 of this paper.

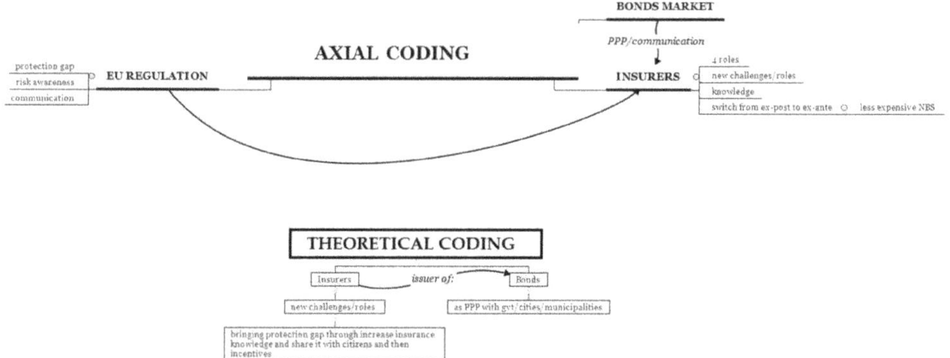

Figure 4. Grounded Theory, methodology explanation from axial coding to theoretical coding (source: Authors' own).

This theory was suitable for the present study to collect and analyze data to generate conceptual theories on the main factors that influence the insurance sector's involvement in climate change adaptation (CCA) and risk prevention through natural infrastructures for five reasons:

(1) it is based on a literature review that provide both background knowledge and interview questions to bridge the research gaps. The combination of the two allows for a deeper theoretical framework and to theorize subcategories, presented within the results, more easily;

(2) Grounded Theory was chosen because it gives room to emergence from the findings from insurance industry and stakeholders involved around the sector as research that will benefit science and new knowledge. The basis was also that this research would be of use to the sector itself and potentially be actionable or transferable to their business model;

(3) it is a useful method to analyze qualitative textual data from semi-structured interviews that target new research topics that are either in their infancy or non-existent. This is particularly suitable to the NAIAD project because research on linkages between risks-NBS-DRR-insurance industry are being developed;

(4) this method is a suitable approach for managing voluminous qualitative data; during the interviews, an hour and a half long interview generated on average 12 pages of text;

(5) the categories and the developed theories in the results are based on the data only and are not developed from researchers' hypotheses. The objective to emphasize the current elements of knowledge, feelings, main questions and understanding from the European insurance industry on CCA, loss prevention and NBS.

3. Results

This section presents the main results from the NAIAD Insurance interviews, the results are presented using the development of categories which include: namely (1) the insurance industry and climate change; (2) the insurance industry's understanding of ecosystem-based DRR and NBS; (3) the role of insurance in eco-DRR; (4) the different roles insurance could play in relation to eco-DRR. The elements within these categories and subcategories (i.e., the different roles of the insurance industry) were thus theorized. Finally, the categories are considered together as a whole in the discussion.

3.1. The (Re)Insurance Industry's Vision of Climate Change

Climate change challenges are being integrated in the insurance industry scenarios, although with different viewpoints. Climate change impacts are considered in three subcategories (1) as a challenge for affordable insurance; (2) as a challenge for risk modelling; and (3) as a potential opportunity to generate a range of innovative services. The first subcategory theorized highlights that participating insurance companies are reinforcing their internal on-going research to assess climate change impacts in combination with increasing exposure of assets and geographical concentration of wealth. The interviewees raised concerns on the fact that natural hazards are expected to become more frequent and intense with climate change (physical risks). The companies also declared, that in recent years, insured damages due to natural hazards have largely increased and are expected to further increase. Only two of the participants shared their reports, the others did not give precise figures.

Also, the interviewees had the opportunity to rank the main natural hazards threats. Meteorological hazards are the main hazards perceived as risks by the insurance industry (38%), followed by hydrological risks (31%), geophysical risks (16%), climate (14%) and pollution risks (1%).

This poses specific challenges for risk modelling based on catastrophic risk modelling expertise (second subcategory). Some companies are developing their own in-house models using historical data and other companies rely on models developed by private consulting companies. The main difference between reinsurance and insurance companies is the development of CAT models. All surveyed reinsurance companies had their own CAT models, on the contrary, most of the insurance companies were using models developed by private consulting companies.

For those companies that have their own CAT models a total of 65 percent of the participating companies declared to have started mapping and understanding the impact of climate change. In contrast, 35% of companies stated that they had not included climate change scenarios due either to the lack of data or to the lack of development of their own models. The survey indicated that not all companies are using projected future data and are assuming stationary scenarios in which underlying assumptions on boundary conditions do not change over time. During the discussions with insurance companies, the participants explained that there is a recent development of in-house CAT models by insurers to improve their knowledge on hazards, vulnerability and damage assessments. In addition, during the interviews, the uncertainties related to determining where and what new type of risks may be appearing with climate change have also be raised. The main barrier identified by the interviewees were the differences between insurers' models and scientists' models, which can limit the integration and knowledge exchange between these two communities. Indeed, it has been justified that non-life insurance have a short perspective for defining the premium each year and the investment terms are about 1–5 years. Thus, this leads to the third subcategory that improvements in sharing risk management expertise could help to address future physical risks and increase the knowledge on climate change potential impacts on institutional investments (assets side). Raising the insurance's sector own awareness and developing targeted communication for their customers is both considered as insurers' responsibilities.

3.2. The (Re)Insurance Industry—Understanding of New Concepts: "Eco-Drr", "Nature-Based Solutions (Nbs)" and "Insurance Value of Ecosystems" (IVE)

In this part, two subcategories were highlighted during the interviews: (1) knowledge and understanding of the sector on NBS and IVE; and (2) reflections on NBS/IVE integration into insurers' models. The analysis of the interviews does not emphasize differences in the state of knowledge between reinsurers and insurers. Rather, it provides an up-to-date review on the current perspectives on these topics for the sector. The interviews conducted illustrate how loss prevention and mitigation are important areas, where the insurance industry is engaging increasingly. Examples provided included: research projects, early warning, and encouraging build back better approaches (resilience), etc. A total of 44% of the respondents claimed to know the eco-DRR concept as related to prevention, even if most did not have a very precise definition, 28% of them gave a comprehensive definition of the concept, while the remaining 28% of the respondents did not know the term. Therefore, given this level of knowledge and awareness on eco-DRR, it is safe to conclude that the insurance industry is currently in a process of increasing its own awareness on NBS. Meanwhile, in relation to the concept "insurance value of ecosystems", 70% of the interviewed expressed unfamiliarity. The remaining 30% had a precise definition and understanding. However, what emerged as one of the main results of the survey is the reluctance of the insurance sector to adopt the IVE concept. Survey participants indicated that IVE concept was a confusing and/or inaccurate concept for the insurance industry. The general feeling was that the concept narrows the value of ecosystems to the avoided damage and would therefore miss the multiple co-benefits of ecosystems. Some interviewees preferred the "resilience dividend of nature" concept instead of the IVE, in order to integrate both avoided damage and the co-benefits of protecting nature. Nevertheless, interviewees had knowledge on the ecosystem's role in risk reduction. Thus, 38% of the participants have knowledge on eco-DRR but are still challenged by its integration into their models. Companies that had their own models commented on the possibility to use their models to measure the effectiveness of NBS in terms of avoided damages due to the reduction of hazard (e.g. extension or water heights). The same participants underlined that currently the assessment of conventional civil engineering measures such as dykes and dams is still at the research project stage. The participants commented how the ability to estimate damages or preventive measures is a nascent field for the sector.

The natural hazards from ecosystems increasing the risks (obstruction of hydraulic structures related to woody debris or vegetation growth) have been raised by 19% of the respondents. The vulnerability of NBS to climate change was considered as a potential reason that could further restrict the successful integration of NBS into insurance business models (by 5% of them). For 39% of the participants, the lack of knowledge and lack of practical demonstrations on the NBS role on hazard reduction were also posed as reasons for the limited action on the topic. When asked which knowledge and tools would be required to integrate NBS into catastrophe loss risk models, 37% of people interviewed commented on the need for more exchanges with the scientific community; 41% of them asked for new studies and data related to NBS; 11% of them considered enough knowledge was already available; and 11% of them had limited suggestions on what would be needed. These responses are linked to the non-life insurers' requirements to have the best understanding on their portfolio exposure. The integration of preventive measures into catastrophe loss risk models is at its infancy for most of the companies interviewed.

3.3. (Eco)-Disaster Risk Reduction and the (re)Insurance Industry

Based on the experiences of the participants, a number of challenges exist to get insurers to participate in eco-DRR. These challenges fall into four subcategories: (1) affordable coverage for weather-related hazards; (2) increasing interest in DRR; (3) changing business models; (4) changing people's perception on the industry.

The assumption was posed that the insurance industry is raising concerns on its liability side (risk providers), with the increasing extreme weather-related hazards and society's vulnerability in a changing climate. The affordability of insurance contracts was raised. The findings indicate that risk management expertise is being developed by the industry to better understand often poorly known natural hazards such as coastal flooding, urban surface runoff or land subsidence.

The interviews confirmed that natural disaster prevention is something relatively new for the (re)insurance industry with a different nuances. The interviews revealed that insurers have taken time to consider natural hazard prevention as an area where insurance could intervene. Indeed, in comparison to fire/thefts prevention, the insurance industry's involvement in natural hazards prevention and mitigation is still small.

The awareness of climate change as a threat to biodiversity and to engineered DRR measures was also raised. The findings highlight that the sector considers NBS assessment even more challenging than mitigation measures. The latter are located at the property level, as compared to NBS which are generally operational at the collective level. It poses specific challenges for integration into the business models (see Section 3.4.1). So, 67% of the interviewees have no specific strategies to incorporate NBS in catastrophe loss risk models. 15% of them have it under consideration and 18% of them have indicated to have the methodology to do so. Results thus show that eco-DRR is considered as an opportunity for the insurance sector to support preparedness for the anticipated impacts from climate change. Yet, the interviews revealed the need for evidence-based knowledge and a high level of confidence on the effectiveness of these proposed green measures and NBS on risk reduction and on the co-benefit generation. The evidence and guidance elements could be provided by both engineering offices which are expert in the area, or by in-house experts in risk assessments. The subcategory "changing business model" was theorized since it is a central theme in the survey. It frequently appears in the data and almost all the participants could relate to the concept. This theme made the participants raise a very important point: a paradigm change from response-based measures towards ex-ante prevention, before a disaster occurs. In addition, for market actors, it could also lead to a strengthening of their position in the market as early movers, by improving the industry's image and the communication of its role in disasters. The insurance industry is currently facing a citizens' risk awareness gap. To bridge this gap, the industry is now promoting better risk communication, climate change, sustainable development and prevention.

3.4. Roles of the (Re)Insurance Industry in Supporting (Eco)-Disaster Risk Reduction

The theorization of the categories, subcategories and of memos offered new insights in the roles of the insurance industry as a driver for natural hazards resilience based on eco-DRR. Two of the four roles: as risk transfer provider and as investors, are the core of the insurers' business with emerging uses of these roles related to eco-DRR. The two other roles and their uses: innovators and partners roles have also been codified and thus analysed.

3.4.1. (Re)Insurers as Providers

The interviewed reinsurers and insurers both stressed the importance of offering affordable insurance coverage, and of reducing the costs from natural disasters. Therefore, the main concern for the interviewees was the affordability of insurance coverage as a societal issue, as mentioned in Section 3.1.

The interviews have confirmed the differences in the national natural hazards insurance schemes to underwrite risks, as well as the current integration of mitigation measures into premiums calculation. For example, for market-based insurance schemes (UK or Germany), the assessment of collective or individual preventive measures are integrated within the risk-based premiums. On the contrary, for mandatory insurance schemes (France or Spain), reinsurance and insurance companies are directly linked to prevention objectives (i.e., Barnier Funds in France), therefore these companies have performed or have required studies to assess the effectiveness of prevention since these companies participate in the financing of those measures. Interviewees from countries with a current low penetration rate on property insurance (Romania, or Slovenia) expressed an interest to have better knowledge on natural hazards and preventive measures to help bridge the current protection gap.

The interviewed insurers also presented the new services developed for their customers on early warning (automatic SMS alerts, etc.) and post-event recovery process (recommendations for reconstruction, etc.). The interviews investigated the current understanding of the insurance industry on the potential integration of scientific evidences of natural infrastructures into insurance business model. The participants stated some key opportunities or barriers to this integration. The first point to be considered is that most of the protective measures are not insured, governments or local authorities insure the structures themselves. Natural areas (forests or marsh areas) owned by landowners are insured for civil liability or for business interruption and are not considered as NBS for their role in reducing risks. This underlines the current lack of knowledge with regards to NBS effectiveness when assessing natural hazard protection measures and highlights the current predominance of grey measures.

Interviewees recognized the need for limiting ecosystems degradation, with some of them highlighting the potential of NBS for insuring degradation or damage as a hazard reduction potential. Regarding the multiple benefits of NBS, the insurance industry expressed interest to link co-benefits' impacts on health insurance.

3.4.2. (Re)insurers as Investors

Results demonstrate that for European insurers to act as institutional investors, further evidence is required on the disaster risk reduction benefits. Viable business models—e.g. like the natural assurance scheme developed under the NAIAD project—where these benefits are demonstrated could play a significant role to increase the financing for the development of NBS. Behaviors towards sustainable and responsible investments are emerging to decrease risks and to diversify investments through the development of loss prevention which could eventually include NBS. Some of the interviewees exemplified the avoided damages from natural infrastructures implementation with a positive return on investments from their in-house research. Participating companies mentioned the current offer of financial products such as green or cat bonds to help finance conservation, restoration, the implementation, maintenance and monitoring of NBS projects. It has been argued that NBS projects, at different scales, could help to support a diversification of risks and to help develop a larger portfolio of return on investments. Some barriers to the integration of NBS into insurance products lie in the current difficulties to assess who would pay and benefit from these measures. Liability of failure and investments that could be beneficial for other insurance companies are some of the other barriers mentioned. Another issue raised during the interviews was the lack of appropriate NBS labels which could impact investors' confidence. Although some industry entities have integrated investments in low-carbon projects and improved environmental performance in their building assets and their governance, in those cases, it has however been highlighted by some participants that the Solvency II Directive [40] penalizes long-term investments as a contradiction to the on-going debates for longer term investments.

3.4.3. (Re)Insurers as Innovators

In light of the current CCA and DRR objectives, the insurance industry has been engaged in the identification of innovative communication tools to increase the awareness of future risks.

The interviews indicated the importance of the insurer´s role to act also as a prevention-advisor to help limit disaster impacts. Some of the participants exemplified this with the role insurance can play for example to foster a more sustainable land-use planning, better building codes, encourage the building of resilient protective infrastructures and on build back-better (BBB) measures. In addition, some of the interviewees highlighted that this sharing of experience could take the form of e.g., performing risk exposure analysis for an area, or a cost-benefit analysis of natural (green/blue) infrastructures by assessing the insurance losses reduction for various scenarios.

This main theme can be theorized on the role that the insurance industry can offer rooted on its wide expertise in risk management for society at large, and on also advice to specific actors based on the requirement of their natural hazards insurance scheme. Some of the insurers, during interview discussions, compared the different schemes to identify the best practices from different schemes. However, some of the respondents questioned whether this was/should be the role of the industry.

Innovation is strongly linked to policy and regulations. The interviews revealed the potential for example for an NBS Floods Directive based on the same principle as the Floods Directive (risk maps integrating protective measures, greener risk financing). Findings revealed the need for clear European and national roadmaps for sustainable insurance. This survey also demonstrates that insurers consider the scale of the European Union as a key area to scale-up socio-economic resilience to natural hazards through policy/regulatory frameworks, and to help foster a greener economy. This underlines the current trigger point on the worldwide natural disaster insurance protection gap [41,42] and the European objective to bridge the gap in the EU through innovative insurance products [5].

3.4.4. (Re)Insurers as Partners

Increasingly, insurers see themselves as institutional partners to help build up societies' resilience after and before a disaster strikes. This can be exemplified in the engagement in ex-ante risk management, developed by the industry for societal and general interest benefits. There was a clear momentum during the interviews were interviewees raised who to foster further collaboration with: the European Commission; national governments; local authorities; scientists; citizens; for loss prevention, CCA, damage estimation or data sharing.

The participants stated that these exchanges are needed to increase knowledge on natural disasters in terms of costs and vulnerability. The willingness for collaboration between reinsurance, insurance companies and scientists on the damage reduction effectiveness of NBS has been pointed out. In that context, the participants emphasize that scientists could provide relevant expertise on hazard assessment while the insurers undertake the damage part. To analyze NBS and their co-benefits on economic and human levels, the catchment scale has been suggested as a good operational level for analysis and implementation. The provided justifications were: it is a relevant scale for land use planning (risk prevention plans); for the maintenance of those measures (water sewage systems, dikes, dams, water retention basin etc.); interesting scale for the possibility of public-private partnerships. In that context, the examples of the France natural hazards insurance scheme, the mechanisms linking compensation (insurance) to prevention (Barnier Fund) and data exchange (ONRN) were raised as good practice. The scheme thus brings together insurers, French State and local governments to manage risks and to foster preventive measures (structural or non-structural ones). Concerning other European schemes, it is the privileged relationship with the insured people that has been raised by insurers, as a support to participate and to define their roles as partners in prevention. Finally, participants from voluntary insurance schemes argued on the need to work with institutional banks. The latest are both change drivers for DRR and CCA and are often linked to insurance contracts. For instance, in some countries it is mandatory to subscribe to natural hazards insurance coverage to obtain mortgages (UK, Sweden).

4. Discussion

Nature-based solutions are seen as a new paradigm for loss prevention to deal with natural hazards in a changing climate [10,27,43–50]. Yet, despite positive debates and discussion, little research has been carried out and published in journals on how exactly the insurance industry can facilitate the operationalization of eco-DRR [16,19,21,51,52]. Natural hazards management requires an integrated approach towards risk management, involving different stakeholders to share the burden of preventive measures. In this study, it was found that the eco-DRR concept, i.e., DRR using NBS, is gaining importance within the insurance industry. In the context of the NAIAD Insurance Survey, the NBS and IVE concepts have been discussed with the insurance industry. The IVE concept seems to create a level of confusion in the insurance industry and for other actors, which could lead to misunderstandings. On the one side, this is due to the lack of knowledge in the industry about these concepts which could result in different understanding from one actor to another, without an agreed definition. On the other side, terminologies already exist for the same or very similar concepts, e.g., resilience dividend of ecosystems, natural infrastructure for green infrastructure, green infrastructures. In this paper, we recommend to use the term *"natural assurance value"*, as a metaphor for the capacity of ecosystems to reduce risk, as compared to the natural insurance value which reflects the possibility of insuring NBS for their risk reduction function. It is important to find ways to mainstream and support NBS implementation, while avoiding the multiplication of terms related to one concept through policy changes.

This survey indicated that the insurance industry is concerned on how climate change could have an impact on natural hazards and ecosystem services. Findings indicate the need for the sector to improve its knowledge on NBS functioning for its potential to help generate avoided damages and co-benefits. This findings is coherent with other studies [18,48,52,53]. Nevertheless, during the interviews it was complex to get access, or to discuss the monetary side. The costs of insured damages are shared annually, or after large damaging events, and mostly by large reinsurance companies. Generally speaking insurance companies do not communicate on the damages, this is more the role of reinsurers or of national federation of insurers. Also, the preventive measures assessment performed by the sector have been developed only twice. One explanation is that during the interview's performance, there was a terminological confusion on the topics. In addition, some interviewees declared not to share information on their on-going research on NBS effectiveness and its integration into their business models due to business and competitiveness confidentiality requirements. The main findings from the survey are that the insurance industry is moving towards more ex-ante actions, in addition to actions in compensation. As found, the sector requires clear demonstration of DRR benefits for each and every dividend. This is aligned with the *"triple dividend of resilience"*; the first dividend is the compensation of losses and avoiding long-term negative impacts of disasters; the second dividend is the simulation of economic activity through risk reduction; the third dividend is the importance of socio-eco-environmental co-benefits [33]. Currently, the benefits of disaster risk management investments are underestimated, and the common opinion is that investing in disaster resilience will be beneficial only once a disaster strikes [54,55]. The triple dividend of resilience concept highlights good reasons for the insurance industry to move towards embracing natural hazards protections. Table 4 synthetizes the triple dividend for DRR, including its counterpart for the insurance sector and how to integrate the different roles of the insurance industry.

Table 4. Triple dividend of risk reduction of insurances, adapted from Surminski, S. & Tanner, T. [56].

What Type of Good Reasons?	Benefits	Role of Insurances
Compensation of losses, reduction of negative impacts from disasters	Availability and affordability of reinsurance and insurance contracts	Providers
Investment in prevention and mitigation	Decrease (non)insured losses, i.e., secured portfolio	Investors
Money saved can be reinvested	Investments dedicated to innovation, research and development, policies and regulation	Innovators
Economic security and co-benefits	Portfolio diversification, valuing co-benefits	Partners/Providers

According to the findings of the NAIAD Insurance Survey, a range of potential uses and new roles of the insurance sector to face natural hazards.

In the case of the interviews presented here, the involvement of the insurance sector is also influenced and dependent by the peculiarities of national insurance schemes, with different relations and visions for the role of the sector on prevention and eco-DRR [36]. The investor role is also particularly important, even if the industry (along with other potential impact investors) requires a strong evidence-based and set of examples that bring a significant return on investment [57]. Their investments can help mainstream the use of NBS in prevention. As shown through the NAIAD Insurance Survey, approval towards financing natural infrastructures can increase the number of NBS projects at catchment scale that support the long-term affordability of insurance. The EU taxonomy work was not yet published when the interviews were performed [58]. This work on sustainable finance has a large impact on what is defined as green investments (EU green bond standards) and green adaptation measures by companies. The report also mentions the importance of better information/data sharing. Under the Non-Financial Reporting Directive [59], banks and insurance companies are targeted to share "non-financial" climate-related information. The findings of the survey present also a theorization on two roles for the industry: innovators and partners. The uses of these two roles can help the sector to develop further advancements in catastrophe risk models and the generation of knowledge and evidence to assess the avoided damage attributable to NBS. This is coherent with the literature review, notably the engagement with different actors through (public-private) partnership mechanisms [60–63]. Some of the participating companies stressed the need to work within research projects. Proactively, the sector can engage governments at different levels and create partnerships to help guide large scale risk management activities, increasing the territorial protection and resilience [34,35].

The findings indicate that the Grounded Theory approach has the potential to reveal a rich and deep understanding of the (re)insurance industry experiences, including the ways that the sector interacts with climate change, their understanding of the NBS and IVE concepts, the specific challenges posed by the eco-DRR, and the different uses of their roles in supporting loss prevention.

Grounded Theory is a systematic method, to deal with interviews, collected self-reported knowledge, opinions and perceptions [36]. The subject of the survey was relatively new and complex for the insurance sector. Another benefit of Grounded Theory is the continuous process of gathering information without preconceptions, helping to provide new information. For example, if literature becomes available then comparison between the current survey results and further investigations will also be possible in the future. Furthermore, if other surveys are conducted, information could be added in the same manner using Grounded Theory to capture specific elements resulting from the interviews. Alternatively, if future updates are made to the underlying NAIAD Insurance Survey questionnaire, the set of interviewees could be extended to capture and to compare these additional results. An important area for further research would be to conduct the survey with other public and private actors involved in risk management (i.e., institutional banks, water agencies and utilities, local authorities, etc.), to better understand their perspectives, actions, and constraints. In short, how to understand better the overall roles in risk management of those actors to mainstream NBS remains a critical issue for further research. Lastly, while the Grounded Theory here is focused on questions raised by the NAIAD project, the theory could be applied to other topics, scales and actors.

5. Conclusions

This paper provides an up-to-date explanation of how the reinsurance and insurance sector is one of the important actors for addressing natural disasters and loss prevention. The use of Grounded Theory, applied to 61 interviews, facilitated the identification of key elements on the challenges related to eco-DRR, IVE and NBS for the insurance industry. The sector is moving from ex-post to ex-ante preventative roles. The main conclusions from the analysis developed are explained below:

(i)   The findings highlight the current understandings of eco-DRR, NBS and IVE concepts by the insurance industry. The IVE concept can be misunderstood (insurance that can be applied to a natural system in case it is damaged by a natural hazard vs. the reduction of risks that a natural system can provide (and hence, act as a natural insurance). To avoid such a misunderstanding, we propose the use of "natural assurance value" to define the reduction of risks that natural systems can produce.

(ii) European natural hazards insurance systems have different considerations of risk management and adaptation strategies, as well as considering ecosystems as reliable means to reduce risks. The generalization of the use of NBS as tools to reduce the impact of natural hazards may take a different time frames in European countries, depending on the existing policies, political frameworks and insurance schemes. Moreover, there is no universal solution for business integration, due to Member States' specificities, different insurance penetration rates and natural hazards/protective measures portfolios. To achieve the European Commission's objective to mainstream the use of NBS to mitigate the effect of natural hazard, differences among insurance schemes, opportunities and barriers, presented in this paper, need to be considered.

(iii) Regarding the roles of the insurance industry (insurance providers, investors, innovators and partners in resilience to natural hazards), it is important to highlight that the industry has still great potential to deliver knowledge, evidence tools and technologies throughout transdisciplinary partnerships. The insurance industry plays a strong role in risk perception (sensibilization, alerts, prevention, incentives, etc.) which could lead to drawing up intelligent coverage concepts and new products to mitigate natural disasters losses. Therefore, the insurance industry has a big role in the mainstreaming of NBS as useful and valuable tools for DRR. The insurance industry can as well have a role to develop new investment strategies (e.g., sustainable investments with green/blue/cat bonds, etc.); to create innovative green ways; to mainstream new environmental partnerships; and to reduce its own environmental impacts. Clarifying, defining and integrating ecosystem co(st)-benefits into the core insurance business is an emerging way to tackle the impact of natural disasters in the future. Thus, NBS may be especially suited for coping with specific hazards and possibly are more scalable.

(iv) The insurance industry requires a clear demonstrations of the DRR benefits as well as viable business models to invest in natural infrastructures. The industry needs clear data that supports that NBS are in good combination: economically viable, financially attractive for investment and measures of the positive effects of NBS on risk reduction. The growing role of the insurance industry in DRR recognizes these new challenges to be addressed. The insurance industry can play a key leading part on the necessary research to help generate new supportive facts and knowledge, always if possible, working in partnerships with the scientific community.

(v) The regulatory frameworks are crucial for the functioning of the insurance systems and, therefore the European Union has an opportunity to stimulate new roles from the insurance industry. As raised during the interviews, the insurance industry is asking for clear European and national roadmaps leading to sustainable insurance systems that include NBS as valuable DRR tools. There are as well divergent opinions on the idea of an EU top down leading role.

(vi) The information gathered during the interviews is aligned with the large work performed by the EU on sustainable finance. Contributions from projects like NAIAD are useful to fill the knowledge gaps since it includes scaling-up experimentations, testing and demonstrating the applicability and the limits of NBS for eco-DRR.

Supplementary Materials: The following are available online at http://www.mdpi.com/2071-1050/11/22/6212/s1.

Author Contributions: Writing, R.M.; Review: E.L.-G., G.P., P.v.d.K., M.A.A., P.P., K.W.J.D., P.Z.-M. and J.H.M. Interviews performance: R.M., N.G., P.Z.-M., M.A.A., J.H.M., J.J., F.N., I.G., K.W.J.D., K.P., P.P., B.C., P.v.d.K., S.B., E.L.-G. Conceptualization, R.M., G.P., E.L.-G., P.Z.-M. and J.-M.T.; Data curation, R.M., E.L.-G., P.Z.-M., K.W.J.D., P.P., N.G., M.A.A., J.J., F.N., I.G., K.P., B.C. and S.B.; Formal analysis, R.M.; Investigation, R.M., G.P., E.L.-G., P.Z.-M., P.v.d.K., K.W.J.D., P.P., J.-M.T., N.G., M.A.A., J.J., F.N., I.G., K.P., B.C. and S.B.; Methodology, R.M. and J.J.; Project administration, E.L.-G.; Supervision, G.P., E.L.-G., J.-M.T. and D.M.; Validation, G.P., E.L.-G., J.-M.T. and D.M.; Visualization, R.M. and G.P.; Writing—original draft, R.M. and G.P.; Writing—review & editing, R.M., G.P., E.L.-G., P.Z.-M., P.v.d.K., K.W.J.D., P.P. and J.H.M.

Funding: This research was funded by the European commission under the H2020 research and innovation programme under grant agreement No 730497.

Acknowledgments: Special gratitude goes to the many international and European reinsurers and insurers that participated in the NAIAD survey. We would like to kindly thankful: Oliver Hauner (German Insurance Association), Flood Re, Juan Duan and James Orr (Bank of England, Prudential Regulation Authority), Marjorie Breyton (Unipol), Pietro Negri (ANIA), Finance Norge, Marlene Lisa Eriksen (Forsikring & Pension), Munich Re, Michael Szoenyi (Zurich Insurance Company, Flood Resilience Program), Fabrice Renaud (United Nations University, now at the School of Interdisciplinary Studies, University of Glasgow), Swenja Surminski (London School of Economics), Justin Butler and Daniel Cook (Ambiental), Daniel Stander (RMS), Emma Raven (JBA), Anaïs Blasco (WBCSD), Markus Feltscher and Reto Stockmann (GVG Switzerland), De Vereende, Dutch Water Authorities, PGGM, Delft University of Technology, Verbond van Verzekeraars, Ignacio Machetti (AgroSeguros), Artur Reñé (Guy Carpenter), Javier Sánchez Martínez (Subdirección General de Gestión Integrada del Dominio Público Hidráulico, Dirección General del Agua, Ministerio para la transición ecológica), Inés Isabel La Moneda (ENESA: Entidad Estatal de Seguros Agrarios, Ministerio de Agricultura, Pesca y Alimentacion), Christopher Bunzl (Mutua de Propietarios), Javier Calatrava (UPCT), Miguel Llorente (IGME), Ministry of Waters and Forests Romania, Ministry of Agriculture and Rural Development Romania, National Administration Apele Romane, National Union of Insurance and Reinsurance Companies Romania (UNSAR), XPRIMM (Specialized media company for global re/insurance industry focused on Central and South-Eastern European regions, as well as CIS region states), Association Natural Park Vacaresti Romania, Expert Romania, Fisheries Research Institute of Slovenia, Institute for spatial policies, Ludovic Faytre (IAU-IDF), XLB Assurances, Syndicat des Etangs de la Dombe, Groupe MMA, AXA, FFA, Romain Marteau (Generali, now at Groupe Covéa), Roland Nussbaum (Mission Risques Naturels), Laurent Montador (Caisse Centrale de Réassurance). The authors would like to thank all the NAIAD partners that have contributed to the design of the survey and that have performed the interviews. Special gratitude goes to Mia Ebelsoft (Finans Norge) for the external review and her advice. The authors would also like to acknowledge the financial support of the European Union under the H2020 research and innovation programme under the grant agreement No 730497.

Conflicts of Interest: The authors declare no conflict of interest.

Appendix A Full Questionnaire

Section 1 Risk-assessment related with natural hazards

1. Have you made a risk assessment of your networks and key assets/Demo/Region? Which assets are more at risk?
2. What are the main natural hazards you perceive as risks for your country/Demo/Asset?
3. Are Climate-driven extreme events one of them?
4. If so, how: do you integrate climate change scenarios in risk evaluation? How is the climate change included in your models? How do you generate climate scenarios?
5. What are the physical parameters describing risk at your country/Demo/Asset? (e.g., wind speed, water level, etc.)
6. How do you cope with (often imperfect) information and knowledge? How are uncertainties taken into account?
7. Is the information/data available to everyone? Where people can find data and other information on natural risks? Can this information help people to have a better understanding of the risk?
8. Who are your peers? How do you seek advices?
9. Who carries the risks of extreme events? You or your (public) client?

Section 2 The role of the protective measures against natural hazards

1. How do you manage the risks (list mentioned above) under your responsibility? We would value concrete examples.

    (a) Taking prevention measures: risk mitigation- risk reduction measures, structural measures, if so, which?
    (b) Preparedness and Crisis Management protocols?
    (c) Insurance/ transferring the residual risk or pooling/compensation.

2. How risks are managed along the project cycle of infrastructure investments (where water management is relevant) and how these risks are shared with the private sector? Give an indication of:

    (a) Which sector—public or private—is responsible for managing these risks and therefore willing to invest in DRR measures?
    (b) Gaps within the investment system that need to be solved with DRR and system understanding expertise.

3. Who maintains and operates protective (Mitigation/adaptation) structures? How are they funded? Who carries the risk of failure?
4. Are insurance contracts available for those protective structures, or are they considered in insurance contracts?

Section 3 Development of a new European legal framework for implementation of the IVE

1. Does the current European legal and market framework permit to harmonize insurance premiums at the EU level? (Develop and discuss your opinion)
2. What are the main differences between legal and market frameworks in each country?
3. Is it possible to create a regional risk-pooling scheme at the EU level? (Develop and discuss your opinion)
4. How are climate risks assigned, for different economic sectors, by European insurance companies? Are these policies under any degree of reconsideration?

Section 4 (re)insurance framework descriptions

1. Can the insurer refuse to cover a property because of the future expected natural hazard (in your country)? Does everybody have access to insurance? If insurance coverage is not possible, what is done in this case?
2. How do you economically assess the indirect costs/effects of a disaster? (e.g., road/airport closure consequences; business interruption)
3. Which insurance options do you have for your key assets and per type of risk?
4. Could you give us an impression of the comparative size of insurance premiums you pay for natural hazard risks, versus other risks as safety of employers, fires, burglary, etc.?

   How are these insurance policies aggregated? (Which subcategories are there and their relative size)

5. How do you evaluate your own insurance system?

Section 5 Economic issues related to insurances

1. What is the basic knowledge for premiums calculation and how are they calculated?
2. Could your insurance company create financial incentives (e.g., premium reduction) for policyholders who have implemented prevention/mitigation measures and what is provided to whom they have not implemented that before any (flooding) natural hazard?
3. Are risk reduction measures considered to modify premiums? If yes, which type of measures are they? Please rank these different resilient measures based on which criterion?

Section 6 Risk awareness risk perception: current knowledge and gaps

1. Does the insurance industry have mechanisms in practice or under development to deal with variance in perception of risk?
2. How far do the member state's risk culture aspects have to be considered by your insurance company in your proposal of risk mitigation measures? What are the main risk culture differences in the EU countries?
3. How would you most likely communicate with your customer about flood/droughts/other climate related hazard resilience? (e.g., standards for reconstruction of their houses, etc.)
4. How has your risk perception changed with the climate change? (brokers: about the perception of their clients)

Section 7 Linkages between the insurance sector and NBS, ecosystems

1. What are the new challenges for your insurance company to cope with climate-related risks?
2. What are tangible examples of projects to address existing or potential climate risks your organization has pursued?
3. What is the insurance company conceptual understanding of ecosystem-based disaster risk reduction?
4. What is the insurance company conceptual understanding of insurance value of ecosystem?
5. What is your current knowledge of positive or negative effects of ecosystems?
6. Do you recognize any particular ecosystem as a defense or resilient measure to face natural hazards?
7. Is there any insurance contract for risks related to ecosystem effects? (e.g., wildfires, woody debris jamming on bridges, etc.)
8. Do you have a specific strategy to incorporate nature-based solution or insurance value of ecosystems in risk assessment strategies? (e.g., key green policies, monetary choices for integration into green solutions portfolios)

9.   How your insurance company can develop models for calculating scenarios of risk reduction for different types of ecosystem services?

10. Which knowledge and tools would be required? Do you know of tools being developed? (Please list them)

11. How your insurance companies can/could develop methods to implement the concept of insurance value of ecosystem?

12. To what extent has the concept of nature-based solution been integrated in your policy framework? Are there pioneer examples around the world?

13. How could you do partnerships with insurance companies/clients? To design insurance schemes and NBS standards that strengthen and incentive DRR?

14. Tell us more about how could that work? (e.g., who pays what, required regulatory changes or economic instruments?) Have you seen such example elsewhere, in EU or outside Europe?

15. What do you know and think of Resilience/Green/Cat Bonds? Could they work to finance NBS? Under which conditions? Which other similar products in development you know of?

16. In addition to risk reduction role of ecosystems, can the environment preservation be an additional motivation for your insurance company? (e.g., for your image? if so, which options would have your preference?)

17. Do you ever finance risk reduction measures? As "insurance" company? And/or as institutional investor? How do you separate these two roles?

    (A)   If you do invest as insurance company, would you be willing to finance Nature-Based Solutions for DRR? Why?

    (B)   And as Institutional Investor? If you are investing, could you give us examples? If not, specify the reasons and/or bottlenecks?

18. Do you ever reimburse risk reduction investments made by your clients?

In example, health insurance policies sometimes reimburse the costs of a gym subscription. Are there plans to do so in this field?

19. What can be the minimal amount of your (re)insurance companies' investments, that will be required to fund those measures? How much of loss can your (re)insurance companies avert, with investments in what grey, green/blue, hybrid measures? (calculating benefit/investment ratios)

20. How do your models take into account the green/NBS options for DRR taken by your client? Why?

21. Concerning the barriers in hindering the uptake of nature-based solutions in practice into insurance policies: How important are the following barriers? Could you please give concrete examples per category?

Natural (physical barriers)
Social (social validation)
Human (cultural)
Financial
Legal

Section 8 Funding DRR measures and ecosystem services

1.   What are the most important sources of Funding for Disaster Risk Reduction Measures? Taxes, Tariffs or Transfer? (Explain more in detail) and who collects them and is responsible for their budgeting? And procurement?

2.   Are there any additional important sources of funds (e.g., Structural funds) and/or strategic partners for the implementation, funding and/or financing of DRR measures?—and for NBS specifically?

3. Are Public-Private Partnerships contracts being used for the procurement of DRR measures? If so, could you give us examples?

4. Which innovative Green/Climate finance (Urban and Rural NBS, Watershed conservation, Natural Foreshores) Funding strategies, financing mechanisms and innovative business models have been applied in your country?

   We would value concrete examples and/or contact persons.

5. Which economic and regulatory instruments do you consider key in incentivizing private sector and society to opt for NBS, for Resilience and Water Security?

6. Do you think a new model would be needed to consider NBS or should we only adapt the current scheme?

7. Do you think a preliminary marketing survey would be needed to check that a market really exists? (a real demand from possible customers)

8. Do you think using NBS could provide a competitive advantage for your company?

9. Do you think those new business models could be created by companies or should it be imposed by European regulations?

References

1. Prudential Regulation Authority. *The Impact of Climate Change on the UK Insurance Sector*; Prudential Regulation Authority, 2015; p. 87. Available online: https://www.bankofengland.co.uk/-/media/boe/files/prudential-regulation/publication/impact-of-climate-change-on-the-uk-insurance-sector.pdf (accessed on 23 March 2018).

2. Surminski, S.; Bouwer, L.M.; Linnerooth-Bayer, J. How insurance can support climate resilience. *Nat. Clim. Chang.* **2016**, *6*, 333. [CrossRef]

3. Faivre, N.; Fritz, M.; Freitas, T.; de Boissezon, B.; Vandewoestijne, S. Nature-based solutions in the EU: Innovating with nature to address social, economic and environmental challenges. *Environ. Res.* **2017**, *159*, 509–518. [CrossRef] [PubMed]

4. Cohen-Shacham, E.; Walters, G.; Janzen, C.; Maginnis, S. *Nature-Based Solutions to Address Global Societal Challenges*; IUCN: Gland, Switzerland, 2016.

5. European Commission. *Towards an EU Research and Innovation policy agenda for Nature-Based Solutions & Re-Naturing Cities*; European Commission: Brussels, Belgium, 2015.

6. Strosser, P.; Delacámara, G.; Hanus, H.; Williams, H.; Jaritt, N. *A Guide to Support the Selection, Design and Implementation of Natural Water Retention Measures in Europe*; Capturing the Multiple Benefits of Nature-Based Solutions; Publications Office of the European Union: Brussels, Belgium, 2015.

7. Daigneault, A.; Brown, P.; Gawith, D. Dredging versus hedging: Comparing hard infrastructure to ecosystem-based adaptation to flooding. *Ecol. Econ.* **2016**, *122*, 25–35. [CrossRef]

8. United Nations World Water Assessment Programme. *Nature-Based Solutions for Water*; The United Nations World Water Development: New York, NY, USA, 2018.

9. Somarakis, G.; Stagakis, S.; Chrysoulakis, N. *Thinknature Nature-Based Solutions Handbook*; Grant Agreement, ThinkNature Project: Crete, Greece, 2019.

10. Raymond, C.M.; Berry, P.; Breil, M.; Nita, M.R.; Kabisch, N.; de Bel, M.; Enzi, V.; Frantzeskaki, N.; Geneletti, D.; Cardinaletti, M.; et al. *An Impact Evaluation Framework to Support Planning and Evaluation of Nature-based Solutions Projects*; Report Prepared by the EKLIPSE Expert Working Group on Nature-based Solutions to Promote Climate Resilience in Urban Areas; Centre for Ecology and Hydrology: Cavite, UK, 2017.

11. CCR. *Département Analyses et Modélisation Cat Conséquences du Changement Climatique sur le Coût des Catastrophes Naturelles en France à Horizon 2050*; CCR: Paris, France, 2018.

12. COWI; Deloitte; Rambøll; KU LIFE; DHI; GRAS. *Copenhagen Climate Adaptation Plan*; COWI: Copenhagen, Denmark, 2011.

13. COWI. *Insurance and Pension Estimation of Unit Costs for Flooding*; COWI: Copenhagen, Denmark, 2014.

14. United Nations Framework Convention on Climate Change. *Paris Agreement—COP21*; UNFCCC: Rio de Janeiro, Brazil; New York, NY, USA, 2015.

15. United Nations Office for Disaster Risk Reduction. *Sendai Framework for Disaster Risk Reduction 2015–2030*; UNDRR: Geneva, Switzerland, 2015.

16. Narayan, S.; Beck, M.W.; Reguero, B.G.; Losada, I.J.; van Wesenbeeck, B.; Pontee, N.; Sanchirico, J.N.; Ingram, J.C.; Lange, G.M.; Burks-Copes, K.A. The Effectiveness, Costs and Coastal Protection Benefits of Natural and Nature-Based Defences. *PLoS ONE* **2016**, *11*, e0154735. [CrossRef] [PubMed]

17. Narayan, S.; Beck, M.W.; Wilson, P.; Thomas, C.J.; Guerrero, A.; Shepard, C.C.; Reguero, B.G.; Franco, G.; Ingram, J.C.; Trespalacios, D. The Value of Coastal Wetlands for Flood Damage Reduction in the Northeastern USA. *Sci. Rep.* **2017**, *7*, 9463. [CrossRef] [PubMed]

18. WBCSD. *Incentives for Natural Infrastructure, Review of Existing Policies, Incentives and Barriers Related to Permitting, Finance and Insurance of Natural Infrastructure*; WBCSD: Geneva, Switzerland, 2017.

19. Dow; Swiss Re; Shell; Unilever; The Nature Conservancy. The Case for Green Infrastructure—Joint-Industry White Paper. 2013. Available online: https://www.nature.org/content/dam/tnc/nature/en/documents/the-case-for-green-infrastructure.pdf (accessed on 20 April 2018).

20. The Nature Conservancy; Dow. Working Together to Value Nature, 2016 Summary Report. 2017. Available online: https://www.nature.org/content/dam/tnc/nature/en/documents/2016-Dow-collaboration-report.pdf (accessed on 20 April 2018).

21. Beck, M.W.; Franco, G.; Guerrero, A.; Ingram, C.J.; Narayan, S.; Reguero, B.G.; Shepard, C.; Thomas, C.; Trespalacios, D.; Wilson, P. *Coastal Wetlands and Flood Damage Reduction: Using Risk Industry-based Models to Assess Natural Defenses in the Northeastern USA*; Lloyd's Tercentenary Research Foundation: London, UK, 2016.

22. Colgan, C.S.; Beck, M.W.; Narayan, S. Financing Natural Infrastructure for Coastal Flood Damage Reduction; Lloyd's Tercentenary Research Foundation. 2017. Available online: https://conservationgateway.org/ConservationPractices/Marine/crr/library/Documents/FinancingNaturalInfrastructureReport.pdf (accessed on 6 May 2018).

23. Comes, M.; Dubbert, M.; Garschagen, M.; und Yew Jin, L.; Grunewald, L.; Lanzendörfer, M.; Mucke, P.; Neuschäfer, O.; Pott, S.; Post, J.; et al. *World Risk Report 2016*; Bündnis Entwicklung Hilft, United Nations University: Tokyo, Japan, 2016, ISBN 978-3-946785-02-6.

24. Frantzeskaki, N.; Mcphearson, T.; Collier, M.J.; Kendal, D.; Bulkeley, H.; Dumitru, A.; Walsh, C.; Noble, K.; Van Wyk, E.; Ordonez, C.; et al. Nature-Based Solutions for Urban Climate Change Adaptation: Linking Science, Policy, and Practice Communities for Evidence-Based Decision-Making. *BioScience* **2019**, *69*, 455–466. [CrossRef]

25. Denjean, B.; Altamirano, M.A.; Graveline, N.; Giordano, R.; van der Keur, P.; Moncoulon, D.; Weinberg, J.; Manez Costa, M.; Kozinc, A.; Mullingan, M.; et al. Natural Assurance Scheme: A level playing field framework for Green-Grey infrastructure development. *Environ. Res.* **2017**, *159*, 24–38. [CrossRef] [PubMed]

26. Lo, V. Synthesis Report on Experiences with Ecosystem-Based Approaches to Climate Change Adaptation and Disaster Risk Reduction; Technical Series, Secretariat of the Convention on Biological Diversity, 2016, No. 85. Available online: https://www.cbd.int/doc/publications/cbd-ts-85-en.pdf (accessed on 26 June 2018).

27. Renaud, F.; Sudmeier-Rieux, K.; Estrella, M. *The Role of Ecosystems in Disaster Risk Reduction*, 3rd ed.; Renaud, F., Sudmeier-Rieux, K., Estrella, M., Eds.; United Nations University Press: Tokyo, Japan, 2013, ISBN 978-92-808-1221-3.

28. Schwarze, R.; Schwindt, M.; Weck-Hannemann, H.; Raschky, P.; Zahn, F.; Wagner, G.G. Natural hazard insurance in Europe: Tailored responses to climate change are needed. *Environ. Policy Gov.* **2011**, *21*, 14–30. [CrossRef]

29. Gizzi, F.T.; Potenza, M.R.; Zotta, C. The Insurance Market of Natural Hazards for Residential Properties in Italy. *Open J. Earthq. Res.* **2016**, *5*, 35–61. [CrossRef]

30. Marchal, R.; Piton, G.; Tacnet, J.-M.; Zorrilla-Miras, P.; Lopez Gunn, E.; Moncoulon, D.; Altamirano, M.; Matthews, J.; Joyce, J.; Nanu, F.; et al. *European Survey on Insurance Systems and Natural Assurance Scheme (NAS)*; No. 730497; Grant Agreement, NAIAD Project: Valladolid, Spain, 2017.

31. Denzin, N.K.; Lincoln, Y.S. (Eds.) *The Sage Handbook of Qualitative Research*, 5th ed.; SAGE Publications, Inc.: Thousand Oaks, CA, USA, 2017, ISBN 9781483349800.

32. World Wildlife Fund (WWF). Natural and Nature-Based Flood Management: A Green Guide. 2017. Available online: https://www.worldwildlife.org/publications/natural-and-nature-based-flood-management-a-green-guide (accessed on 22 July 2017).

33. Weingärtner, L.; Simonet, C.; Caravani, A.; Overseas Development Institute (ODI). Disaster Risk Insurance and the Triple Dividend of Resilience. 2017. Available online: https://www.odi.org/publications/10926-disaster-risk-insurance-and-triple-dividend-resilience (accessed on 15 October 2017).

34. Tipper, A.W.; Francis, A.; Green Alliance & National Trust. Natural Infrastructure Schemes in Practice How to Create New Markets for Ecosystem Services from Land. 2017. Available online: https://www.green-alliance.org.uk/resources/Natural_infrastructure_schemes_in_practice.pdf (accessed on 28 October 2017).

35. Francis, A.; Brown, S.A.; Tipper, W.A.; Wheeler, N.; Green Alliance & National Trust. New Markets for Land and Nature How Natural Infrastructure Schemes Could Pay for a Better Environment. 2016. Available online: https://www.green-alliance.org.uk/resources/New_markets_for_land_and_nature.pdf (accessed on 28 October 2017).

36. Xavier, L.D.; Matilda, P.; Audrey, B.; Paul, H.; De Ruiter, M.; Laars, D.R.; Kuik, O. *Insurance of Weather and Climate-Related Disaster Risk: Inventory and Analysis of Mechanisms to Support Damage Prevention in the EU*; European Commission: Brussels, Belgium, 2017, ISBN 978-92-79-73173-0.

37. Annells, M. Grounded Theory Method: Philosophical Perspectives, Paradigm of Inquiry, and Postmodernism. *Qual. Health Res.* **1996**, *6*, 379–393. [CrossRef]

38. Saldana, J. (Ed.) *The Coding Manual for Qualitative Researches*, 3rd ed.; SAGE Publications, Inc.: Thousand Oaks, CA, USA; Arizon State University: Tempe, AZ, USA, 2015, ISBN 978-1847875495.

39. Richards, L. (Ed.) *Handling Qualitative Data, a Practical Guide*; SAGE Publications, Inc.: Thousand Oaks, CA, USA; RMIT University: Melbourne, Australia, 2014, ISBN 978-1446276051.

40. European Parliament. Directive 2009/138/EC of the European Parliament and of the Council of 25 November 2009 on the Taking-up and Pursuit of the Business of Insurance and Reinsurance (Solvency II). 2009. Available online: http://data.europa.eu/eli/dir/2009/138/oj (accessed on 5 March 2018).

41. Baur, E.; Schnarwler, R.; Prystav, A.; Sundermann, L.; Swiss, R. Closing the Protection Gap, Disaster Risk Financing: Smart Solutions for the Public Sector. 2018. Available online: https://www.swissre.com/Library/closing-the-protection-gap-disaster-risk-financing.html (accessed on 13 September 2018).

42. University of Cambridge Institute for Sustainability Leadership (CISL). *Investing for Resilience*; Cambridge Institute for Sustainability Leadership: Cambridge, UK, 2016. Available online: https://www.cisl.cam.ac.uk/resources/sustainable-finance-publications/investing-for-resilience (accessed on 5 November 2019).

43. Ozment, S.; Ellison, G.; Jongman, B. *Nature-based Solutions for Disaster Risk Management*; World Bank Group: Washington, DC, USA, 2019. Available online: http://documents.worldbank.org/curated/en/253401551126252092/Booklet (accessed on 4 February 2019).

44. Kabisch, N.; Frantzeskaki, N.; Pauleit, S.; Naumann, S.; Davis, M.; Artmann, M.; Haase, D.; Knapp, S.; Korn, H.; Stadler, J.; et al. Nature-based solutions to climate change mitigation and adaptation in urban areas: Perspectives on indicators, knowledge gaps, barriers, and opportunities for action. *Ecol. Soc.* **2016**, *21*, 39. [CrossRef]

45. Maes, J.; Jacobs, S. Nature-Based Solutions for Europe's Sustainable Development. *Conserv. Lett.* **2015**. [CrossRef]

46. Munang, R.; Thiaw, I.; Alverson, K.; Liu, J.; Han, Z. The role of ecosystem services in climate change adaptation and disaster risk reduction. *Curr. Opin. Environ. Sustain. Sci.* **2013**, *5*, 47–52. [CrossRef]

47. Sudmeier-Rieux, K. Ecosystem Approach to Disaster Risk Reduction—Basic Concepts and Recommendations to Governments, with a Special Focus on Europe; European and Mediterranean Major Hazards Agreement (EUR-OPA). 2013. Available online: https://www.coe.int/t/dg4/majorhazards/ressources/pub/Ecosystem-DRR_en.pdf (accessed on 11 October 2017).

48. Renaud, F.G.; Sudmeier-Rieux, K.; Estrella, M.; Nehren, U. (Eds.) Ecosystem-Based Disaster Risk Reduction and Adaptation in Practice. In *Advances in Natural and Technological Hazards Research*; Springer International Publishing: Berlin/Heidelberg, Germany, 2016, ISBN 978-3-319-43631-9.

49. O'Mara, C.; Carmilani, S. We Can't Control Where Storms Hit, but We Can Harness Nature to Better Protect Us. The Hill 2018. Available online: https://thehill.com/opinion/energy-environment/396738-we-cant-control-where-storms-hit-but-we-can-harness-nature-to (accessed on 28 January 2018).

50. Browder, G.; Ozment, S.; Rehberger Bescos, I.; Gartner, T.; Lange, G.-M. *Integrating Green and Gray Creating Next Generation Infrastructure*; World Bank Group: Washington, DC, USA, 2019. Available online: https://www.worldbank.org/en/news/feature/2019/03/21/green-and-gray (accessed on 3 March 2019).

51. Cohn, C. Mangroves, Coral Reefs Could Cut Flood Insurance Premiums: Lloyd's. Reuters 2017. Available online: https://www.reuters.com/article/us-insurance-climatechange-research/mangroves-coral-reefs-could-cut-flood-insurance-premiums-lloyds-idUSKBN1932NF?platform=hootsuite (accessed on 25 August 2017).
52. Murti, R.; Buyck, C. (Eds.) *Safe Havens, Protected Areas for Disaster Risk Reduction and Climate Change Adaptation*; IUCN: Gland, Switzerland, 2014.
53. Caisse Centrale de Réassurance. Retour Sur Les Inondations de Janvier et Février 2018, Modélisation des Dommages et Evaluation Des Actions de Prévention. 2018. Available online: https://www.ccr.fr/documents/23509/29230/CCR+Etude+inondations+janvier+f%C3%A9vrier+2018_vf.pdf/ (accessed on 12 October 2018).
54. Tanner, T.; Surminski, S.; Wilkinson, E.; Reid, R.; Rentschler, J.; Rajput, S. Global Facility for Disaster Reduction and Recovery (GFDRR) at the World Bank and Overseas Development Institute (ODI) The Triple Dividend of Resilience: Realising Development Goals through the Multiple Benefits of Disaster Risk Management. 2015. Available online: https://www.odi.org/publications/9599-triple-dividend-resilience-development-goals-multiple-benefits-disaster-risk-management (accessed on 5 November 2018).
55. Tanner, T.; Rentschler, J.; Surminski, S.; Mitchell, T.; Wilkinson, E.; Peters, K.; Overseas Development Institute (ODI); International Bank for Reconstruction and Development; International Development Association. Unlocking the "triple dividend" of Resilience: Why Investing in Disaster Risk Management Pays off. 2015. Available online: https://www.odi.org/tripledividend (accessed on 5 November 2018).
56. Surminski, S.; Tanner, T. (Eds.) Realising the 'Triple Dividend of Resilience': A New Business Case for Disaster Risk Management. In *Climate Risk Management, Policy and Governance*; Springer International Publishing: Berlin/Heidelberg, Germany, 2016, ISBN 978-3-319-40693-0.
57. Sudmeier-Rieux, K.; Masundire, H.; Rietbergen, S. (Eds.) Ecosystems, Livelihoods and Disasters: An Integrated Approach to Disaster Risk Management. In *Ecosystem Management Series*; IUCN: Gland, Switzerland, 2006, ISBN 978-2-8317-0928-4.
58. EU Technical Expert Group on Sustainable Finance. Financing a Sustainable European Economy. 2019. Available online: https://ec.europa.eu/info/sites/info/files/180131-sustainable-finance-final-report_en.pdf (accessed on 12 August 2019).
59. European Commission. Guidelines on Non-Financial Reporting: Supplement on Reporting Climate-Related Information. 2019. Available online: https://ec.europa.eu/finance/docs/policy/190618-climate-related-information-reporting-guidelines_en.pdf (accessed on 12 August 2019).
60. Crick, F.; Jenkins, K.; Surminski, S. Strengthening insurance partnerships in the face of climate change—Insights from an agent-based model of flood insurance in the UK. *Sci. Total Environ.* **2018**, *636*, 192–204. [CrossRef] [PubMed]
61. Mysiak, J.; Pérez-Blanco, C.D. Partnerships for disaster risk insurance in the EU. *Nat. Hazards Earth Syst. Sci.* **2016**, *16*, 2403–2419. [CrossRef]
62. UNISDR. Disaster Risk Reduction Private Sector Partnership. 2016. Available online: https://www.unisdr.org/files/42926_090315wcdrrpspepublicationfinalonli.pdf (accessed on 26 October 2018).
63. Walker, G.R.; Mason, M.S.; Crompton, R.P.; Musulin, R.T. Application of insurance modelling tools to climate change adaptation decision-making relating to the built environment. *Struct. Infrastruct. Eng.* **2016**, *12*, 450–462. [CrossRef]

Article

Planning Nature-Based Solutions for Urban Flood Reduction and Thermal Comfort Enhancement

Abdul Naser Majidi [1,2,*], **Zoran Vojinovic** [1,2,3,4], **Alida Alves** [1], **Sutat Weesakul** [2,5], **Arlex Sanchez** [1], **Floris Boogaard** [6,7] **and Jeroen Kluck** [7,8]

[1]   IHE Delft, Institute for Water Education, Westvest 7, 2611 AX Delft, The Netherlands; z.vojinovic@un-ihe.org (Z.V.); a.alves@un-ihe.org (A.A.); a.sanchez@un-ihe.org (A.S.)
[2]   School of Engineering and Technology, Asian Institute of Technology, P.O. Box 4 Klong Luang, Pathumthani 12120, Thailand; sutat@ait.asia
[3]   College of Engineering, Mathematics and Physics, University of Exeter, Exeter EX4 4QF, UK
[4]   Faculty of Civil Engineering, Department of Hydraulic and Environmental Engineering, University of Belgrade, Bulevar kralja Aleksandra 73, 11000 Beograd, Serbia
[5]   Hydro and Agro Informatics Institute (HAII), Khwaeng Thanon Phaya Thai, Khet Ratchathewi, Krung Thep Maha Nakhon, Bangkok 10400, Thailand
[6]   NoorderRuimte, Centre of Applied Research and Innovation on Area Development, Hanze University of Applied Sciences, Zernikeplein 7, P.O. Box 3037, 9701 DA Groningen, The Netherlands; floris@noorderruimte.nl
[7]   TAUW, Zekeringstraat 43g, P.O. Box 20748, 1001 NS Amsterdam, The Netherlands; j.kluck@hva.nl
[8]   Amsterdam University of Applied Sciences, Urban Technology, Weesperzijde 190, 1097 DZ Amsterdam, The Netherlands
*   Correspondence: st117444@alumni.ait.asia

Received: 20 September 2019; Accepted: 29 October 2019; Published: 12 November 2019

Abstract: As a consequence of climate change and urbanization, many cities will have to deal with more flooding and extreme heat stress. This paper presents a framework to maximize the effectiveness of Nature-Based Solutions (NBS) for flood risk reduction and thermal comfort enhancement. The framework involves an assessment of hazards with the use of models and field measurements. It also detects suitable implementation sites for NBS and quantifies their effectiveness for thermal comfort enhancement and flood risk reduction. The framework was applied in a densely urbanized study area, for which different small-scale urban NBS and their potential locations for implementation were assessed. The overall results show that the most effective performance in terms of flood mitigation and thermal comfort enhancement is likely achieved by applying a range of different measures at different locations. Therefore, the work presented here shows the potential of the framework to achieve an effective combination of measures and their locations, which was demonstrated on the case of the Sukhumvit area in Bangkok (Thailand). This can be particularly suitable for assessing and planning flood mitigation measures in combination with heat stress reduction.

Keywords: nature-based solutions; flood risk reduction; thermal comfort enhancement; microclimatic simulations; Mike Urban; ENVI-met

---

1. Introduction

There is an increasing awareness that the interplay between the supposed effects of climate change and global warming combined with rapid and uncontrolled urbanization can lead to serious challenges to urban water managers and city planners. Since the vegetation coverage and green areas are decreasing significantly, the imperviousness rate in different urban areas is increasing [1,2]. As a consequence, many cities will deal with less reliable drainage systems, more flooding, extreme heat stress and droughts.

Urban flooding leads to numerous direct and indirect impacts, and it causes high social, environmental and financial damages to the more vulnerable and less prepared cities around the world [3,4]. Heat stress is considered to be a phenomenon induced by a hot atmospheric condition, implying an increase of heat-related mortality and morbidity [5]. The increase in urban air temperature can affect human well-being and energy consumption due to the need for extra cooling. Therefore, in order to reduce the vulnerability and increase the capacity of cities to cope with these effects, a paradigm shift in the management and design of urban water systems is required. In this new management approach, multifunctional designs will deal with multiple hazards, meaning that the hazards are not targeted individually, and therefore that urban water systems will now deal with multiple challenges at the same time [6,7].

In urban drainage management, similar structures are named differently. For instance, green infrastructure (GI), best management practices (BMP), low impact development (LID), water sensitive urban design (WSUD), sustainable drainage systems (SuDS), ecosystem-based adaptation (EbA) and nature-based solutions (NBS), are broadly used. In this work we use the term nature-based solutions. NBS is a relatively new concept; it comprises solutions inspired and supported by nature, which provide multiple benefits and help society to adapt to climate change [8,9].

Several studies have investigated the effectiveness of NBS for different aspects in urban areas. In particular, several works focus upon the application of NBS to achieve multiple benefits at the same time, e.g., [10–13]. Furthermore, several studies have assessed the effectiveness of NBS measures separately on either urban flooding, e.g., [2,14–16], or on heat stress, e.g., [17–22]. However, urban flooding and heat stress frequently occur simultaneously, and NBS have the potential to be effective in mitigating both. To the best of our knowledge and the literature review to date, there are no reports of an integrated (combined) assessment using quantitative effectiveness of NBS measures for both flooding and heat stress mitigation. Moreover, a limited number of works studied the effectiveness of these measures in a highly dense urban area of a tropical environment [23–25].

Further to the above, there is a need to undertake more studies towards the understanding of interactions between different hazards and how they shape vulnerabilities and risk. This can help city planners to make better decisions, and to gain a better understanding of how urban development on one site can influence vulnerability on the other site, and how both of them can happen within the same urban area. Such understanding can lead towards a better identification of locations where mitigation strategies can contribute more efficiently in achieving sustainable urban conditions.

The present work provides a contribution in this direction, and it presents a novel framework for the selection and location of NBS to achieve both urban flood reduction and heat stress mitigation. This framework was applied in a case study area in Bangkok (Thailand) through the application of a macro scale model for urban flooding, and a micro scale microclimatic model for human thermal comfort.

2. Methods and Application

2.1. Framework Description

The framework presented in this study includes three parts. The first part evaluates hazards to identify flood and heat stress problems areas. The second part includes site selection and a feasibility analysis of Nature-Based Solutions (NBS) measures, and the third part applies numerical modeling to quantitatively assess the effectiveness of these measures. Figure 1 illustrates different steps of the proposed framework.

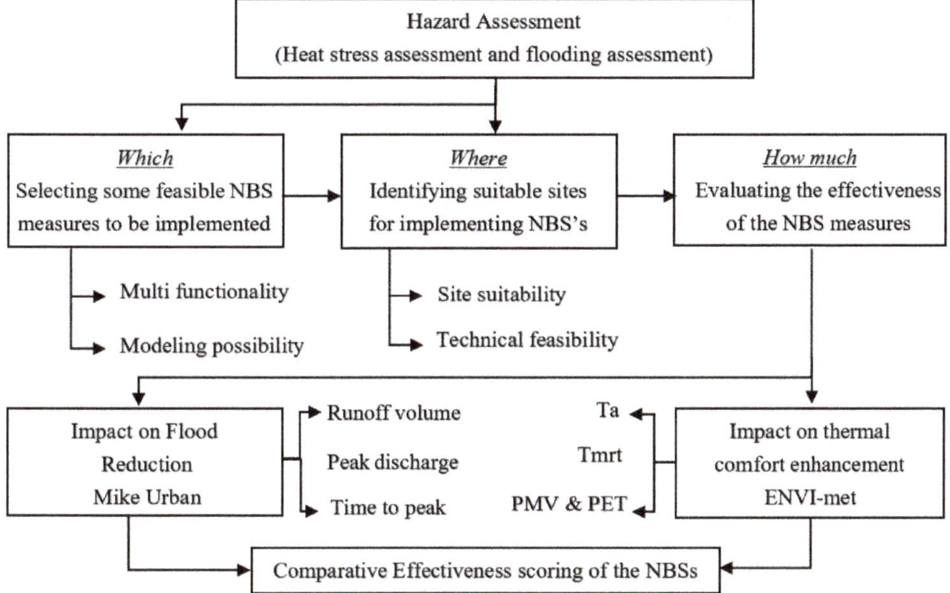

Figure 1. Methodological framework.

The results obtained from hazard assessment are used in the selection of small-scale urban NBS ("Which") and the identification of suitable areas for their implementation ("Where"). Best management practices (BMP) Sitting and ArcGIS tools are used to identify the locations suitable for the implementation of the selected NBS measures. The last part ("How much") evaluates the effectiveness of the selected measures for flood reduction and thermal comfort enhancement. Hydrodynamic and microclimatic modeling are used to assess flood and heat stress mitigation effectiveness. The use of micro-climatic models is often used for urban assessments [26,27].

Six parameters are considered for the effectiveness assessment, three related to flood mitigation and three oriented to evaluate heat stress mitigation. Finally, according to the results obtained for these parameters, scores are given to the measures for their comparison. The following sections introduce the case study area framework.

2.2. Study Area

The framework was applied in the Sukhumvit area, located in central Bangkok (Figure 2), which is a highly dense urban area of approximately 23 Km^2. According to The World Bank [28], urban growth in Thailand is mostly situated in the Bangkok urban area, which is among the twenty largest cities in Asia in terms of population, approaching 10 million people. Bangkok is a growing city located in a tropical area, and it is facing many extreme climatic conditions, which will also be intensified in the future as a result of climate change [29].

The annual rainfall in the city is 1651 mm, which mainly takes place in the wet season (from May to October). According to Rehan et al. [30] the number of annual rainy days was increased from 90 to 110 days in the last 30 years. Additionally, the study done by Sheikh [31] shows that the average mean temperature in Bangkok was increased by 0.6 °C between 1985 and 2014. Arifwidodo and Tanaka [32] studied the effect of the Urban Heat Island (UHI) in Bangkok, showing that there is a mean maximum of 5 °C UHI intensity (UHII) between semi urban and urban areas, and a mean maximum UHII of 2 °C between dry and in rainy seasons. In addition, there is a maximum night time UHII of 7 °C during January in Bangkok [29,32].

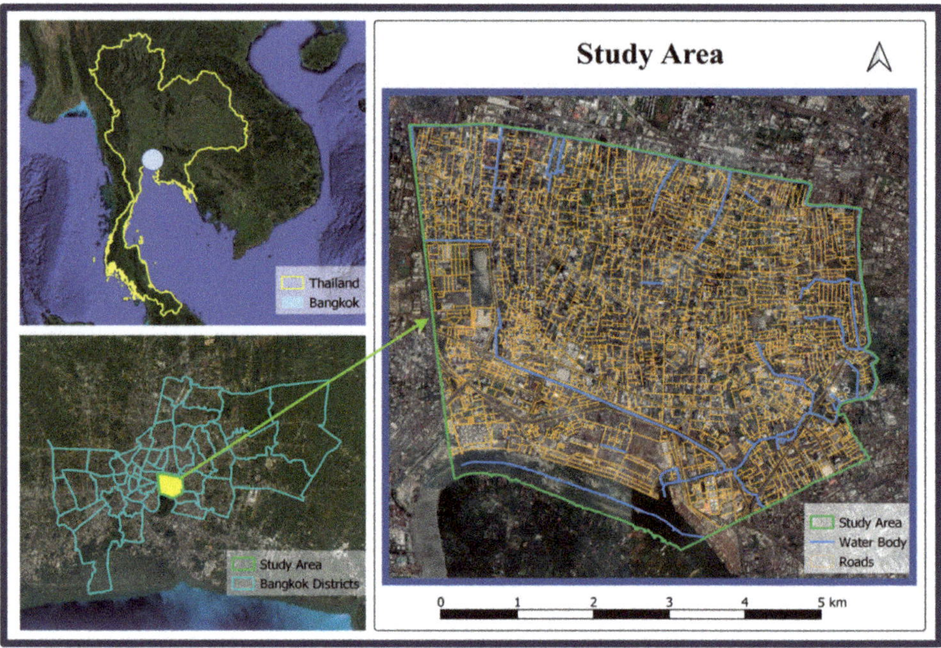

Figure 2. Location of the study area.

Flooding is also a severe problem in Bangkok, causing important economic and health-related problems. This problem has been aggravated in the last years due to urbanization, which has generated land use changes in the city. According to Srivanit et al. [29] the urban/built-up land in Bangkok and its metropolitan area increased by almost three times between 1994 and 2009, growing from 15% in 1994, to 42% in 2009. In contrast, a pronounced decrease in the vegetated area was observed from 1994 (72%) to 2009 (40%). The land use in the Study Area of Sukhumvit is presented in Appendix A.

Flooding is caused by excessive local rainfalls or by the overtopping of embankments due to a high water level in the Chao Phraya River [4,33]. Even though there are numerous pumping stations inside the city to pump the excess of storm water to the river, the city is still highly vulnerable to flooding. The problem is aggravated by over extraction of ground water, which has caused land subsidence of up to 15 cm in many locations [34].

2.3. Hazard Assessment

Hazard assessment was undertaken to gain a better understanding of existing conditions, and to identify locations which are more hazardous in terms of urban flooding and heat stress. The choice to undertake a macro scale approach for urban flood modeling and a micro scale approach for the thermal comfort modeling was based on field observations and relevant literature review.

For assessing urban flood hazards, the existing sewer system was modeled with a hydrodynamic model, considering several scenarios with particular emphasis on areas with a higher frequency of flooding. In terms of hazards due to heat stress, this assessment aimed to identify and evaluate the effects from different urban land uses on heat stress and human thermal comfort. For this purpose, field data was collected from both fixed weather stations and mobile weather measurements (using the instrument Kestrel 5400 Heat Stress Tracker). Five different urban land uses were considered for the heat stress assessment with mobile weather measurements (Table 1), and categorized following the work of Stewart and Oke [35].

Table 1. Mobile measurements in five different urban land uses (M1-5 refer to measuring locations).

Sn	ID	Installed Height	Type of Measurement	Characteristic of the Location
1	M1	1.4 m	Mobile	Water Body
2	M2	1.4 m	Mobile	Highly dense urban area (high vehicle traffic and buildings construction)
3	M3	1.4 m	Mobile	Compact high-rise
4	M4	1.4 m	Mobile	Urban Green (Park)
5	M5	1.4 m	Mobile	Open low-rise

The assessment of heat stress variation was based on measurements of different weather parameters taken at five different locations. These weather parameters were: Air temperature (Ta), mean radiant temperature (Tmrt), predicted mean vote (PMV) and physiological equivalent temperature (PET). The rationale behind this assessment was to analyze the effect of different urban land uses on heat stress and thermal comfort.

In terms of the selection of parameters, according to Coccolo et al. [36], "the Tmrt is considered to be as an artificial measure to express the degree of exposure to the environmental radiation. The radiant temperature is related to the amount of radiant heat transferred from a surface, and it depends on the material's ability to absorb or emit heat, or its emissivity". As Höppe [37] stated, "PET is defined as the air temperature at which, in a typical indoor setting (without wind and solar radiation), the heat budget of the human body is balanced with the same core and skin temperature as under the complex outdoor conditions to be assessed". Predicted mean vote (PMV) is one of the most used thermal indices by researchers. It is initially based on Fanger's heat balance model, and is an index ranging from −3 for cold weather and +3 for hot weather). This index is an outcome of A result of the perceived sensation of the thermal environment of a group of people, and was initially developed for indoor environments [38].

2.4. Selection of NBS Types and Their Suitable Sites: Which and Where

Four measures were selected based on several factors, including: Recommendations in the literature [39–41], feasibility of implementation in the case study area and the capability to be modeled in Mike Urban (for urban flood analysis) and the ENVI-met microclimatic simulation model (for thermal comfort analysis). The selected measures are: green roofs, pervious pavements, bio retentions and rain gardens. Details about each measure and the parameters used in the models are provided in Appendix B.

For suitability analysis, these measures were categorized in two groups: Green roof (GR) and pervious pavement (PP) in the first category, and bio-retention (BR) and rain garden (RG) in the second one. The analysis of possible locations and the maximum application of these measures was done using satellite images and the geographic information system (GIS) data. For the second category, the BMP sitting tool Sustain [42] was also used (see Appendix C).

The boundary of the case study area (macro scale) is depicted with a green line in Figure 3 (left). Two distinct locations inside the case study area of Sukhumvit were chosen for microclimatic simulations, and they are shown in Figure 3 (within the red rectangles). These locations were categorized as open low-rise (A) and compact high-rise (B), and are examples of two most common urban configurations in the case study area: A site with low-rise buildings and a site with dense high-rise buildings. This selection was made to achieve a comprehensive analysis of how the measures will be effective in each of these conditions. The two selected micro scale sites represent two different urban configurations for urban climate zones based on the research done by Stewart and Oke [35] (see Appendix D).

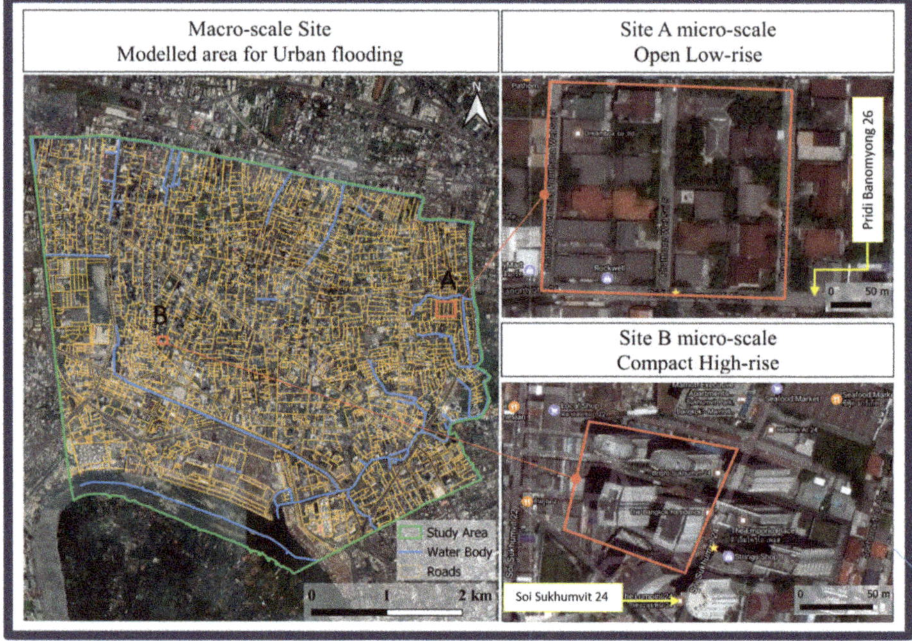

Figure 3. Study area showed as a green boundary on the left image, which was the modeled area for flooding. Two representative urban configurations showed as areas with red boundaries and were used for thermal comfort modeling: Site A: Open low-rise (**right up** image) & Site B: Compact high-rise (**right bottom** image).

2.5. Evaluating Effectiveness: How Much

2.5.1. Model Development and Data Analysis

In order to quantify the effectiveness of NBS for urban flood reduction, a flood hazard map was produced by applying a 1D/2D modeling approach within Mike Urban software [43] (see Appendix C), see also Vojinovic and Tutulic [44]. The model was run for different rainfall return periods and different cases of NBS measures applications in the study area. The input data to create the model was collected from the Bangkok Metropolitan Administration office, Hydro and Agro Informatics Institute and the Department of Drainage and Sewerage office in Bangkok. The flooding reduction in the study area was assessed considering storm water runoff reduction, peak flow reduction and time to peak delay.

In terms of the heat stress, the ENVI-met v4.1.3 model [45] was used for assessing the effectiveness of measures in relation to thermal comfort. ENVI-met is a three-dimensional computational fluid dynamics non-hydrostatic S.V.A.T. (soil, vegetation, atmosphere, and transfer) model (Appendix C). This software is commonly used for modeling surface-plant-air interactions in urban environments, and it can also simulate flows around buildings, heat and vapor transfer at urban surfaces, turbulence and exchanges of energy and mass between the vegetation and its surroundings, and simple chemical reactions [46–49].

The input data and parameters used in each of the models are presented in Table 2. The changes made in each model in order to represent the application of NBS are presented as inputs, while the outputs explain the type of results obtained from each model. The results are compared with the case that no such measures are applied (also referred to as a 'baseline scenario' or 'business as usual') in order to evaluate the change of conditions obtained from each alternative.

Table 2. Model input data and parameters.

NBS ID	Mike Urban Model (Scenarios X & Y)		ENVI-Met Model (Scenarios A & B)	
	Inputs	**Outputs**	**Inputs**	**Outputs**
PP (Pervious Pavement)	Surface, Pavement, Storage and Drain parameters	The amount of change in volume, flow and time to peak for each of the X and Y Scenarios and for each of the variables.	The surface Albedo and emissivity of the PP is changed from 0.4 to 0.8.	The average of change in Ta, MRT and PMV for each of the scenarios A and B and for each of the variables.
GR (Green Roof)	Surface, Soil and Drainage mat parameters		Grass on top of the buildings. The characteristics of the grass are LAD, Albedo, Cell size and intensity.	
BR (Bio-Retention)	Surface, Soil, storage and Underdrain parameters		The Green area percentage is increase by 5%. The inputs are Number of trees, LAD, RAD, plant height, Albedo and Leaf type.	
RG (Rain Garden)	Surface and Soil parameters		The Green area percentage is increase by 5%. The inputs are Number of trees, LAD, RAD, plant height, Albedo and Leaf type.	

2.5.2. Scenarios Development

Further to the above, ten scenarios were considered in each case, to evaluate urban flooding and thermal comfort (macro and micro scales respectively). Table 3 shows the scenarios for urban flood simulations, which include two different precipitation return periods for each of the four selected NBS, in addition to the baseline scenario.

Table 3. Nature-Based Solutions (NBS) implementation scenarios for urban flood assessment.

Measures and Their Implemented Scenarios for Assessment of Urban Flood Reduction (F)—Macro Scale			
Implemented Measures	**Description of Measures**	**Scenarios According to Rainfall Return Periods**	
		2 Year	**20 Year**
Business as usual	Business as usual	X-B	Y-B
PP (all str. and pavements) (implementing area: 15%)	Pervious Pavements (with high albedo material)	X-PP	Y-PP
GR (all feasible roofs) (implementing area: 27%)	Green roof (extensive vegetation)	X-GR	Y-GR
BR (alongside the streets) (implementing area: 4%)	Bio-retention (with shrub/bush)	X-BR	Y-BR
RG (alongside the streets) (implementing area: 4%)	Rain garden (with street trees)	X-RG	Y-RG

Table 4 shows scenarios for microclimatic thermal comfort simulations, which include two different site characteristics and the four selected NBS, in addition to the baseline scenario.

Figure 4 shows the overall framework for assessment of effectiveness from NBSs in relation to urban flood and thermal comfort. For each case, the models used, scenarios, variables and outputs, are presented.

Table 4. NBS implementation scenarios for thermal comfort assessment.

Scenarios for Thermal Comfort Effectiveness Assessment (T)—Micro Scale			
Implemented Measures	**Description of Measures**	**Scenarios According to Site Characterizes**	
		Low Rise	**High Rise**
Business as usual	Business as usual	A-B	B-B
PP (all str. and pavements) (implementing area: 25%)	Changing the albedo from 0.4 to 0.8	A-PP	B-PP
GR (all feasible roofs) (implementing area: 35%)	Adding 50 cm height grass on top of the roofs	A-GR	B-GR
BR (alongside the street) (implementing area: 5%)	Planting shrubs (1.2 m height) alongside the street edges	A-BR	B-BR
RG (alongside the street) (implementing area: 5%)	Planting trees (6.0 m height) alongside the street edges	A-RG	B-RG

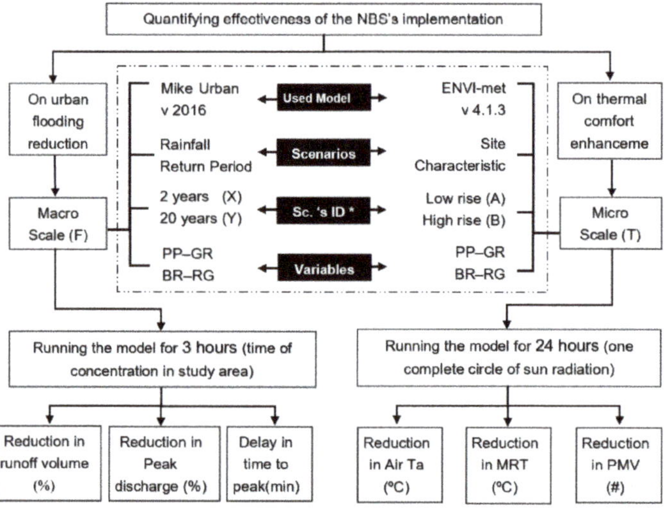

Figure 4. Overall framework for effectiveness assessment of NBSs.

2.5.3. Comparative Effectiveness of NBSs

The main purpose of the comparative performance scoring performed here is to identify the most effective measure, taking into consideration all parameters and scenarios. The effectiveness of measures for flooding reduction and thermal comfort enhancement are evaluated using the six parameters shown in Figure 4. According to the results obtained, the measures are scored from 1 to 4 according to the relation to their performance compared with other measures for the same scenario. This implies that the measure with the highest effectiveness will be scored as 4, and the one with the least effectiveness (comparably) will be scored as 1.

3. Results and Discussion

3.1. Hazard Assessment

The initial heat stress assessment was performed from mobile weather measurements. These measurements were performed in five different urban land uses within the case study area (M sites in Figure 5). Values of air temperature, wind speed, glob temperature and humidity were collected; and some thermal comfort indices including PMV and PET were calculated. The Kestrel 5400 Heat Stress Tracker was used for this purpose. This procedure showed that urban parks (M4) and open low-rise sites (M5) can be respectively 1.4 °C and 1.1 °C cooler than a highly dense urban site in the study area (M2 and M3). Furthermore, the results showed that the PMV and PET can be also lower in these two sites compared to highly dense urban sites. Notice that M3 and M5 are the sites chosen to perform the thermal modeling, presented as areas B and A, respectively, in Figure 3.

Additionally, by overlaying the flood hazard map with the buildings data (Figure 5), it was observed that less floods and more heat stress are likely to occur in the highly dense upstream area (A1 in Figure 5), when compared other parts of the study area. However, less urbanized areas (A2) located downstream of this high-rise area present more of the flood-related issues. Since in a highly dense urban area there are more impervious surfaces, it can be expected that this can have a great impact upon flood-related issues in downstream areas.

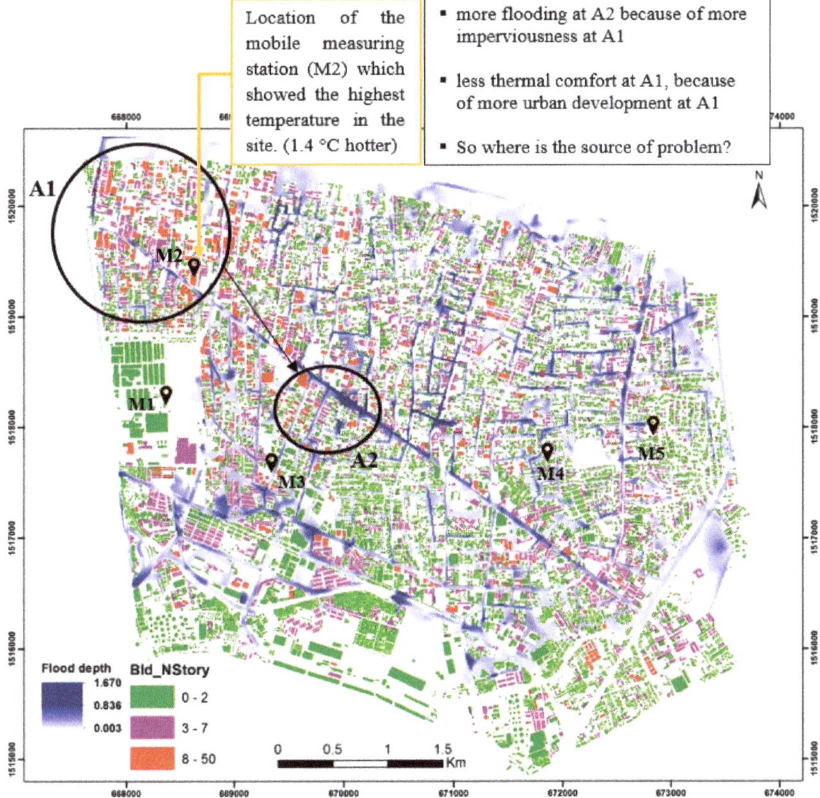

Legend: M1: Water Body; M2: Highly dense urban area; M3: Compact high-rise; M4: Urban Green; M5: Open low-rise.

Figure 5. Relation of urban flooding with heat stress and urbanization, shown through the overlay of the flood inundation map (for a 20-year return period rainfall) and variation in real estate.

3.2. NBS Types Selection and Suitability Analysis (Which and Where)

Four different small-scale NBS measures were selected from the analysis of local characteristics, namely the availability of possible locations for green roofs implementation, or low slope and low traffic pavements, which could be changed into pervious pavements:

1. Green roof (GR), with extensive vegetation
2. Pervious pavement (PP), with high albedo construction material
3. Bio-retention (BR), with shrubs as topping vegetation at a height of 1.2 m
4. Rain garden (RG), with street trees and lawn as topping vegetation at the height of 6 m

From the suitability analysis, maximum rates of measures application where obtained. The results of these analysis showed that for the macro scale simulations (urban flooding), green roofs and pervious pavements can be implemented as an average on 27% and 15% of the whole study area, respectively. In addition, bio-retention and rain gardens can both be implemented as an average on 4% of the whole study area. For the micro scale simulations (thermal comfort), green roofs and pervious pavements can be implemented within a maximum of 36% and 27% of the selected micro scale study areas, respectively. While bio-retentions and rain gardens can both be implemented as a maximum of 5% of the selected microscale study area.

The two different results from suitability analysis are due to two different scales used in this study. Urban flood assessment and analysis requires a macro scale simulation study. However, the microclimatic thermal comfort assessment requires a microscale simulation study. The maximum possible application of the measures will be different when studying a macro scale site as a whole, and when studying the selected micro scale sites within the whole.

3.3. Effectiveness of NBS's on Urban Flooding (How Much Impact on Flood Reduction)

Figure 6 shows the effectiveness of each NBS for flood mitigation according to the results obtained from the hydrodynamic model for two rainfall scenarios. The parameters presented are runoff volume reduction (Figure 6a) and peak discharge reduction (Figure 6b). From the analysis of results we can observe that the effectiveness of the measures is reduced when the rainfall return period increases. Additionally, it appears that 'green roofs' is the most efficient NBS type for this case study area, having effectiveness of up to 39% and 40% in reduction for total runoff volume and peak discharges, respectively, for a two-year return period rainfall. On the other hand, 'pervious pavements' was found to be the least effective NBS type. The main reason for green roofs for being the most effective is the relatively large suitable area for its implementation. According to the suitability analysis, around 27% of the area was considered suitable for implementing green roofs.

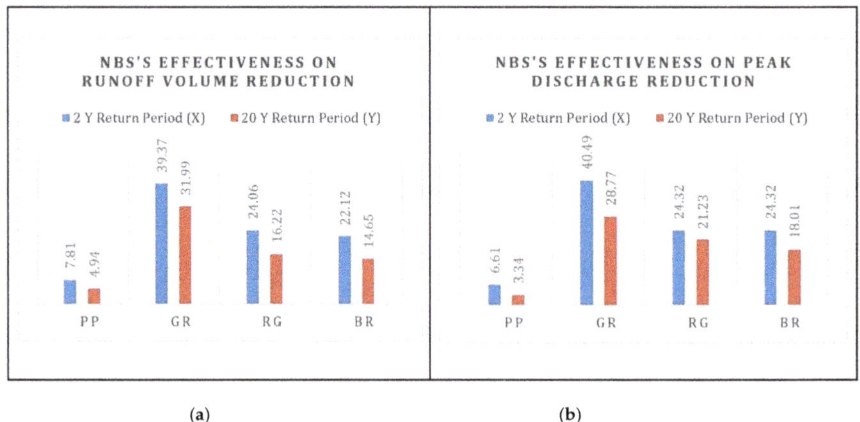

(a)   (b)

Figure 6. (a) NBS effectiveness on runoff volume reduction, (b) NBS effectiveness on peak discharge reduction.

The results also show that there was no significant change in the delay for time to peak in most of the cases; thus this result was not plotted in Figure 6. One reason for this could be the existence of numerous small catchments in the case study area.

3.4. The Effectiveness of NBS on Thermal Comfort Enhancement (How Much Impact on Thermal Comfort)

Regarding the microclimatic situation, model results for the base case and for the case of implementing rain gardens in open low-rise sites are shown as an example in Figure 7. The effectiveness of NBS was measured by comparing the results of variation in Air temperature (Ta) and Mean radiant temperature (Tmrt) with the base case scenario. In addition, variations of Predicted mean vote (PMV) in relation with the base case scenario were considered. The obtained results show that the cooling effect of trees, which were used in rain gardens, was widely dispersed if we analyze the reduction in air temperature. Whereas, Tmrt and PMV were significantly reduced, but only in the shaded areas of the trees, as a result of the prevention of direct sun radiation.

Figure 7 shows the distribution of the air temperature at 4:00 p.m. and at the height of 1.0 m from the ground. According to the American Society of Heating, Refrigerating and Air-Conditioning Engineers (ASHRAE) [50], the thermal comfort measurements have to be done at the center of the human body, which internationally is established at 1.10 m of height. However, a vertical grid size configuration was used in this model, and the software only provided the vertical elevation data at the height of 1.0 m and 1.4 m. Therefore, the results in this work are presented for an elevation of 1.0 m, which is considered the pedestrian level.

The results from the microclimatic modeling for this case established that 'rain gardens' was the most effective measure in the open low-rise buildings (site A), with a maximum reduction of 0.66 °C in air temperature (Ta) (see Table 5). The reason why rain gardens performed best is that trees were considered as vegetation for this measure. Similar studies on the effect of different vegetation on thermal comfort also revealed that 'trees' was the most effective measure in providing outdoor thermal comfort, and this effectiveness can be enhanced by increasing the LAD (Leaf Area Density) and height of the trees [51–53]. In contrary, in a compact high-rise building (site B) rain gardens did not have the same effect. In site B, even though this measure was still the most effective in air temperature reduction, it had an impact of only 0.25 °C maximum reduction in Ta. This difference on the impact of the same measure in sites A and B is because more paved surfaces were exposed to sunlight in the low-rise site compared to the high-rise site. As a result, the application of trees prevented the sunlight to reach the paved surfaces in site A, which consequently lead to a higher decrease in the sensible heat fluxes. In other words, in site B the shadow provided by the tall buildings did not give the chance for the trees to further decrease the temperature by providing their own shadow.

Regarding bio-retention, this measure was the least effective in the open low-rise building or site A, with a maximum reduction capacity of 0.16 °C. This result confirmed the role of the tree's height and leaf area density (LAD) upon thermal comfort, since by reducing these parameters for the bio-retention case compared to the rain gardens case, the air temperature reduction was significantly less. On the other hand, pervious pavements had a good impact of 0.41°C on air temperature reduction in the site A. The implementation of pervious pavements in the model was represented by changing the albedo of pavements from 0.4 to 0.8, in order to reduce the absorption of sun short wave radiation. It was observed that this measure had a better performance during the time that the sun was shining almost vertically, around 1:00 p.m. However, the other measures had their highest performance during the heat stress peak daytime, at 4 p.m.

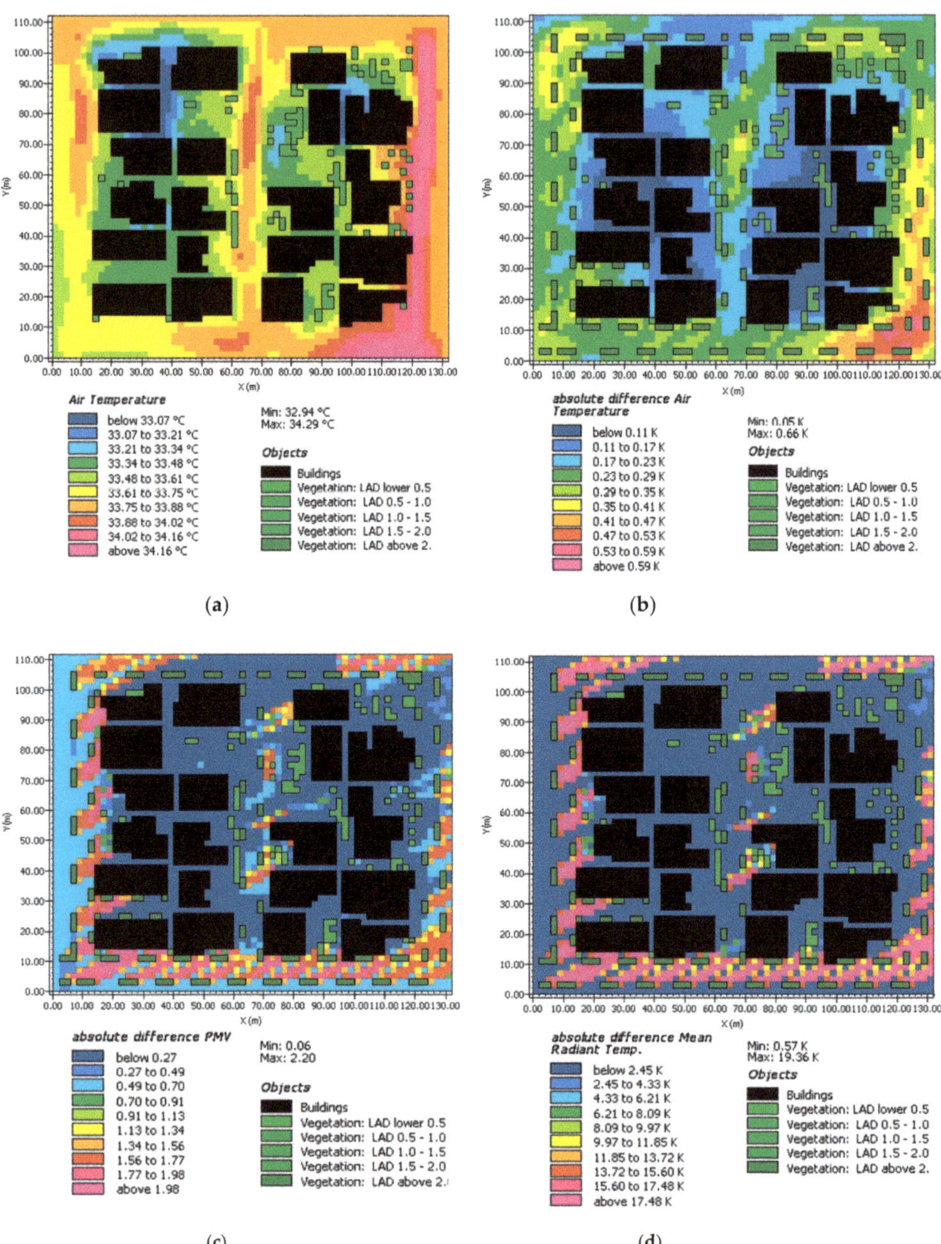

(**a**) (**b**)

(**c**) (**d**)

Figure 7. (**a**) Model of the base case scenario in ENVI-met for scenario A-B (see Table 3); (**b**) Effectiveness of RG implementation on Ta reduction for scenario A-RG; (**c**) Effectiveness of RG implementation on PMV reduction for scenario A-RG; (**d**) Effectiveness of RG implementation on Tmrt reduction for scenario A-RG.

Table 5. Effectiveness of the NBS measures on reduction of Ta, Tmrt and PMV.

Simulated Sub-Scenario (NBS's Variation)	Description of Measures	Max Reduction in Ta 4:00 p.m. (°C)	Max Reduction in Tmrt at 4:00 p.m. (°C)	Max Reduction in PMV at 4:00 p.m. (−5 to 5)
Open Low rise buildings (Site A)				
A-PP	Changing the albedo from 0.4 to 0.8	0.41	−0.6 (from 51.18)	0.68 (from 4.67)
A-GR	Adding 50 cm height grass on top of the roofs	0.17	17.81 (from 51.18)	0.8 (from 4.67)
A-BR	Planting shrubs (1.3 m height) alongside the street edges	0.16	16.2 (from 51.18)	1.52 (from 4.67)
A-RG	Planting trees (6.0 m height) alongside the street edges	0.66	19.36 (from 51.18)	2.21 (from 4.67)
Compact high-rise buildings (Site B)				
B-PP	Changing the albedo from 0.4 to 0.8	0.10	−0.69 (from 52.31)	−0.05 (from 4.09)
B-GR	Adding 50 cm height grass on top of the roofs	0.00	0.10 (from 52.31)	0.01 (from 4.09)
B-BR	Planting shrubs (1.3 m height) alongside the street edges	0.07	15.64 (from 52.31)	1.07 (from 4.09)
B-RG	Planting trees (6.0 m height) alongside the street edges	0.25	19.26 (from 52.31)	1.52 (from 4.09)

Table 5 also shows that the effectiveness of the measures on Ta reduction was quite different when the site characteristic changed from open low-rise (A) to compact high-rise buildings (B). For instance, 'green roofs' had almost no effect on the reduction of Ta in scenario B, at 4:00 p.m. and at the pedestrian level. However, in site A, 'green roofs' had an effectiveness of 0.17 °C on reduction of air temperature at the pedestrian level. In fact, as the buildings in site B are very high (around 70 to 104 m), the effect of vegetation on top of such buildings did not reach the pedestrian level. There is previous research that revealed that the green roofs on top of high-rise buildings do not have any significant effect on air temperature for pedestrians [18,54–56].

An interesting finding of this research was that in a site of high-rise buildings, tree plantation and vegetation can be more effective during the night time. The results clearly show that both measures, bio-retention and rain gardens at site B, had better effectiveness at 1:00 a.m., with 0.43 °C and 0.15 °C reduction in air temperature, respectively. Therefore, it can be derived that vegetation planting (especially in the high-rise building site) is the best practice for the urban heat island control, which is at its peak during night time. High-rise buildings absorb sun energy during the day and release it during the night. As a result, the temperature does not get reduced enough during the night time in these sites, and it causes an urban heat island (see also [55]), which can be controlled relatively by implementing NBS measures.

Regarding other parameters, the effectiveness of bio-retention and rain gardens on the reduction of Tmrt were almost the same in both site characteristics. The results show that these two measures had a maximum reduction in Tmrt of around 16 °C and 19 °C, respectively (at 4:00 p.m.) in both sites.

Whereas, 'green roofs' had also a reasonable contribution to Tmrt reduction in site A (Max 17.81 °C from 51.18 °C), it had a very low effect on Tmrt reduction at pedestrian level at site B (0.10 °C from 52.31 °C), which is logical, according to the previous discussion on the height of the buildings. Similarly, its effect on PMV max reduction was also 0.8 at site A and 0.01 at site B, which again does not show any significant change at site B.

The effect of pervious pavements on reduction of Tmrt is quite controversial. The results show that pervious pavements not only had no contribution in reduction of the Tmrt at any of the study areas, but it even increased slightly the value of Tmrt by 0.6 °C and 0.69 °C at sites A and B, respectively. This result shows that by changing the albedo of the streets from 0.4 to 0.8, the sun radiation was being more reflected to the atmosphere. Therefore, the pedestrians experienced more radiation, which led to a growth in Tmrt. Similarly, the impact on PMV for pervious pavements application at site B was slightly negative. However, there was a positive impact in site A, showing a change of 0.68 from 4.67 on PVM.

3.5. Discussion on NBS's Performance

In general, the simulations of the base case scenario for both sites (A and B) on thermal comfort show that during the day, and specially in the peak temperature time of the day (4:00 p.m.), site B (compact high-rise) provided a better thermal comfort at the pedestrian level, compared to site A (open low-rise). The main reason for this was related to the shadow provided during the day by the high-rise buildings. On the other hand, during the nights this issue reversed, and the site A had a better thermal comfort, as there was more air ventilation and less structures to release the heat accumulated during the day, to the environment during the night time. A research done by Hedquist and Brazel [57] on a case study of Arizona, U.S.A., presented a similar result.

An overall comparison of the measure's effectiveness in both scenarios (A and B) interestingly shows that the NBS measures had a better performance in the low-rise buildings (A) site during the daytime and a better performance in the high-rise buildings (B) site during the night time. This is because the more direct sunshine in site A during the day provided a better opportunity for NBS measures to improve the existing thermal comfort of the site. This was achieved by either providing more shadow or reradiating the sun's shortwaves. However, in site B, the shadow provided by the buildings during the day gave less opportunity for the NBS to show their effectiveness. On the other hand, as during the night, site B was hotter compared to site A, which is a result of the urban heat island [58], the implementation of NBS was more effective at site B during the night, which could contribute to the reduction of urban heat island intensity at these locations.

In conclusion, the effectiveness assessment of NBS measures using a microclimatic modeling clearly showed that the effectiveness of the measures on thermal comfort enhancement depends on several factors. These factors are: The characteristics of the site, the type of NBS, the coverage and location of the measures and the time of the day.

3.6. Comparative Effectiveness Scoring for the NBS

3.6.1. Comparative Effectiveness for Urban Flood Reduction

Analyzing the effectiveness scoring of the measures in relation to urban flood reduction (Table 6), it can be observed that 'green roofs' represent the most effective measure for this purpose while 'pervious pavements' appears as the least effective measure. This is in line with other studies, e.g., Carpenter and Kaluvakolanu [59] and Berardi et al. [60]. However, this effectiveness scoring does not consider the differences on percentages of implementation area for each measure. It is to be noted that one of the main reasons why 'green roofs' is the most effective NBS for control of urban flooding, is the fact that this measure is the most feasible to be implemented at such an urbanized part of the city. Therefore, again it is that site characteristics determine the performance of the measures.

3.6.2. Comparative Effectiveness for Thermal Comfort Enhancement

In relation to thermal control, 'rain gardens' is found to be the most effective measure in order to provide more thermal comfort (Table 7). Besides, 'pervious pavements' followed by 'green roofs' are the least effective measures for providing thermal comfort at the pedestrian level. Similar results were obtained in previous studies [61,62].

Table 6. Comparative effectiveness scoring of the measures in urban flooding.

Effectiveness Aspect	Scenarios	Criteria	Comparative Effectiveness Scoring of NBS Measures			
			PP	GR	BR	RG
Reduction in urban flooding	2 years (X)	Runoff volume	1	4	2	3
		Peak discharge	1	4	3	3
	Performance score for scenario (X)		2	8	5	6
	20 years (Y)	Runoff volume	1	4	2	3
		Peak discharge	1	4	2	3
Performance score for scenario (Y)			2	8	4	6
Total comparative performance score			4	16	9	12

Table 7. Comparative effectiveness scoring of the measures on thermal comfort.

Effectiveness Aspect	Scenarios	Criteria	Comparative Effectiveness Scoring of NBS Measures			
			PP	GR	BR	RG
Thermal comfort enhancement	Low rise (A)	Ta	3	2	1	4
		Tmrt	1	3	2	4
		PMV	1	2	3	4
	Performance in scenario (A)		5	7	6	12
	High rise (B)	Ta	3	1	2	4
		Tmrt	1	2	3	4
		PMV	1	2	3	4
Performance in scenario (B)			5	5	8	12
Total comparative score			10	12	14	24

3.6.3. Overall Analysis of Effectiveness and Recommendation for NBS Application

From the comparative rankings of the measures for flood reduction and thermal comfort enhancement in the Sukhumvit area in Bangkok (Thailand), it can be observed that 'rain gardens' are likely to be the most effective NBS type with respect to both criteria, flood reduction and thermal comfort enhancement (Table 8). It is interesting to observe that trees (with average height of six meters), which were part of the design of rain gardens in this case, represent the most influencing factor for urban microclimate conditions. As such, for this particular case study area this should be maintained in order to achieve better thermal comfort. Furthermore, design of rain gardens also included greater depth of soil zone for infiltration (e.g., 80 cm) purposes and the inclusion of a storage zone (e.g., 18 cm depth), which is likely to contribute towards higher effectiveness in flood volume reduction. For the Sukhumvit area, this measure should play an important role in urban planning activities. Green roofs and rain gardens are found to be as the second and third most effective NBS types, respectively.

Regardless of the overall effectiveness results, it is important to consider that different measures performed best on the two different criteria. As a result, to achieve the best performance in both objectives for the Sukhumvit case study area, it is recommendable to implement a combination of different NBS types.

Table 8. The overall effectiveness scoring of the measures' performance.

Effectiveness Aspect.	Scenarios	Comparative Effectiveness Scoring for each of the NBS Measures			
		PP	GR	BR	RG
Reduction in urban flooding (F)	2 year (X)	2	8	5	6
	20 year (Y)	2	8	4	6
Thermal comfort enhancement (T)	Low rise (A)	5	7	6	12
	High rise (B)	5	5	8	12
Overall comparative score		14	28	23	36

This is possible, since the NBS types analyzed here do not compete for free spaces. For example, green roofs and rain gardens can be applied in a site simultaneously, since one uses roofs and the other one uses spaces alongside the streets.

Additionally, the location of the site as well as its land use type play a role on the performance of each NBS type, and as such, they should be considered for implementing the right measure. In this case, measures with higher effectiveness in flood reduction should be applied in the high-rise site, since it is at the upstream area, and has higher imperviousness. Reducing runoff through providing more disconnections in this case study area is likely to have a positive impact downstream, where flood impacts are higher. For Sukhumvit, the NBS type which showed best flood reduction performance is 'green roofs'. This can be explained due to its enhanced suitability that results on increased disconnection area; that leads to a better effectiveness for urban flood reduction when compared to other NBS types. However, this NBS type has low effectiveness on heat stress reduction at this compact high-rise site. Therefore, the best alternative for the site B is the combination of green roofs and rain gardens, which together would have a good performance on both criteria. Moreover, even though green roofs are less effective in relation to outdoor thermal comfort enhancement, their application may have many other benefits, such as energy consumption reduction, air pollution mitigation, storm water management, sound absorption and ecological preservation [63].

4. Conclusions

In this paper, a novel framework that can be used for the selection of small-scale urban nature-based solutions to reduce flooding and enhance human thermal comfort has been presented. The framework has been applied in the Sukhumvit area in Bangkok (Thailand). The obtained results show that the combined implementation of different NBS types is likely to have a good potential to make this area more resilient and sustainable to cope with future challenges related to climate change and the high rate of urbanization. By applying this novel framework, it was possible to identify the most promising NBS types that can be applied in different parts of the area to effectively achieve both objectives at the same time.

Several interesting findings were obtained from the present work. For instance, green roofs are likely to achieve better performance in the reduction of urban flooding when compared to the other NBS types studied. However, this particular NBS type is not effective in thermal comfort enhancement in sites with compact, high-rise buildings. Regarding the effectiveness of the NBS measures for thermal comfort, the results showed that this is mainly related to the provision of shadows from trees. Therefore, rain gardens with street trees as covering vegetation would have the best performance for the open low-rise scenario. Although, the results also showed that the effectiveness of different NBS types changes according to the site characteristics and time of the day. Therefore, a combined application of green roofs and rain gardens is recommended in compact high-rise building areas. We conclude then that this method is very helpful to identify adequate measures according to local characteristics, and to choose a combination of measures that is best for each particular site.

The importance of combing micro scale and macro scale effectiveness assessment of NBS was demonstrated through the present work. The use of microclimatic modeling in this framework showed that the effectiveness of NBS for thermal comfort enhancement depends on several factors. These factors are: The characteristics of the implementation site, the type of NBS, the coverage and location of the measures, and the time of day. Moreover, the results illustrate the usefulness of macro scale urban flood modeling for an assessment of the effectiveness of different NBS types for the reduction of flood impacts. Consequently, the present work proves the effectiveness of different NBS types taken from a micro and macro scale perspectives.

The outcome of the present research aims to provide some additional knowledge to city planners and decision makers in gaining better understanding of the effectiveness of different NBS measures for different sites and local conditions.

Author Contributions: Conceptualization, A.N.M., A.A., Z.V., A.S. and S.W.; Data curation, F.B. and J.K.; Formal analysis, A.N.M., A.A. and A.S.; Funding acquisition, Z.V. and S.W.; Investigation, A.N.M.; Methodology, A.N.M.; Project administration, Z.V.; Supervision, Z.V., S.W., A.S. and A.A.; Validation, F.B. and J.K.; Writing—original draft, A.N.M.; Writing—review & editing, A.A.

Funding: We are thankful to the Canadian Department of Foreign Affairs, Trade and Development, for providing financial support through Asian Development Bank (ADB) for performing this research. Furthermore, this research was also partially supported by the European Union Seventh Framework Programme (FP7/2007–2013) under Grant agreement n° 603,663 for the research project PEARL (Preparing for Extreme And Rare events in coastal regions) and from the European Union's Horizon 2020 Research and Innovation Programme under grant agreement No 776866 for the research project RECONECT. The study reflects only the authors' views and the European Union is not liable for any use that may be made of the information contained herein.

Acknowledgments: We thank Wendy Wuyts for improving earlier drafts of this manuscript.

Conflicts of Interest: The authors declare no conflict of interest.

Appendix A

Figure A1. Land use in the study area and details of land use in the two sites chosen to model thermal comfort.

Appendix B

<u>Pervious Pavements</u>

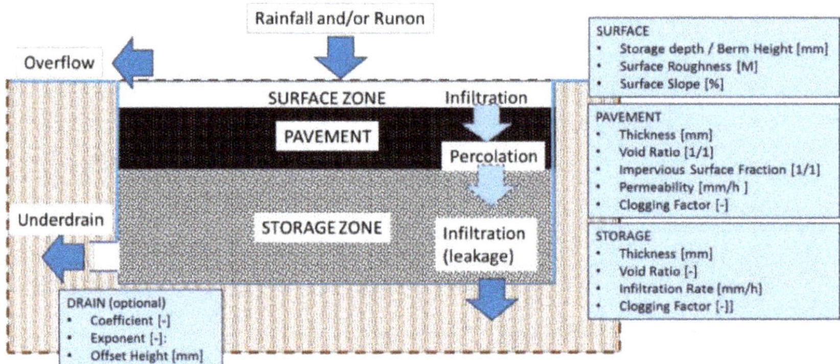

Figure A2. Schematic view of a pervious pavement system.

Table A1. PP Input parameters for PP modeling in ENVI-met.

Parameter	Unit	Value
Z0 Roughness	m	0.01
Albedo	fraction	0.8
Emissivity	fraction	0.9
Surface irrigated	-	No

Table A2. Input parameters for PP modeling in Mike Urban.

Parameter (Units)	Value	Source
Surface		
Storage height (mm)	0	[43]
Vegetation volume (fraction)	-	Assumption
Surface Roughness (Manning's m)	20	[43]
Surface Slope (%)	1	[64,65]
Pavement		
Thickness (mm)	150	[64,65]
Void Ratio (voids/solids)	0.15	[64,65]
Impervious Surface Fraction (fraction)	0	[64,65]
Permeability (mm/h)	200	[64,65]
Clogging Factor	300	Formula based
Storage		
Height (mm)	300	[43,66]
Porosity (fraction)	0.70	[64,65]
Infiltration capacity of surrounding soil (mm/h)	10	[64,65]
Clogging Factor	0	Assumed no clogging
Drain		
Drain Capacity (mm/h)	0	[43]
Drain Exponent	0.5	[64,65]
Drain Offset Height (mm)	0	[64,65]

Green roofs

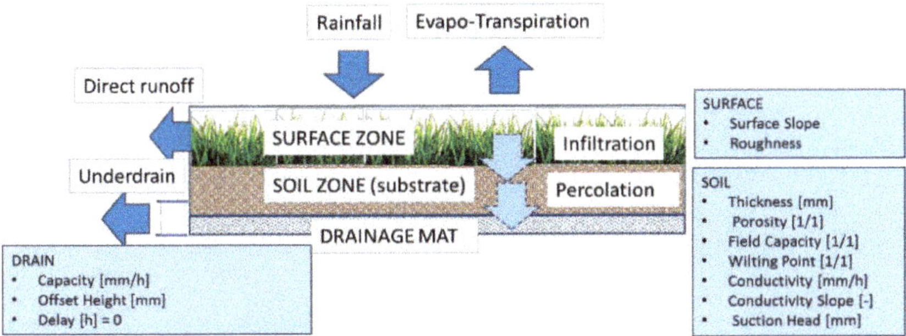

Figure A3. Schematic view of a green roof system.

Table A3. Input parameters for GR modeling in ENVI-met (The toping grass).

Parameter	Unit	Value
Leaf Type	-	Grass
Albedo	fraction	0.2
Plant height	m	0.5
Root zone height	m	0.5
LAD (Leaf area density) profile	-	Default
RAD (Root area density) profile	-	Default

Table A4. Input parameters for GR modeling in Mike Urban.

Parameter (Units)	Value	Source
Surface		
Storage Depth (mm)	20	
Vegetative Volume (fraction)	0.1	[65]
Surface Roughness (Manning's m)	5	[65]
Surface Slope (percent)	1	[65]
Soil		
Thickness (mm)	150	[66]
Porosity (volume fraction)	0.5	[66]
Field Capacity (volume fraction)	0.20	[66]
Wilting Point (volume fraction)	0.1	[66]
Conductivity (mm/h)	12.7	[66]
Conductivity Slope	10	[66]
Suction Head (mm)	88.9	[65]
Drainage Mat		
Thickness (mm)	25	[64,65]
Void fraction	0.5	[64,65]
Roughness (Manning M)	5	[64,65]

Bio-retention and Rain garden

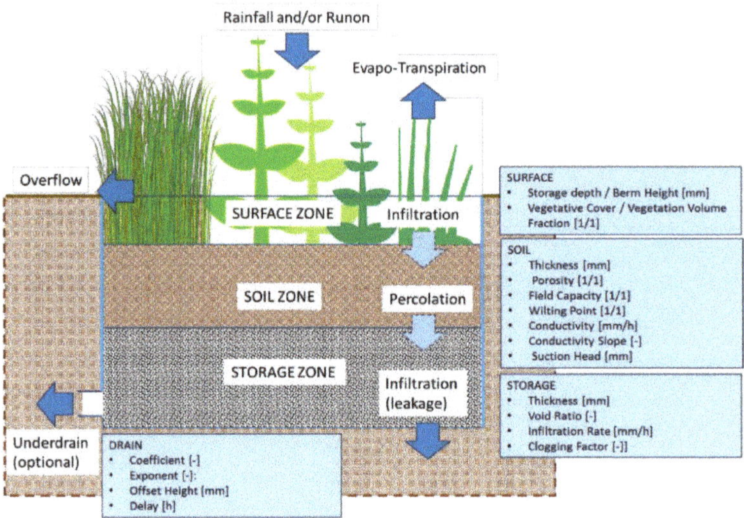

Figure A4. Schematic view of a Bio-retention system.

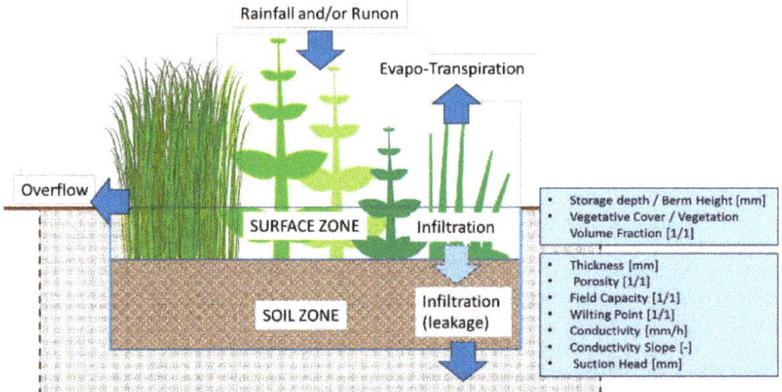

Figure A5. Schematic view of a Rain Garden system.

Table A5. Input parameters for BR modeling in ENVI-me.

Parameter	Unit	Value
Leaf Type	-	Deciduous
Albedo	fraction	0.2
Plant height	m	1.2
Root zone height	m	1
LAD (Leaf area density) profile	-	Default
RAD (Root area density) profile	-	Default

Table A6. Input parameters for RG modeling in ENVI-met.

Parameter	Unit	Value
Leaf Type	-	Deciduous
Albedo	fraction	0.2
Plant height	m	6.0
Root zone height	m	1
LAD (Leaf area density) profile	-	Default
RAD (Root area density) profile	-	Default

Table A7. Input parameters for modeling BR and RG in Mike Urban.

Parameter (Units)	(RG) Value	(BR) Value	Source
Surface			
Storage Depth (mm)	180	150	[43]
Vegetative Volume (fraction)	0.10	0.15	[64,65]
Surface Roughness (Manning's m)	5	2.5	[64,65]
Surface Slope (percent)	1	1	[64,65]
Soil			
Thickness (mm)	800	550	[66]
Porosity (volume fraction)	0.5	0.5	[66]
Field Capacity (volume fraction)	0.20	0.20	[66]
Wilting Point (volume fraction)	0.10	0.10	[66]
Conductivity (mm/h)	12.7	12.7	Default; [64]
Conductivity Slope	10	10	Default; [64]
Suction Head (mm)	88.9	88.9	Default; [64]
Storage			
Height (mm)		250	[66]
Void Ratio (voids/solids)		0.70	[66]
Infiltration capacity of surrounding soil (mm/h)		5	[66]
Clogging Factor		0	Assumed no clogging
Underdrain			
Drain Capacity (mm/h)		0	Default; [64]
Drain Exponent		0.5	Default; [64]
Drain Offset Height (mm)		0	Default; [64]

Appendix C

Site selection assessment for implementing the measures

A GIS extension tool called the Sustain-BMP siting tool, developed by the United States Environmental Protection Agency (US EPA), was used to analyze possible locations of NBS measures in the area. This tool was used only in the case of BR and RG, because the implementation of these two measures needs more detailed site feasibility assessment than in the case of GR and PP. This tool has been used for similar studies in several previous cases [67–69]. Figure A6 shows the overall methodology for the suitability analysis of the four selected NBS measures.

The input data for this tool, such as land use and a two meters' resolution DEM, were provided for the Bangkok Metropolitan Administration office. Data about soil types, imperviousness and ground water level were obtained from the Department of Drainage and Sewage in Bangkok.

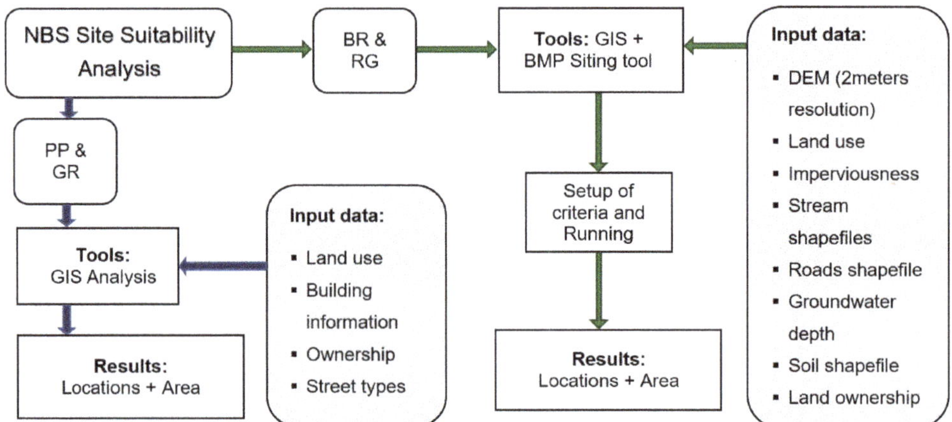

Figure A6. Flow chart for suitability analysis methodology.

Mike Urban simulation

A 1D/2D modeling simulation for producing the flood hazard map had been run by Mike Urban in different scenarios of implementing the multifunctional measures in the case study area. This is a well-recognized software for hydrodynamic modeling [70,71]. Additionally, the existing sewer network of Sukhumvit was also initially modeled using Mike Urban, which facilitated the building of the model for this study. Two modifications to the previously built model of Sukhumvit were introduced: Change in runoff routing and change in the simulation engine. Figure A7 shows the overall methodology followed in Mike Urban simulations to study the effectiveness of NBS measures on urban flooding.

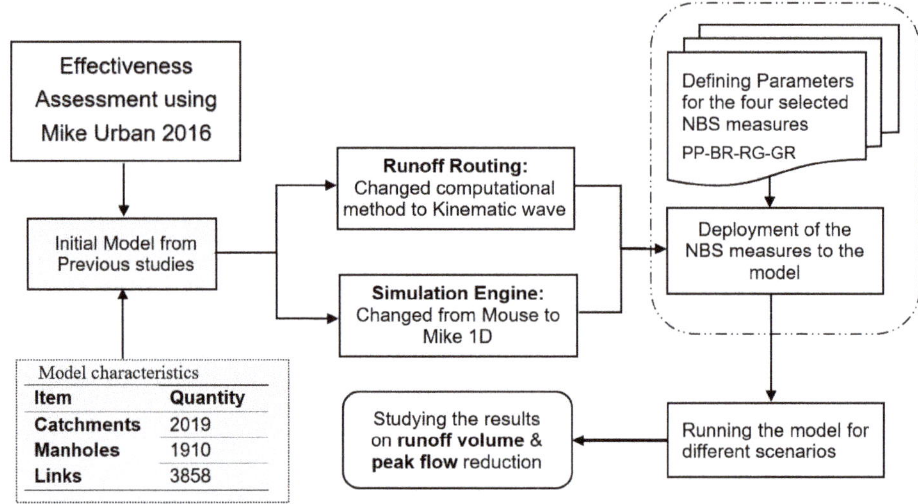

Figure A7. The Mike Urban simulation methodology.

ENVI-met microclimatic simulation

ENVI-met is a three-dimensional non-hydrostatic climate model for the simulation of surface-plant-air interactions, especially for conditions inside urban environments. Since the model is designed for the microscale, the resolution output is high, ranging from 0.5 to 10 m horizontally, with a temporal resolution of 10 sec and the ability to simulate timeframes from 24 to 48 h. The model requires the user to input certain parameters, such as defining the model area (area input file) and configuring the initial atmospheric conditions, surfaces (including soils), vegetation, and time intervals [72]. Figure A8 shows the conceptual framework describing the required procedure and steps for running the microclimatic simulations in ENVI-met and visualizing and analyzing the results using LEONARDO 2014 and Biomet, respectively.

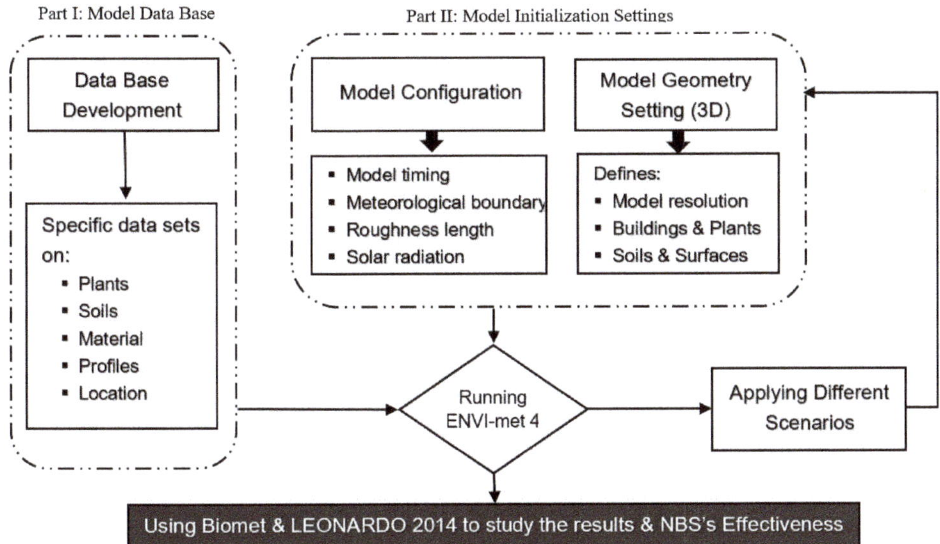

Figure A8. Procedures and steps for running the microclimatic simulations.

Appendix D

The selection of urban climate zones follows the urban climate zone categorization proposed by Stewart and Oke [35]. The zone numbers 1 & 6 of this categorization (see Figure A9) were used as sites B & A in microclimatic simulation.

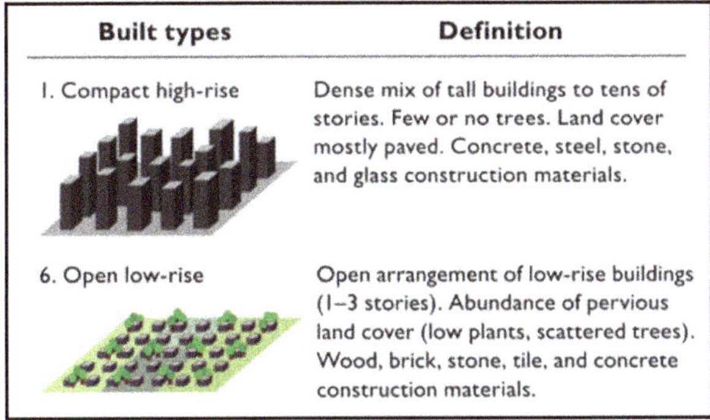

Built types	Definition
1. Compact high-rise	Dense mix of tall buildings to tens of stories. Few or no trees. Land cover mostly paved. Concrete, steel, stone, and glass construction materials.
6. Open low-rise	Open arrangement of low-rise buildings (1–3 stories). Abundance of pervious land cover (low plants, scattered trees). Wood, brick, stone, tile, and concrete construction materials.

Figure A9. Characteristics of zones chosen for microclimatic simulation (modified from [35]).

References

1. Goonetilleke, A.; Thomas, E.; Ginn, S.; Gilbert, D. Understanding the role of land use in urban stormwater quality management. *J. Environ. Manag.* **2005**, *74*, 31–42. [CrossRef]

2. Liu, W.; Chen, W.; Peng, C. Assessing the effectiveness of green infrastructures on urban flooding reduction: A community scale study. *Ecol. Model.* **2014**, *291*, 6–14. [CrossRef]

3. Andersson-Sköld, Y.; Thorsson, S.; Rayner, D.; Lindberg, F.; Janhäll, S.; Jonsson, A.; Moback, U.; Bergman, R.; Granberg, M. An integrated method for assessing climate-related risks and adaptation alternatives in urban areas. *Clim. Risk Manag.* **2015**, *7*, 31–50. [CrossRef]

4. Hilly, G.; Vojinovic, Z.; Weesakul, S.; Sanchez, A.; Hoang, D.; Djordjevic, S.; Chen, A.; Evans, B. Methodological Framework for Analysing Cascading Effects from Flood Events: The Case of Sukhumvit Area, Bangkok, Thailand. *Water* **2018**, *10*, 81. [CrossRef]

5. Kovats, R.S.; Hajat, S. Heat Stress and Public Health: A Critical Review. *Annu. Rev. Public Health* **2008**, *29*, 41–55. [CrossRef]

6. Derkzen, M.L.; van Teeffelen, A.J.A.; Verburg, P.H. Green infrastructure for urban climate adaptation: How do residents' views on climate impacts and green infrastructure shape adaptation preferences? *Landsc. Urban Plan.* **2017**, *157*, 106–130. [CrossRef]

7. Jia, Z.; Tang, S.; Luo, W.; Li, S.; Zhou, M. Small scale green infrastructure design to meet different urban hydrological criteria. *J. Environ. Manag.* **2016**, *171*, 92–100. [CrossRef]

8. Nesshöver, C.; Assmuth, T.; Irvine, K.N.; Rusch, G.M.; Waylen, K.A.; Delbaere, B.; Haase, D.; Jones-Walters, L.; Keune, H.; Kovacs, E.; et al. The science, policy and practice of nature-based solutions: An interdisciplinary perspective. *Sci. Total Environ.* **2017**, *579*, 1215–1227. [CrossRef]

9. Raymond, C.M.; Frantzeskaki, N.; Kabisch, N.; Berry, P.; Breil, M.; Nita, M.R.; Geneletti, D.; Calfapietra, C. A framework for assessing and implementing the co-benefits of nature-based solutions in urban areas. *Environ. Sci. Policy* **2017**, *77*, 15–24. [CrossRef]

10. Alves, A.; Gersonius, B.; Sanchez, A.; Vojinovic, Z.; Kapelan, Z. Multi-criteria Approach for Selection of Green and Grey Infrastructure to Reduce Flood Risk and Increase CO-benefits. *Water Resour. Manag.* **2018**, *32*, 2505–2522. [CrossRef]

11. Alves, A.; Gómez, J.P.; Vojinovic, Z.; Sánchez, A.; Weesakul, S. Combining Co-Benefits and Stakeholders Perceptions into Green Infrastructure Selection for Flood Risk Reduction. *Environments* **2018**, *5*, 29. [CrossRef]

12. Ossa-Moreno, J.; Smith, K.M.; Mijic, A. Economic analysis of wider benefits to facilitate SuDS uptake in London, UK. *Sustain. Cities Soc.* **2017**, *28*, 411–419. [CrossRef]

13. Demuzere, M.; Orru, K.; Heidrich, O.; Olazabal, E.; Geneletti, D.; Orru, H.; Bhave, A.G.; Mittal, N.; Feliu, E.; Faehnle, M. Mitigating and adapting to climate change: Multi-functional and multi-scale assessment of green urban infrastructure. *J. Environ. Manag.* **2014**, *146*, 107–115. [CrossRef] [PubMed]

14. Ahiablame, L.M.; Engel, B.; Chaubey, I. Effectiveness of low impact development practices in two urbanized watersheds: Retrofitting with rain barrel/cistern and porous pavement. *J. Environ. Manag.* **2013**, *119*, 151–161. [CrossRef]

15. Zahmatkesh, Z.; Burian, S.; Karamouz, M.; Tavakol-Davani, H.; Goharian, E. Low-Impact Development practices to mitigate climate change effects on urban stormwater runoff: Case study of New York City. *J. Irrig. Drain. Eng.* **2015**, *141*, 04014043. [CrossRef]

16. Garcia-Ayllon, S. Long-term GIS analysis of seaside impacts associated to infrastructures and urbanization and spatial correlation with coastal vulnerability in a mediterranean area. *Water (Switzerland)* **2018**, *10*, 1642. [CrossRef]

17. Hendel, M.; Gutierrez, P.; Colombert, M.; Diab, Y.; Royon, L. Measuring the effects of urban heat island mitigation techniques in the field: Application to the case of pavement-watering in Paris. *Urban Clim.* **2016**, *16*, 43–58. [CrossRef]

18. Wang, Y.; Berardi, U.; Akbari, H. Comparing the effects of urban heat island mitigation strategies for Toronto, Canada. *Energy Build.* **2016**, *114*, 2–19. [CrossRef]

19. Marando, F.; Salvatori, E.; Sebastiani, A.; Fusaro, L.; Manes, F. Regulating Ecosystem Services and Green Infrastructure: Assessment of Urban Heat Island effect mitigation in the municipality of Rome, Italy. *Ecol. Model.* **2019**, *392*, 92–102. [CrossRef]

20. Nastran, M.; Kobal, M.; Eler, K. Urban heat islands in relation to green land use in European cities. *Urban For. Urban Green.* **2019**, *37*, 33–41. [CrossRef]

21. Coccolo, S.; Pearlmutter, D.; Kaempf, J.; Scartezzini, J.L. Thermal Comfort Maps to estimate the impact of urban greening on the outdoor human comfort. *Urban For. Urban Green.* **2018**, *35*, 91–105. [CrossRef]

22. Amani-Beni, M.; Zhang, B.; Xie, G.D.; Shi, Y. Impacts of urban green landscape patterns on land surface temperature: Evidence from the adjacent area of Olympic Forest Park of Beijing, China. *Sustainability* **2019**, *11*, 513. [CrossRef]

23. Peng, L.L.H.; Jim, C.Y. Green-roof effects on neighborhood microclimate and human thermal sensation. *Energies* **2013**, *6*, 598–618. [CrossRef]

24. Herath, H.M.P.I.K.; Halwatura, R.U.; Jayasinghe, G.Y. Evaluation of green infrastructure effects on tropical Sri Lankan urban context as an urban heat island adaptation strategy. *Urban For. Urban Green.* **2018**, *29*, 212–222. [CrossRef]

25. Galagoda, R.U.; Jayasinghe, G.Y.; Halwatura, R.U.; Rupasinghe, H.T. The impact of urban green infrastructure as a sustainable approach towards tropical micro-climatic changes and human thermal comfort. *Urban For. Urban Green.* **2018**, *34*, 1–9. [CrossRef]

26. Tsoka, S.; Tsikaloudaki, A.; Theodosiou, T. Analyzing the ENVI-met microclimate model's performance and assessing cool materials and urban vegetation applications—A review. *Sustain. Cities Soc.* **2018**, *43*, 55–76. [CrossRef]

27. Crank, P.J.; Sailor, D.J.; Ban-Weiss, G.; Taleghani, M. Evaluating the ENVI-met microscale model for suitability in analysis of targeted urban heat mitigation strategies. *Urban Clim.* **2018**, *26*, 188–197. [CrossRef]

28. World Bank. *East Asia's Changing Urban Landscape: Measuring a Decade of Spatial Growth*; World Bank: Washington, DC, USA, 2015.

29. Srivanit, M.; Hokao, K.; Phonekeo, V. Assessing the Impact of Urbanization on Urban Thermal Environment: A Case Study of Bangkok Metropolitan. *Int. J. Appl. Sci. Technol.* **2012**, *2*, 243–256.

30. Rehan, M.M.; Weesakul, S.; Chaowiwat, W.; Charoensukrungruang, W. Development of Design Storm Pattern with Climate Change in Monsoon Asia. In Proceedings of the THA 2017 International Conference on "Water Management and Climate Change Towards Asia's Water-Energy-Food Nexus", Bangkok, Thailand, 25–27 January 2017.

31. Sheikh, Z.A. *Farmer's Perceived Agricultural Adaptation to Climate Change Impact in Rangsit Canal Area of Nong Sua District, Thailand*; Asian Institute of Technology: Bangkok, Thailand, 2016.

32. Arifwidodo, S.D.; Tanaka, T. The Characteristics of Urban Heat Island in Bangkok, Thailand. *Procedia-Soc. Behav. Sci.* **2015**, *195*, 423–428. [CrossRef]

33. Mark, O.; Weesakul, S.; Apirumanekul, C.; Aroonnet, S.B.; Djordjević, S. Potential and limitations of 1D modelling of urban flooding. *J. Hydrol.* **2004**, *299*, 284–299. [CrossRef]

34. Price, R.K.; Vojinovic, Z. *Urban Hydroinformatics: Data, Models and Decision Support for Integrated Urban Water Management*, 1st ed.; IWA Publishing: London, UK, 2011.

35. Stewart, I.D.; Oke, T.R. Local climate zones for urban temperature studies. *Bull. Am. Meteorol. Soc.* **2012**, *93*, 1879–1900. [CrossRef]

36. Coccolo, S.; Kämpf, J.; Scartezzini, J.L.; Pearlmutter, D. Outdoor human comfort and thermal stress: A comprehensive review on models and standards. *Urban Clim.* **2016**, *18*, 33–57. [CrossRef]

37. Hoppe, P. The physiological equivalent temperature—A universal index for the biometeorological assessment of the thermal environment. *Int. J. Biometeorol.* **1999**, *43*, 71–75. [CrossRef] [PubMed]

38. ISO 7726. *Ergonomics of the Thermal Environment, Instruments of Measuring Physical Quantities*; ISO: Geneva, Switzerland, 1998.

39. Ahiablame, L.M.; Engel, B.; Chaubey, I. Effectiveness of Low Impact Development Practices: Literature Review and Suggestions for Future Research. *Water Air Soil Pollut.* **2012**, *223*, 4253–4273. [CrossRef]

40. Connop, S.; Vandergert, P.; Eisenberg, B.; Collier, M.J.; Nash, C.; Clough, J.; Newport, D. Renaturing cities using a regionally-focused biodiversity-led multifunctional benefits approach to urban green infrastructure. *Environ. Sci. Policy* **2015**, *62*, 99–111. [CrossRef]

41. Sharma, R.; Joshi, P.K. Mapping environmental impacts of rapid urbanization in the National Capital Region of India using remote sensing inputs. *Urban Clim.* **2016**, *15*, 70–82. [CrossRef]

42. Tetra Tech. *BMP Siting Tool: Step-by-Step Guide*; Tetra Tech: Pasadena, CA, USA, 2013.

43. DHI. *Storm Water Runoff from Green Urban Areas—Modellers' Guideline*; DHI: Hørsholm, Danmark, 2015.

44. Vojinovic, Z.; Tutulic, D. On the use of 1D and coupled 1D-2D approaches for assessment of flood damages in urban areas. *Urban Water J.* **2009**, *6*, 183–199. [CrossRef]

45. ENVI-met. Available online: https://www.envi-met.com/ (accessed on 1 December 2016).

46. Berardi, U. The outdoor microclimate benefits and energy saving resulting from green roofs retrofits. *Energy Build.* **2016**, *121*, 217–229. [CrossRef]

47. Bruse, M. Simulating microscale climate interactions in complex terrain with a high-resolution numerical model: A case study for the Sydney CBD Area (Model Description). In Proceedings of the International Conference on Urban Climatology & International Congress of Biometeorology, Sydney, Australia, 8 November 1999.

48. Salata, F.; Golasi, I.; de Lieto Vollaro, R.; de Lieto Vollaro, A. Urban microclimate and outdoor thermal comfort. A proper procedure to fit ENVI-met simulation outputs to experimental data. *Sustain. Cities Soc.* **2016**, *26*, 318–343. [CrossRef]

49. Yang, X.; Zhao, L.; Bruse, M.; Meng, Q. Evaluation of a microclimate model for predicting the thermal behavior of different ground surfaces. *Build. Environ.* **2013**, *60*, 93–104. [CrossRef]

50. ASHRAE. *Standard 55—Thermal Environmental Conditions for Human Occupacy*; ASHRAE: Atlanta, GA, USA, 2013.

51. Lee, H.; Mayer, H.; Chen, L. Contribution of trees and grasslands to the mitigation of human heat stress in a residential district of Freiburg, Southwest Germany. *Landsc. Urban Plan.* **2016**, *148*, 37–50. [CrossRef]

52. Perini, K.; Magliocco, A. Effects of vegetation, urban density, building height, and atmospheric conditions on local temperatures and thermal comfort. *Urban For. Urban Green.* **2014**, *13*, 495–506. [CrossRef]

53. Vailshery, L.S.; Jaganmohan, M.; Nagendra, H. Effect of street trees on microclimate and air pollution in a tropical city. *Urban For. Urban Green.* **2013**, *12*, 408–415. [CrossRef]

54. Jamei, E.; Rajagopalan, P. The effect of green roof on pedestrian level air temperature. In Proceedings of the 14th Conference of International Building Performance Simulation Associatiion, Hyderabad, India, 7–9 December 2015.

55. Lobaccaro, G.; Acero, J.A. Comparative analysis of green actions to improve outdoor thermal comfort inside typical urban street canyons. *Urban Clim.* **2015**, *14*, 251–267. [CrossRef]

56. Sharma, A.; Conry, P.; Fernando, H.J.S.; Hamlet, A.F.; Hellmann, J.J.; Chen, F. Green and cool roofs to mitigate urban heat island effects in the Chicago metropolitan area: Evaluation with a regional climate model. *Environ. Res. Lett.* **2016**, *11*, 064004. [CrossRef]

57. Hedquist, B.C.; Brazel, A.J. Seasonal variability of temperatures and outdoor human comfort inPhoenix, Arizona, U.S.A. *Build. Environ.* **2014**, *72*, 377–388. [CrossRef]

58. Rizwan, A.M.; Dennis, L.Y.C.; Liu, C. A review on the generation, determination and mitigation of Urban Heat Island. *J. Environ. Sci.* **2008**, *20*, 120–128. [CrossRef]

59. Carpenter, D.D.; Kaluvakolanu, P. Effect of Roof Surface Type on Storm-Water Runoff from Full-Scale Roofs in a Temperate Climate. *J. Irrig. Drain. Eng.* **2011**, *137*, 161–169. [CrossRef]

60. Berardi, U.; GhaffarianHoseini, A.H.; GhaffarianHoseini, A. State-of-the-art analysis of the environmental benefits of green roofs. *Appl. Energy* **2014**, *115*, 411–428. [CrossRef]

61. Duarte, D.H.S.; Shinzato, P.; Gusson, C.D.S.; Alves, C.A. The impact of vegetation on urban microclimate to counterbalance built density in a subtropical changing climate. *Urban Clim.* **2015**, *14*, 224–239. [CrossRef]

62. Wang, Y.; Akbari, H. The effects of street tree planting on Urban Heat Island mitigation in Montreal. *Sustain. Cities Soc.* **2016**, *27*, 122–128. [CrossRef]

63. Vijayaraghavan, K. Green roofs: A critical review on the role of components, benefits, limitations and trends. *Renew. Sustain. Energy Rev.* **2016**, *57*, 740–752. [CrossRef]

64. Kjølby, M.J. Modellering af LAR Anlæg Samt Modellering af Stoffjernelse i LAR Anlæg og Bassiner. Available online: http://www.evanet.dk/modellering-af-lar-anlaeg-samt-modellering-af-stoffjernelse-i-lar-anlaeg-og-bassiner/ (accessed on 1 December 2016).

65. James, W.; Rossman, L.E.; James, R.C. *User Guide to SWMM5*; CHI Press: Guelph, ON, Canada, 2010.

66. Rossman, L.A. *Storm Water Management Model User's Manual Version 5.1*; US EPA: Cincinnati, OH, USA, 2015.

67. Gao, J.; Wang, R.; Huang, J.; Liu, M. Application of BMP to urban runoff control using SUSTAIN model: Case study in an industrial area. *Ecol. Model.* **2015**, *318*, 177–183. [CrossRef]

68. Mao, X.; Jia, H.; Yu, S.L. Assessing the ecological benefits of aggregate LID-BMPs through modelling. *Ecol. Model.* **2016**, *353*, 139–149. [CrossRef]

69. Maryati, S.; Humaira, A.N.S.; Adianti, P. Green Infrastructure Development in Cisangkuy Subwatershed, Bandung Regency: Potential and Problems. *Procedia-Soc. Behav. Sci.* **2016**, *227*, 617–622. [CrossRef]

70. Elliott, A.; Trowsdale, S. A review of models for low impact urban stormwater drainage. *Environ. Model. Softw.* **2007**, *22*, 394–405. [CrossRef]
71. Zhou, Q. A review of sustainable urban drainage systems considering the climate change and urbanization impacts. *Water* **2014**, *6*, 976–992. [CrossRef]
72. Declet-Barreto, J.; Brazel, A.J.; Martin, C.A.; Chow, W.T.L.; Harlan, S.L. Creating the park cool island in an inner-city neighborhood: Heat mitigation strategy for Phoenix, AZ. *Urban Ecosyst.* **2013**, *16*, 617–635. [CrossRef]

 *sustainability*

Article

Ecological Environment Vulnerability and Driving Force of Yangtze River Urban Agglomeration

Benhong Peng [1,2,*], Qianqian Huang [2,*], Ehsan Elahi [3] and Guo Wei [4]

[1] Binjiang College, Nanjing University of Information Science & Technology, Wuxi 214105, China
[2] School of Management Science and Engineering, Nanjing University of Information Science & Technology, Nanjing 210044, China
[3] School of Business, Nanjing University of Information Science and Technology, Nanjing 210044, China; ehsanelahi@nuist.edu.cn
[4] Department of Mathematics and Computer Science, University of North Carolina at Pembroke, Pembroke, NC 28372, USA; guo.wei@uncp.edu
* Correspondence: 002426@NUIST.EDU.CN (B.P.); hqq19961101@163.com (Q.H.)

Received: 15 September 2019; Accepted: 21 November 2019; Published: 23 November 2019

Abstract: The vulnerability of ecological environment threatens social and economic development. Recent studies failed to reveal the driving mechanism behind it, and there is little analysis on the spatial clustering characteristics of the vulnerability of urban agglomerations. Therefore, this article estimates ecological environment vulnerability in 2005, 2011, and 2017, determines Moran Index (MI) with spatial autocorrelation model, analyzes the spatial-temporal difference characteristics of ecological environment vulnerability of Yangtze River Urban Agglomeration and the spatial aggregation effect, and discusses its driving factors. The study results estimate that the overall vulnerability index of the Yangtze River Urban Agglomeration is in a mild fragile state. However, most fragile and slightly fragile cities are developing in the direction of moderate to severe vulnerability. The spatial agglomeration effect of the ecological environment vulnerability of the Yangtze River Urban Agglomeration is not obvious, and the effect of mutual ecological environment influence among cities is not obvious. Moreover, the driving factors of ecological environment vulnerability of Yangtze River city group changed from natural factors to social economic factors and then to policy factors. It is necessary to develop an ecological economy, coordinate the spatial agglomeration of urban agglomerations, and make balance the internal differences of urban agglomerations.

Keywords: ecological security; driving force; yangtze river urban agglomeration

1. Introduction

Urban agglomeration is an important form of urban regionalization development [1]. However, in the process of urban agglomeration development, the construction of ecological environment is neglected. Consequently, contradiction between ecology and urban agglomeration development gradually developed, and the vulnerability of ecological environment has become more prominent. In the early 20th century, Clement, an American scholar, proposed the concept of "ecological transition zone" [2]. The ecological environment vulnerability refers to the weak ability of the ecological environment that can resist when the regional environment is externally interfered [3]. The ability to recover after being disturbed is low, and it is difficult to change the current vulnerability status. Currently, ecological and environmental problems such as global warming has reduced per capita arable land, forest resources, supply of fresh water, and biological species [4]. This destruction has increased the vulnerability of the ecological environment.

China's urban ecological environment has low carrying capacity, is sensitive and difficult to recover, and has weak anti-interference ability. Barry and Wandel (2006) [5] studies the ecological adaptability of communities but have not focused on urban agglomerations. However, in the mature stage of urban development, the collection of highly integrated cities forms urban agglomerations [6]. Under the rapid development of the Yangtze River Urban Agglomeration, it faces problems such as over-exploitation, lagging ecological governance, and imperfect environmental governance mechanisms. These restrict the realization of high-quality development of the ecological environment and the upgrading of urban agglomerations [7]. From the perspective of ecological environment vulnerability, this article studies the ecological environment of the Yangtze River Urban Agglomeration, solves outstanding environmental problems, examines its ecological environment vulnerability, and explores the driving factors behind it [8].

At present, scholars at home and abroad have carried out a wealth of research on the vulnerability of ecological environment and achieved fruitful results. From the perspective of research objects, most of them are concentrated in river basins [9,10], extremely arid desert climate areas [11], developed cities [12], and areas rich in ecological resources [13], but for the newly divided Yangtze River city group with high level of economic development, high intensity of industrial agglomeration and opening up in recent years. There are few studies on environmental vulnerability. In terms of research methods, the current research on ecological vulnerability evaluation mainly includes SRP evaluation model [12], analytic hierarchy process [14], fuzzy cloud model [15], comprehensive index method [16], and principal component analysis method [12]. At the same time, combining MI and Lisa cluster map [17], geographic information system, and remote sensing data image [18], quantitative analysis of the driving forces affecting the ecological vulnerability of the study area is carried out. For example, Li et al. (2016) [12] used the SRP evaluation model to obtain the spatiotemporal dynamics of ecological environment vulnerability in Chaoyang County, Beijing, China. Li et al. (2016) [12] used principal component analysis to explore the main driving factors of ecological environment vulnerability. However, the existing research methods have the following problems. First, the selection of evaluation factors will have the influence of human subjective factors, which is difficult to reflect the objectivity of evaluation indicators, and the subjective weighting of evaluation factors will also affect the objectivity of results. Second, the repeatability of evaluation indicators, whether it is natural indicators, social and economic indicators, or policy indicators, will have internal indicators. In the evaluation process, there will be redundancy and repeated conclusions. From the perspective of research content, most of them are from the analysis of ecological environment vulnerability, such as biological quantitative characteristics, spatial distribution and differences. The analysis of driving factors and spatial agglomeration characteristics of ecological environment vulnerability in different years is insufficient.

According to the above analysis, the work of this paper is as follows. First, to build the evaluation index of ecological environment vulnerability of Yangtze River Urban Agglomeration. Second, to calculate the index of ecological environment vulnerability, to use the spatial principal component analysis method and Moran index analysis, and to get the interaction relationship between ecological environment vulnerability regions. Finally, to analyze the driving mechanism of ecological environment vulnerability of Yangtze River Urban Agglomeration. This paper summarizes the temporal and spatial evolution law of the ecological environment vulnerability of the Yangtze River Urban Agglomeration and puts forward suggestions and Countermeasures for the green development of the ecological environment of the Yangtze River Urban Agglomeration.

2. Research Methodology

2.1. Study Area

The Yangtze River Urban Agglomeration is located in the southeastern part of China, including Jiangsu Nanjing, Zhenjiang, Changzhou, Wuxi, Suzhou, Yangzhou, Taizhou, and Nantong along the Yangtze river [19]. Figure 1 illustrates the key ecological protection area for the ecological environment of the Yangtze River Urban Agglomeration. Within the red circle is the Yangtze River city group, and each red circle is a city. It is quite necessary to work on environmental vulnerability to create a good ecological environment and improve the protection mechanism of the Yangtze river basin [20].

Figure 1. Location map of Yangtze River Urban Agglomeration.

2.2. Data Source

The data of elevation, slope, and land use degree were obtained from remote sensing maps and DEM analysis of elevation, slope, and land use degree of each city. Landscape diversity is based on land related data. Lithology and soil type data were mainly obtained by digitizing geological maps on the website of China geological bureau. The normalized difference vegetation index (NDVI) was sourced from the Resource and Environmental Science Data Center of the Chinese Academy of Sciences (www.resdc.cn). The annual precipitation, annual temperature population density, per capita GDP, and road network density were mainly derived from the annual statistical yearbook and statistical bulletin of the municipal statistics bureau of the Yangtze River Urban Agglomeration.

2.3. Construction of Ecological Environment Vulnerability Assessment Model

Referring to the index selection method of Tian (2012) [21] genetic model, 12 indicators such as elevation and slope were selected to construct the evaluation index system (Table 1). Among them, the slope and elevation are positive indicators, reflecting the characteristics of topography and geomorphology. When the slope and elevation are larger, the stability is worse, and it is easy to be eroded by rainstorm. The soil type and lithology are literal data, different types of soil reflect the intensity of erosion, and the lithology reflects the geological conditions of the area, reflecting the weathering resistance. NDVI is a negative indicator, reflecting the growth of vegetation, and the larger the value is, the more stable the ecosystem is, while the more moderate the annual average precipitation is, the more abundant the annual average precipitation is. The more negative the annual average temperature is, the more stable the ecosystem is. The population density, per capita GDP, and road network density are positive indicators, which reflect the stress of the social and economic development on the ecological environment. The landscape diversity and land use degree reflect the environmental pressure brought by unreasonable land use. Land use degree is a positive indicator while landscape diversity is a negative indicator.

Table 1. The index system for assessing the vulnerability of the Yangtze River Urban Agglomeration based on the "genesis-result" model.

Sub Indicator	Unit	Nature of the Indicator	Source of Indicators
Elevation	Meter	Negative indicator	Li (2016) [12]
Landscape diversity	——	Positive indicator	
Annual average temperature	Celsius	Positive indicator	
Lithology	——	Negative indicator	Du (2016) [22]
slope	degree	Negative indicator	Li (2016) [12]
Soil type	——	Negative indicator	
NDVI	——	Positive indicator	
Average annual precipitation	Millimeter	Negative indicator	
The population density	People per square kilometer	Negative indicator	
Per capita GDP	Ten thousand yuan/person	Negative indicator	Ma (2015) [23]
Road network density	Km/km^2	Negative indicator	Xu (2012) [24]
Land use	%	Negative indicator	Tian (2012) [21]

2.4. Standardization of the Data

Following Li (2016) [12] 12 indicators such as elevation, slope, lithology, soil type, NDVI, annual precipitation, annual temperature, population density, per capita GDP, road network density, landscape diversity, and land use degree in 2005, 2011, and 2017 were selected. The original digital data need to input, but the lithology and soil type are written data which is not convenient for data analysis. Therefore, following Lin (2018) [17] on the basis of the cause analysis, the standardized valuation method of was used and two indexes of lithology and soil type were normalized to 2, 4, 6, 8, and 10 to ensure the analysis of data and make it unified (Table 2).

Table 2. Assignment and standardization of ecological environment vulnerability assessment indicators.

Evaluation Index	Standardized Assignment				
	2	4	6	8	10
Lithology	Mudstone, limestone, clay rock, shale	Conglomerate, breccia, siltstone	Schist, quartzite, marble, amphibolite	Andesite, fluke, tuff	Granite, granite porphyry, monzonite
Soil type	Paddy soil, gray tidal soil	Saline soil, wind sand	Brick red soil, red soil	Yellow soil, coarse soil, lime soil	Purple soil

According to the grading standard of Du (2016) [22] ecological environment vulnerability assessment, the ecological environment vulnerability index in 2005 was graded by natural breaks method, and the grading of ecological environment vulnerability index is shown in Table 3 below.

Table 3. Ecological environment vulnerability and its ecological characteristics.

Vulnerability Level	Vulnerability Index	Ecological Characteristics
Micro-fragility	<-0.75	Good ecological environment, reasonable structure, strong anti-interference ability, stable ecosystem, and high ecological security.
Mildly fragile	$(-0.75)–(-0.02)$	The ecological environment is relatively good, the structural configuration is relatively reasonable, the anti-interference ability is relatively strong, the ecological system is relatively stable, and the ecological security degree is relatively high.
Moderately vulnerable	$(-0.02)–0.79$	The ecological environment is general, the structure is general, the anti-interference ability is general, the ecosystem is unstable, and the ecological security is general.
Heavier and more fragile	>0.79	Poor ecological environment, unreasonable structural configuration, poor anti-interference ability, unstable ecosystem, and low ecological security.

Following Ord and Arthur (2010) [25], Local Indicators of Spatial Association (LISA) clustering map was obtained by spatial clustering on the calculation results of the local MI. The meaning of different spatial aggregation modes is defined in Table 4.

Table 4. Connotation of different LISA aggregation modes.

Aggregate Type	Meaning
High-High accumulation (H-H)	Areas with high observations are surrounded by high-value areas space agglomeration.
High-Low accumulation (H-L)	High-value areas around low-value areas space agglomeration.
Low-High accumulation (L-H)	High-value areas around low-observation areas space agglomeration.
Low-Low accumulation (L-L)	Areas with low observations are surrounded by low-value areas spatial agglomeration feature.
Not obvious	There are no significant spatial agglomeration features.

3. Analytical Methods

3.1. Principal Component Analysis

The principal component analysis can simplify through dimensionality reduction analysis.

$$F_i = \alpha_1 x_1 + \alpha_2 x_2 + \alpha_3 x_3 + \cdots\cdots + \alpha_n x_n \tag{1}$$

where F_i is the i-*th* principal component, α is a feature vector, and $x_i, i \in \{1, 2, 3 \cdots\cdots n\}$ is the selected evaluation index.

3.2. MI Analysis

The current spatial autocorrelation statistical analysis methods mainly include MI and Geary's C index. Following Cliff and Ord (1981) [26], the global and local MI were used to evaluate and analyze the ecological vulnerability. The spatial distribution diagram and LISA cluster diagram of vulnerability were drawn to facilitate the analysis of spatiotemporal differences of ecological environment vulnerability and its driving forces in the following sections.

The grading standard of global MI can be calculated as:

$$I = \frac{\sum\limits_{i=1}^{n} \sum\limits_{j=1}^{n} w_{ij}(x_i - \bar{x})(x_j - \bar{x})}{\sum\limits_{i=1}^{n} \sum\limits_{j=1}^{n} w_{ij} \sum\limits_{i=1}^{n} (x_i - \bar{x})^2} \tag{2}$$

The calculation formula for the local MI is given as:

$$I = \frac{(x_i - \bar{x})}{s^2} \sum_{j} w_{ij}(x_j - \bar{x}) \tag{3}$$

where I represents the MI, x_i, x_j represents the mean of the vulnerability index in the i-*th*, j-*th* evaluation unit, $\bar{x}$ refers to the mean of the vulnerability of all evaluation units, W_{ij} refers to the spatial weight matrix, and S represents the sum of the elements of the spatial weight matrix.

3.3. Ecological Environment Vulnerability Index

The ecological environment vulnerability index of the Yangtze River Urban Agglomeration can be determined by using Equation (4):

$$EVI = r_1 y_1 + r_2 y_2 + r_3 y_3 + \ldots + r_n y_n \tag{4}$$

where EVI is the eco-vulnerability composite index, r_i is the contribution rate of the i-*th* spatial principal component, and y_i is the value of the i-*th* spatial principal component.

4. Results and Discussion

4.1. Ecological Environment Vulnerability and its Ecological Characteristics

The Eco-environment Vulnerability Index (EVI) was used to evaluate the ecological environment vulnerability of the Yangtze River Urban Agglomeration. Principal component analysis was used to analyze 12 evaluation indicators to eliminate the overlap and correlation in the indicator information. The four principal components of 2005, 2011, and 2017 were determined according to the cumulative contribution rate of the principal component of 85% or more (Table 5). Among them, the cumulative contribution rate in 2005 was 90.64%, the cumulative contribution rate in 2011 was 88.36%, and the cumulative contribution rate in 2017 was 92.14%, respectively.

Table 5. Characteristic values of each principal component, contribution rate, and cumulative contribution rate.

Years	Principal Component Coefficient	Main Ingredient			
		PC1	PC2	PC3	PC4
2005	Characteristic value λ	4.2	3.19	2.25	1.23
	Contribution rate%	34.97	26.62	18.77	10.28
	Accumulated contribution rate%	34.97	61.59	80.36	90.64
2011	Characteristic value λ	3.84	2.62	2.19	1.96
	Contribution rate%	32.02	21.79	18.24	16.32
	Accumulated contribution rate%	32.02	53.81	72.05	88.36
2017	Characteristic value λ	3.56	3.5	2.2	1.8
	Contribution rate%	29.67	29.12	18.41	14.94
	Accumulated contribution rate%	29.67	58.79	77.2	92.14

Using data of Table 5, the Equations (5)–(7) can be estimated.

$$EVI_{2005} = 0.3497 \times Y_1 + 0.2662 \times Y_2 + 0.1877 \times Y_3 + 0.1028 \times Y_4 \tag{5}$$

$$EVI_{2011} = 0.3202 \times Y_1 + 0.2179 \times Y_2 + 0.1824 \times Y_3 + 0.1632 \times Y_4 \tag{6}$$

$$EVI_{2017} = 0.2967 \times Y_1 + 0.2912 \times Y_2 + 0.1841 \times Y_3 + 0.1494 \times Y_4 \tag{7}$$

where Y_1–Y_4 are the normalized values of the first four principal components extracted by spatial principal component analysis. Due to the excessive number of charts related to normalized values, EVI in 2005, 2011, and 2017 are mainly shown.

Table 6 depicts that only Nanjing was severely vulnerable in 2005 and 2011. In 2017, both Nanjing and Suzhou were severely vulnerable. Nantong has been in a state of mild vulnerability, and Nantong city ranked first in the province in terms of its green development index in 2016. Recently, with the implementation of ecological protection related policies, the ecological environment vulnerability index of the Yangtze River Urban Agglomeration has gradually become better, but the economic development would always be moderately and severely fragile with the development and utilization of the ecological environment.

Table 6. The EVI (2005–2017).

CITY	2005 EVI	2011EVI	2017EVI
Nanjing	0.7948	0.6950	0.4443
Zhenjiang	0.1580	−0.2163	0.2825
Changzhou	0.1131	−0.1328	0.0284
Wuxi	−0.2294	−0.1513	0.1455
Suzhou	0.3948	0.5973	0.3643
Yangzhou	−0.1242	0.1357	0.0192
Taizhou	−0.4868	−0.2596	−0.2264
Nantong	−0.6203	−0.6680	−1.0580

4.2. Characteristics of Time Difference of Ecological Environment Vulnerability

The time distribution characteristics of the ecological environment vulnerability of the Yangtze River Urban Agglomeration from 2005 to 2017 were obtained (Figure 2). Results found that in 2005, Nanjing was relatively fragile, while Zhenjiang, Changzhou, and Suzhou were moderately vulnerable. Taizhou and Wuxi were slightly fragile, and the rest of the cities were slightly vulnerable. In 2011, Nanjing and Suzhou were relatively weak and vulnerable. Yangzhou was in the middle vulnerability, Nantong was slightly fragile, and other cities were slightly vulnerable. In 2017, Nanjing and Suzhou were relatively weak, while Zhenjiang and Wuxi were moderately vulnerable. Nantong was slightly fragile, and the rest of the cities were slightly vulnerable. Overall, the ecological environment of the Yangtze River Urban Agglomeration is moderately and slightly fragile. Our results are in line with [27]. (In Figure 2, "horizontal ordinate 1" represents 2005, "horizontal ordinate 2" represents 2011, and "horizontal ordinate 3" represents 2017. The vertical coordinate represents the EVI.)

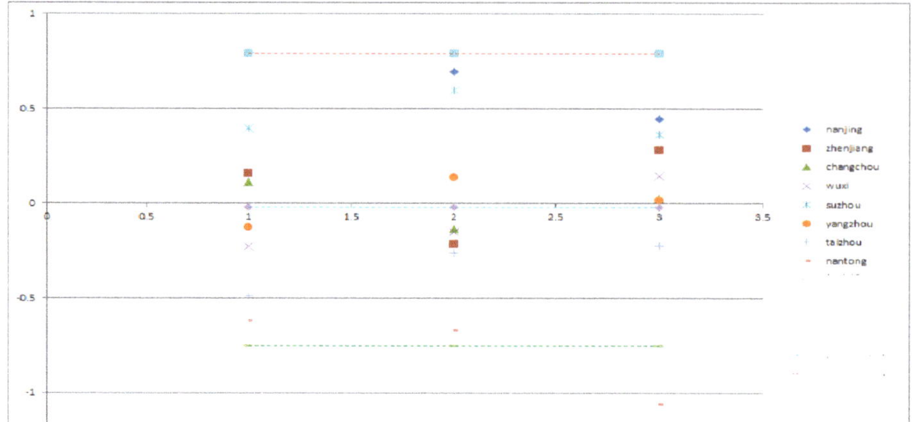

Figure 2. Distribution of ecological environment vulnerability index of Yangtze River Urban Agglomeration from 2005 to 2017.

Figure 3 illustrates that the vulnerability index of Nanjing has declined, but it is still in a severe vulnerability. The vulnerability index of Zhenjiang, Changzhou, Suzhou, and Yangzhou has changed repeatedly, showing a trend of first rise and then fall or, first fall and then rise. Suzhou and Yangzhou have improved gradually, but other regions were more serious. The severity in Taizhou was increased year by year but the vulnerability index was always low. The vulnerability index of Nantong was decreased year by year and the ecological environment was excellent.

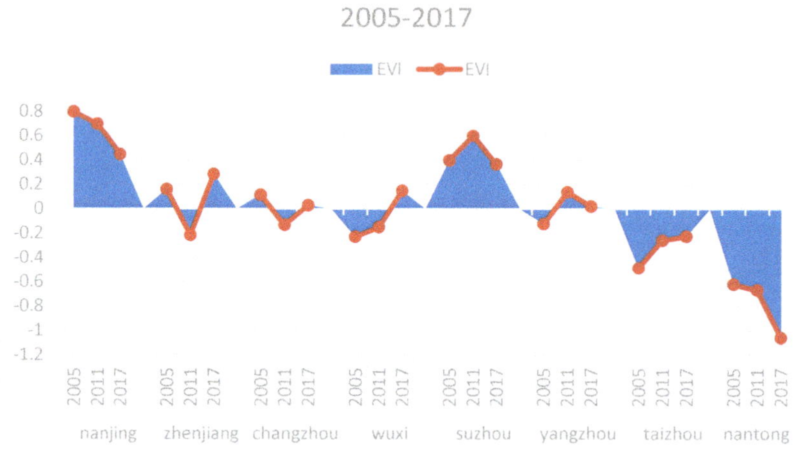

Figure 3. The EVI trend map of Yangzijiang urban agglomeration from 2005 to 2017.

4.3. Spatial Difference Characteristics of Ecological Environment Vulnerability

The distribution map of the ecological environment vulnerability of the Yangtze River Urban Agglomeration illustrates that during 2005 and 2017, the ecological environment fragility of the Yangtze River Urban Agglomeration showed an increasing trend from the central to the northwest. The east and west were more vulnerable and the middle was weaker. Moreover, between 2005 and 2011, the ecological environment vulnerability of Nanjing and Suzhou were basically relatively fragile, which was closely related to the rapid development of industrial economy and over-exploitation in recent years (Figure 4). Overall, between 2005 and 2017, the vulnerable areas of the Yangtze River Urban Agglomeration were mainly distributed in the southeast and west and the vulnerability of the central and northeastern parts did not change significantly.

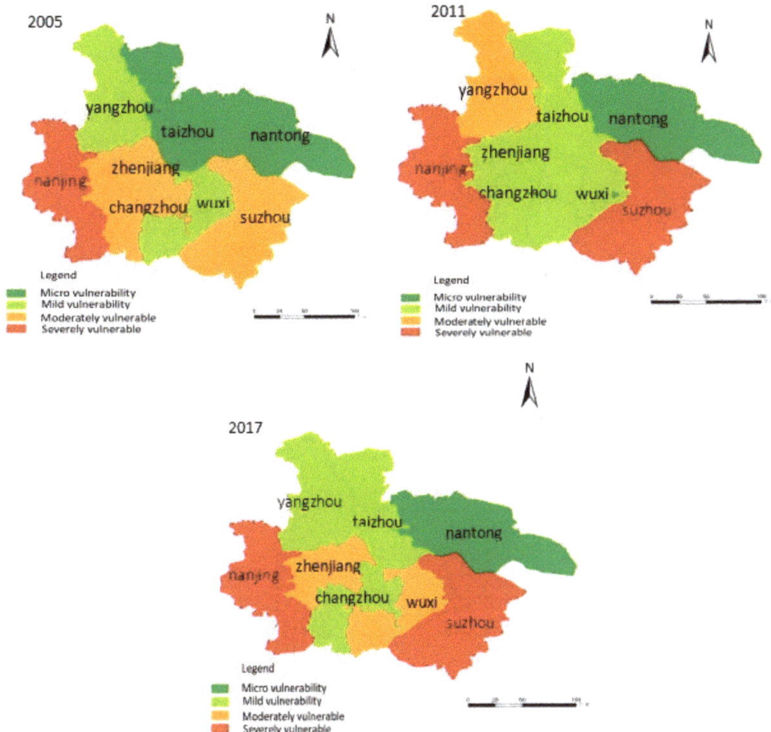

Figure 4. Spatial distribution of ecological environment vulnerability of the Yangtze River Urban Agglomeration from 2005 to 2017.

4.4. Characteristics of Spatial Clustering of Ecological Environment Vulnerability

4.4.1. Global MI

Based on the evaluation results of ecological environment vulnerability in 2005, 2011, and 2017, the global MI of ecological environment vulnerability was calculated (Table 7). Moran Index is calculated as −0.0567, −0.2636, and −0.0745, indicating that there is a negative correlation, which is not significant. It is speculated that this is related to the selected indicators.

Table 7. Spatial autocorrelation parameters of ecological environment vulnerability from 2005 to 2017.

Year	Moran I	Expected Value	Z Value	*p* Value
2005	−0.0567	−0.1429	0.1888	0.2700
2011	−0.2636	−0.1429	0.6540	0.2740
2017	−0.0745	−0.1429	0.1424	0.2740

4.4.2. Local MI

On the basis of the global MI, the local MI was analyzed and a LISA clustering map is drawn as to explore the vulnerability aggregation state (Figure 5). Since the analysis in 2017 showed no correlation, the LISA maps in 2005 and 2011 did not change, and only the ecologically vulnerable LISA cluster map in 2005 was retained. From the perspective of space, in 2005 and 2011, the ecological environment vulnerability of the Yangtze River Urban Agglomeration has showed a small spatial agglomeration, especially in Suzhou, showing high-low concentration and indicating that there was no obvious spatial agglomeration. The low-low concentration of the Taizhou area indicates that the ecological environment of the area is better.

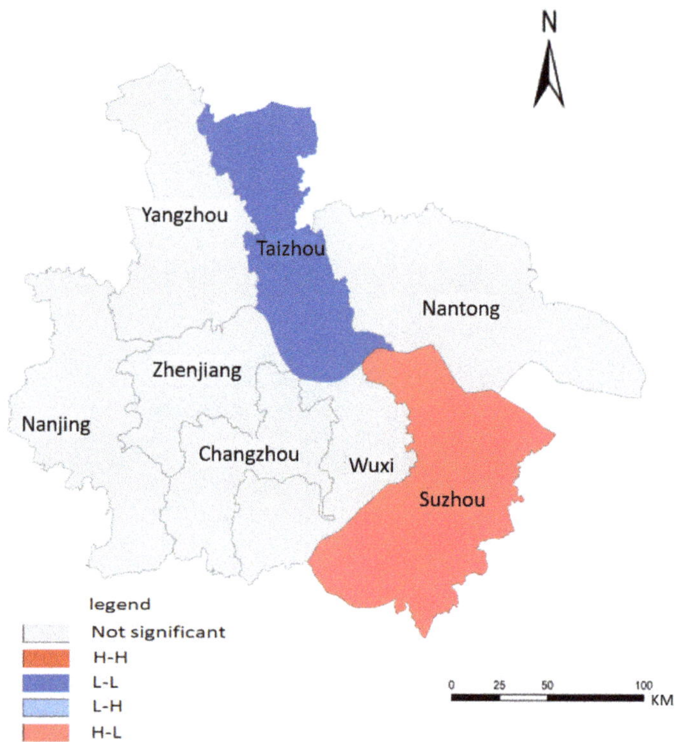

Figure 5. LISA cluster diagram of the ecological environment vulnerability of the Yangtze River Urban Agglomeration.

4.5. Analysis of the Driving Force of Ecological Environment Vulnerability

Using principal component analysis, the Yangtze River Urban Agglomeration spatial and temporal differences of ecological environment vulnerability and its driving force was analyzed for 2005, 2011, and 2017 using the indexes of ecological environment vulnerability.

4.5.1. Analysis of the Driving Force of Ecological Environment Vulnerability (2005)

In 2005, in the PC1, the contribution rate of annual average precipitation and road network density is large; in the PC2, the contribution rate of NDNI is large; in the PC3, the contribution rate of slope is large; in the PC4, the contribution rate of lithology is large (Table 8). This is in line with the rapid development of social economy in the region and the impact of industrial development on the ecological environment has initially appeared, but the main driving factor is the natural factor.

Table 8. Principal component load matrix (2005).

Index	Factor Load Factor			
	PC1	PC2	PC3	PC4
Road network density	0.91	0.04	−0.25	0.266
Average annual precipitation	0.897	−0.081	−0.084	0.335
NDVI	−0.249	0.919	−0.022	0.072
Slope	−0.009	−0.209	0.911	0.173
Lithology	0.582	0.512	0.179	−0.605
The population density	−0.858	−0.027	−0.439	0.063
Landscape diversity	−0.789	−0.005	0.401	−0.026
Annual average temperature	−0.703	0.05	0.201	−0.237
Soil type	0.466	0.658	0.331	−0.464
Land use	−0.232	0.859	0.072	0.41
Elevation	0.162	−0.196	0.901	0.286
Per capita GDP	−0.113	0.907	−0.001	0.352

4.5.2. Analysis of the Driving Force of Ecological Environment Vulnerability (2011)

In the PC1 in 2011, the larger contribution rate is population density and land use degree, indicating that with the development and utilization of land, its impact on environmental vulnerability is gradually increasing. In the PC2, the greater contribution rate is road network density; in the PC3, the greater contribution rate is slope; and in the PC4, the greater contribution rate is lithology (Table 9). Compared with 2005, with the sustained and high-speed development of social economy, more social and economic factors have become the driving factors of the vulnerability of ecological environment. The reason lies in that in 2011, cities first developed industrial economy, and the increase of personnel density indirectly affected the intensity of land development. However, the overall topographic and geomorphic characteristics have not changed greatly, so the slope and lithology are still the main components.

Table 9. Principal component load matrix (2011).

Index	Factor Load Factor			
	PC1	**PC2**	**PC3**	**PC4**
The population density	0.969	−0.156	−0.112	0.135
Land use	0.76	0.361	−0.24	0.152
Road network density	0.009	−0.874	−0.026	0.015
Slope	−0.414	0.55	0.693	−0.043
Lithology	−0.457	0.187	−0.317	0.806
Per capita GDP	0.657	0.713	−0.056	0.151
NDVI	0.632	0.625	0.453	−0.011
Landscape diversity	0.542	−0.168	0.675	0.385
Elevation	−0.476	0.393	0.671	−0.081
Average annual precipitation	−0.416	0.761	−0.289	−0.246
Soil type	−0.351	0.425	−0.24	0.794
Annual average temperature	0.124	0.629	−0.551	−0.481

4.5.3. Analysis of the Driving Force of Ecological Environment Vulnerability (2017)

In 2017, in the PC1, the larger contribution rate is population density and land use degree, indicating that with the development and utilization of land, its impact on environmental vulnerability is gradually increasing. In the PC2, the greater contribution rate is road network density; in the PC3, the greater contribution rate is slope; and in the PC4, the greater contribution rate is lithology (Table 10). Compared with the driving factors of ecological environment vulnerability in 2011, only the population density is still the main component because the government carries out a series of ecological protection leading areas, promotes environmental supervision reform, etc. With the increase of environmental protection and environmental awareness, policy factors become the main driving force of ecological environment vulnerability.

Table 10. Principal component load matrix. (2017).

Index	Factor Load Factor			
	PC1	**PC2**	**PC3**	**PC4**
Per capita GDP	0.823	0.309	0.39	0.038
Soil type	0.77	−0.002	−0.305	0.549
The population density	0.177	0.924	0.311	−0.053
Landscape diversity	−0.214	0.388	0.793	0.292
Annual average temperature	0.749	−0.097	−0.211	−0.596
Road network density	−0.718	0.377	−0.313	0.276
Average annual precipitation	0.694	−0.6	−0.334	0.014
Land use	0.663	0.607	0.109	0.245
Lithology	0.655	0.279	−0.532	0.304
NDVI	0.516	−0.082	0.764	−0.269
Slope	0.14	−0.795	0.519	0.196
Elevation	0.007	−0.779	0.337	0.421

Results found that in 2005, urban development paid more attention to develop rapidly, ignoring the ecological environment protection under the blind development economy. Since 2011, socioeconomic factors have become the main driving force of ecological environment vulnerability, indicating that the ecological environment of the Yangtze River Urban Agglomeration is increasingly affected by human socioeconomic activities. Cities with better development momentum are beginning to realize the importance of environmental protection. These findings are in line with [28].

In 2017, the Yangtze River Urban Agglomeration found negative effects of human activities on the ecological environment. Zhao et al. (2006) [29] conducted a study on the ecological consequences of rapid urban expansion in Shanghai province of China and also found negative interaction of human activities with ecological environment. Implementation of environmental policies are required to protect ecological system [30]. However, due to weak ecological resilience, it takes years of efforts to restore the ecological environment. Once the ecological environment is destroyed, it takes 100 years to recover, but with the changes in driving factors, it also reflects that the environmental behavior of these years is still effective [31].

5. Conclusion and Recommendations

Taking the Yangtze River Urban Agglomeration as the research object, the dynamic ecological environment vulnerability index of each region is calculated and classified. Based on the Moran Index (MI), the spatial agglomeration characteristics were obtained, and the spatial and temporal distribution characteristics and driving forces of the ecological environment vulnerability in the region were explored. From the analysis, the results can be summarized as:

(1) The degree of economic development has a great impact on the ecological vulnerability index. Nanjing and Suzhou have been at the forefront of economic development, facing severely fragile ecological risks. The economic aggregates of Nantong and Taizhou in the northeast are behind Nanjing and Suzhou, but the ecological environment is good. In the past 10 years, the industrial structure of Nanjing and Suzhou has focused on the chemical industry. Air pollution, water pollution, and land pollution have caused certain damages to the ecological environment. In addition, increasing population size and distribution have negative impact on the ecological environment.

(2) The spatial agglomeration of ecological fragility is low and the ecological environment hazards among cities are weak. The spatial agglomeration of the Yangtze River Urban Agglomeration has not changed significantly, and the interaction between cities was little. The spatial agglomeration of Suzhou presents high-low concentration, indicating that Suzhou's ecological environment vulnerability index is large, ecological and environmental issues are significant, and other cities ecological vulnerability index was lower than that of Suzhou. Taizhou, which was a good ecological environment, presents low-low concentration, indicates that the ecological environment of Taizhou and its surrounding cities is good. The spatial agglomeration effect was not significant and it would have a great impact on the economic development of the Yangtze River Urban Agglomeration. In the process of urban development in the new era, the spatial agglomeration effect was significantly more beneficial than disadvantages.

(3) The driving factors of ecological environment vulnerability have changed, and it has been found that from natural factors to social factors to policy factors. In 2005, it was still the initial stage of economic development of the Yangtze River Urban Agglomeration. It is in the stage of economic growth, with good ecological carrying capacity, and can be well digested and treated for human activities. In 2011, the cities within the Yangtze River Urban Agglomeration were gradually developed, and the demand for natural resources was increased, which caused certain damages to the ecological environment. The gradual policy influence factors appear in 2017, and the government has formulated a strategic goal of ecological environmental protection to provide a solid ecological environment for high-quality development. Based on results, the study suggests following recommendations.

The driving factors of the ecological environment vulnerability of the Yangtze River Urban Agglomeration were changed from natural to social economic factors. It is necessary to properly control the population to adapt to the development of the ecological environment, rationally plan to use the land, establish an ecological protection zone, reasonably plan the mileage and location of the railway and highway, and the density of the road network should be consistent with the development of the ecological environment. It is a dire need to strictly control pollution sources, integrate various resources and technologies to rectify contaminated areas, and increase the construction of infrastructure conditions necessary to protect the ecological environment.

Author Contributions: Conceptualization, B.P. and G.W.; Data curation, Q.H.; Formal analysis, Q.H.; Investigation, E.E.; Methodology, B.P.; Project administration, B.P. and G.W.; Software, B.P. and Q.H.; Supervision, E.E.; Writing—original draft, B.P. and Q.H.; Writing—review & editing, B.P., Q.H., E.E. and G.W.

Funding: This research was funded by HRSA, US DHHS, grant number H49MC00068; and was funded by The National Natural Science Foundation of China, grant number 71263040; and was funded by Key Project of National Social and Scientific Fund Program, grant number 18ZDA052; and was funded by Project of National Social and Scientific Fund Program, grant number 17BGL142; and was funded by Open Project of Jiangsu Productivity Society, grant number JSSCL 2018A005; and was funded by Soft Science Project of China Meteorological Administration, grant number 2019ZDIANXM25; and was funded by The Natural Science Foundation of China, grant number 91546117.

Acknowledgments: The authors are grateful to the case company for permitting and supporting this research. This work was financially supported by HRSA, US DHHS (Grant number. H49MC00068), the National Natural Science Foundation of China (No. 71263040), Key Project of National Social and Scientific Fund Program (18ZDA052), Project of National Social and Scientific Fund Program (17BGL142), Open Project of Jiangsu Productivity Society (JSSCL 2018A005), Soft Science Project of China Meteorological Administration (2019ZDIANXM25). The Natural Science Foundation of China (91546117).

Conflicts of Interest: The authors declare no conflict of interest.

References

1. Herold, M.; Goldstein, N.C.; Clarke, K.C. The spatiotemporal form of urban growth: Measurement, analysis and modeling. *Remote Sens. Environ.* **2003**, *86*, 286–302. [CrossRef]

2. Rakotoarisoa, J.-E.; Raheriarisena, M.; Goodman, S.M. A phylogeographic study of the endemic rodent Eliurus carletoni (Rodentia: Nesomyinae) in an ecological transition zone of northern Madagascar. *J. Hered.* **2013**, *104*, 23–35. [CrossRef] [PubMed]

3. Zhao, Y.Z.; Zou, X.Y.; Cheng, H.; Jia, H.K.; Wu, Y.Q.; Wang, G.Y.; Zhang, C.L.; Gao, S.Y. Assessing the ecological security of the Tibetan plateau: Methodology and a case study for Lhaze County. *J. Environ. Manag.* **2006**, *80*, 120–131. [CrossRef] [PubMed]

4. Wackernagel, M.; Onisto, L.; Bello, P.; Linares, A.C.; Falfán, I.S.L.; García, J.M.; Guerrero, A.I.S.; Guerrero, M.G.S. National natural capital accounting with the ecological footprint concept. *Ecol. Econ.* **1999**, *29*, 375–390. [CrossRef]

5. Smit, B.; Wandel, J. Adaptation, adaptive capacity and vulnerability. *Glob. Environ. Chang.* **2006**, *16*, 282–292. [CrossRef]

6. Whitford, V.; Ennos, A.R.; Handley, J.F. "City form and natural process"—Indicators for the ecological performance of urban areas and their application to Merseyside, UK. *J. Arch. Eng.* **2001**, *57*, 91–103.

7. Hongfu, X. Current situation and Countermeasures of sustainable development of water resources in small and medium-sized cities in water shortage areas of China. *Res. Water Conserv. Dev.* **2007**, *11*, 48–50.

8. Collins, S.L.; Carpenter, S.R.; Swinton, S.M.; Orenstein, D.E.; Childers, D.L.; Gragson, T.L.; Grimm, N.B.; Grove, J.M.; Harlan, S.L.; Kaye, J.P. An integrated conceptual framework for long-term social–ecological research. *Front. Ecol. Environ.* **2011**, *9*, 351–357. [CrossRef]

9. Wang, S.Y.; Liu, J.S.; Yang, C.J. Eco-environmental vulnerability evaluation in the Yellow River Basin, China. *Pedosphere* **2008**, *18*, 171–182. [CrossRef]

10. Kong, L.Q.; Zhang, L.; Zheng, H.; Xu, W.H.; Xiao, Y.; Ouyang, Z.Y. Driving forces behind ecosystem spatial changes in the Yangtze River Basin. *Acta Ecol. Sin.* **2018**, *38*, 741–749. [CrossRef]

11. Chen, H.M.; Li, Q. Assessment and analysis on the water resource vulnerability in arid zone based on the PSR model. *Adv. Mater. Res.* **2014**, *955–959*, 3757–3760. [CrossRef]

12. Li, Y.; Fan, Q.; Wang, X.; Xi, J.; Wang, S.; Yang, J. Temporal and spatial differentiation of ecological vulnerability in natural disaster-prone areas based on SRP model: A case study of Chaoyang County, Liaoning Province. *Geogr. Sci.* **2016**, *11*, 1452–1459. (In Chinese)

13. Shi, Y.Z.; Li, F.L.; Fan, M.Y.; Liu, H.J.; Yang, X.F. Applications of variable fuzzy set theory and GIS for water resources vulnerability assessment of Yellow River Delta. *Appl. Mech. Mater.* **2014**, *700*, 501–505. [CrossRef]

14. Nguyen, A.K.; Liou, Y.A.; Li, M.H.; Tran, T.A. Zoning eco-environmental vulnerability for environmental management and protection. *Ecol. Indic.* **2016**, *69*, 100–117. [CrossRef]

15. Jing, S. *Comprehensive Assessment of Ecological Vulnerability and Research on Development Strategies*; North China Electric Power University: Beijing, China, 2018.

16. Yao, X.; Yu, K.Y.; Liu, J.; Yang, S.P.; He, P.; Deng, Y.B.; Yu, X.Y.; Chen, Z.H. Spatial and temporal changes of the ecological vulnerability in a serious soil erosion area, Southern China. *Yingyong Shengtai Xuebao J. Appl. Ecol.* **2016**, *27*, 735–745.

17. Lin, J.; Hu, G.; Yan, X.; Xu, C.; Zhang, A.; Chen, W.; Shuai, C.; Liang, C. The ecological environment vulnerability and its driving force in the urban agglomeration of the Yangtze River Delta. *Acta Ecol. Sin.* **2018**, *38*, 4155–4166. (In Chinese)

18. Kang, H.; Tao, W.; Chang, Y.; Zhang, Y.; Li, X.; Chen, P. A feasible method for the division of ecological vulnerability and its driving forces in Southern Shaanxi. *J. Clean. Prod.* **2018**, *205*, 619–628. [CrossRef]

19. Chen, W.; Cutter, S.L.; Emrich, C.T.; Shi, P. Measuring social vulnerability to natural hazards in the Yangtze River Delta region, China. *Int. J. Disaster Risk Sci.* **2013**, *4*, 169–181. [CrossRef]

20. Gu, C.; Hu, L.; Zhang, X.; Wang, X.; Guo, J. Climate change and urbanization in the Yangtze River Delta. *Habitat Int.* **2011**, *35*, 544–552. [CrossRef]

21. Tian, Y.; Chang, W. Bibliometric analysis of the research progress of ecological vulnerability in China. *Acta Geogr. Sin.* **2012**, *67*, 1515–1525. (In Chinese)

22. Du, Y.; Peng, J.; Zhao, S.; Hu, Z.C.; Wang, Y.L. Ecological risk assessment of landslide hazards in Southwestern China: A case study of Dali Bai Autonomous Prefecture. *J. Geogr. Sci.* **2016**, *71*, 1544–1561. (In Chinese)

23. Ma, J.; Li, C.; Wei, H.; Ma, P.; Yang, Y.; Ren, Q.; Zhang, W. Evaluation of ecological vulnerability in the three gorges reservoir area. *Chin. J. Ecol.* **2015**, *35*, 7117–7129. (In Chinese)

24. Xu, G.; Kang, M.; Metzger, M.; Li, Y. Ecological vulnerability of Xilin Gol League. *Chin. J. Ecol.* **2012**, *32*, 1643–1653. (In Chinese)

25. Ord, J.K.; Getis, A. Local spatial autocorrelation statistics: Distributional issues and an application. *Geogr. Anal.* **2010**, *27*, 286–306. [CrossRef]

26. Cliff, A.D.; Ord, J.K. *Spatial Processes: Models & Applications*; Pion Ltd.: London, UK, 1981; pp. vii+266. ISBN 0-85086-081-4.

27. Gu, Q.; Wang, H.; Zheng, Y.; Zhu, J.; Li, X. Ecological footprint analysis for urban agglomeration sustainability in the middle stream of the Yangtze River. *Ecol. Model.* **2015**, *318*, 86–99. [CrossRef]

28. Liu, H. Comprehensive carrying capacity of the urban agglomeration in the Yangtze River Delta, China. *Habitat Int.* **2012**, *36*, 462–470. [CrossRef]

29. Zhao, S.; Da, L.; Tang, Z.; Fang, H.; Song, K.; Fang, J. Ecological consequences of rapid urban expansion: Shanghai, China. *Front. Ecol. Environ.* **2006**, *4*, 341–346. [CrossRef]

30. McKenzie, D.H.; Hyatt, D.E.; McDonald, V.J. *Ecological Indicators*; Springer: Berlin/Heidelberg, Germany, 2012.

31. Guiltinan, J. Creative destruction and destructive creations: Environmental ethics and planned obsolescence. *J. Bus. Ethics* **2009**, *89* (Suppl. S1), 19–28. [CrossRef]

Article

Design Optimization of Productive Façades: Integrating Photovoltaic and Farming Systems at the Tropical Technologies Laboratory

Abel Tablada [1,*], Vesna Kosorić [1,2], Huajing Huang [1], Ian Kevin Chaplin [1], Siu-Kit Lau [1], Chao Yuan [1] and Stephen Siu-Yu Lau [1]

[1] Department of Architecture, National University of Singapore, Singapore 117566, Singapore; vesna.kosoric@gmail.com (V.K.); huajinghuang@hotmail.com (H.H.); ik.chaplin@gmail.com (I.K.C.); slau@nus.edu.sg (S.-K.L.); akiyuan@nus.edu.sg (C.Y.); akilssy@nus.edu.sg (S.S.-Y.L.)
[2] BauLab GmbH, 4656 Starrkirch-Wil, Switzerland
* Correspondence: abel@nus.edu.sg or abeltablada@yahoo.com; Tel.: +65-6601-2435

Received: 23 September 2018; Accepted: 16 October 2018; Published: 18 October 2018

Abstract: Singapore's high dependence on imported energy and food resources, and the lack of available land requires an efficient use of the built environment in order to increase energy and food autonomy. This paper proposes the concept of a productive façade (PF) system that integrates photovoltaic (PV) modules as shading devices as well as farming planters. It also outlines the design optimization process for eight PF prototypes comprising two categories of PF systems: Window façade and balcony façade, for four orientations. Five criteria functions describing the potential energy and food production as well as indoor visual and thermal performance were assessed by a parametric modelling tool. Optimal PF prototypes were subsequently obtained through the VIKOR optimization method, which selects the optimal design variants by compromising between the five criteria functions. East and West-facing façades require greater solar protection, and most façades require high-tilt angles on their shading PV panels. The optimal arrangement for vegetable planters involves two planters located relatively low with regard to the railing or window sill. Finally, the optimal façade designs were adjusted according to the availability of resources and the conditions and context of the Tropical Technologies Laboratory (T^2 Lab) in Singapore where they are installed.

Keywords: building-integrated photovoltaics; vertical farming; shading devices; design optimisation; low-carbon architecture; multi-criteria decision assessment

---

1. Introduction

Owing to scarcity of land and natural assets in Singapore, more than 95% of resources necessary for electricity generation [1] and over 90% of food [2] consumed in the country are imported. The Singapore government aims to reduce the dependency on imports for vital and strategic products such as food and energy [1,2], but its options are limited. Additionally, as a part of the 2015 Paris Agreement, Singapore has committed to reducing greenhouse gases (GHG) emissions to 36% of their 2005 level by 2030 [3].

With a land area of just 7199 km^2 accommodating a population of 5,638,700 [4], Singapore is the third most densely populated country in the world [5], with 7796 inhabitants per km^2 [4]. The total population is expected to increase to approximately 6.5 million by 2030 [6,7]. On the other hand, the proportion of residents aged 65 years or above against the population aged 20–64 years has increased from 13% to 19.7% in the last decade [8]. The aging of the population is expected to further increase in the coming two decades according to the current age pyramid of resident population with the largest number of residents being aged between 40 and 60 [6,9]. Regarding the residential building stock, of the total 1.29 million dwellings, 94.6% are apartments in multi-story buildings and 79% comprise public housing developed by the Housing and Development Board (HDB). Landed properties comprise only 5.2% [10].

Particular conditions related to population density and structure and scarcity of land and natural resources in Singapore require innovative solutions for reducing its dependency on energy and food imports and decrease of GHG emissions and the overall per capita carbon footprint. Apart from its work-force, Singapore's two main resources are year-round solar irradiation and the manmade built environment. The limited marine area around Singapore archipelago and its current use for fishing and port activities reduce its potential use as a source of solar and wind energy, the latter being not feasible due to average low wind speeds between two and three m/s [11]. While the government, agencies, and communities in Singapore are responding to the abovementioned challenges by promoting research into and the development of solar energy technologies, advanced farming systems, and community farming, such steps are still insufficient.

Singapore has an annual solar irradiance of 1580 kWh/m^2 [12], hence investing in photovoltaic (PV) electricity generation is not only the most practical, but a highly promising solution. At the end of 2017, PV installed capacity in Singapore, mostly rooftop-based, reached 143 MWp [13]. This, however, accounts for only 1% of the total installed power generation capacity in Singapore and even less than 1% of the total electricity generation [1]. Reliance on rooftop PV installations alone is not sufficient to noticeably reduce the dependency on natural gas. Large façade areas of residential buildings may be affected to a larger extent by urban compactness [14] and solar irradiation on vertical surfaces is indeed much lower than on horizontal surfaces (approximately 51.5% (East) up to 35.7% (South orientation)) [15]; however, with the total available surface for rooftop installations being limited [16], façade PV integrations should be exploited as a means of electricity generation from renewable energy sources [17] and may significantly contribute to the PV integration potential of the cityscape [18–20].

According to the annual report on renewables of the International Energy Agency (IEA) [21], PVs comprise the largest annual capacity additions for renewables, well above wind and hydropower energy sources. It is expected that solar and wind energy sources will represent more than 80% of the global renewable capacity growth between 2017 and 2022. Such favorable conditions will contribute to 25% reduction of the average PV module cost between 2015 and 2020 [22]. Therefore, the application of building-integrated photovoltaics (BIPV) is a feasible solution, especially when BIPV devices substitute conventional building materials and elements such as shading devices. Several studies examined the feasibility and benefits of BIPV panels as shading devices. The study performed in Hong Kong by Zhang et al. [23] determined that the energy saving potential of PV shading devices is significantly larger than that of the conventional interior blinds in terms of electricity used for achieving thermal and visual comfort. For the Mediterranean region, Mandalaki et al. [24] and Stamatakis et al. [25] compared a series of BIPV designs acting as shading devices. Multiple criteria decision-making (MCDM) methods were applied to select the preferable designs in terms of visual comfort and energy yield as well as other factors such as aesthetics and outdoor views by applying. A simulation of a dynamic BIPV-shading system for different orientations was performed in Zurich and tested by Jayathissa et al. [26] on a real façade considering multiple façade criteria functions in addition to the energy yield. Regarding the applicability of BIPV systems in Singapore, Luther, and Reindl [27] made an estimation of the potential PV area to be installed on Singapore's building façades resulting in a potential energy yield equivalent to 4 km^2 of roof-top PV panels facing the sun with an optimum angle. On a building scale, Wittkopf et al. [28] reported on the design development and implementation process of BIPV systems, including shading devices, at the Building Construction Authority (BCA) Academy. Saber et al. [29] studied the PV performance and energy yield of the same building, the first zero-energy building in Singapore, through data-collection and simulations and concluded that the most effective tilt angle for shading BIPV panels in terms of electricity generation potential is 30°. The study of Ong and Tablada [30] on residential buildings in Singapore under various sky-view factor conditions obtained multiple optimal shading BIPV designs considering other environmental and performance parameters such as thermal and visual comfort in addition to electricity generation. The most important advances regarding food production in Singapore refer to the application of rooftop and vertical farming technologies both indoors and outside. This is a logical response to the limitations of traditional cultivation methods, which require large areas that pointedly counter the gradually shrinking farming areas in Singapore. Industrialized vertical farming is currently expanding in Singapore. The first commercial tropical vegetable urban vertical farm, Sky Greens, was established in 2012 and uses a method called "A-Go-Gro Vertical Farming", which enables the production of one ton of fresh vegetables every two days [31]. Vertical farming high-tech systems including hydroponics, aeroponics and aquaponics are reshaping the traditional approaches to farming and food production by offering efficient and sustainable methods for city farming that minimize maintenance and maximize yield [32]. Other initiatives developed by the HDB and local communities have focused on using the rooftops of multi-storey carparks and green areas around HDB buildings for installing raised beds for vegetable cultivation. In a study on the potential use of rooftops for vegetable cultivation, Lim and Kishnani [33] determined that the rooftop surface of the existing residential buildings may be sufficient to satisfy approximately 35% of vegetable demand in Singapore. A design prototype of an apartment tower with farming areas on elevated terraces and balconies was proposed by Surbana Jurong Consultants for Singapore. Bay et al. [34] reported the design challenges of the integration of farming areas in high-rise buildings, especially the sunlight availability under potential overshadowing in high density urban areas, which require considering the location and type of crops. The study of Tablada and Zhao [19], focusing on the use of building façades and open areas of residential precincts with different urban densities, determined that all façade orientations in Singapore receive sufficient sunlight for vegetable cultivation. The potential vegetable production could be as high as 50% and 75% for plot ratios of 2.7 and 2.2, respectively, according to the most recent building typology. Song et al. [35] examined the photosynthetically active radiation (PAR) along exposed corridors of HDB residential apartment blocks and determined that, if the façade experienced a minimum of half-day direct insolation, the vegetables requiring moderate to high-light would sufficient light. The application of farming systems on building façades from the construction point of view was reported by Suparwoko and Taufani [36]. However, no other studies addressed the façade arrangement and the potential yield.

The integration of both solar and farming systems on building façades was investigated by Tablada and Zhao [19] and Tablada et al. [37–39] at both urban and façade scales, respectively. Apart from these preliminary studies, no other studies have addressed the combining BIPV, as shading devices and building-integrated agriculture (BIA) on the building façades. Therefore, the purpose of this study was to design and analyse possible solutions for simultaneous food and energy production on vertical façades of residential buildings by means of productive façade (PF) systems that include both BIPV and BIA systems, with a view to reducing the dependency on food and energy imports and considering the social, economic, and environmental benefits of BIPV panels [40–46] and urban farming [47–51]. The paper presents the design optimisation framework applied in the development and optimization of modular PF prototypes and provides clear systematic guidelines through the iterative integrative design process leading to optimal PF solutions for residential buildings in Singapore. These solutions produce both food and electricity and act as passive devices for reduction of solar heat gains, and improve visual and thermal comfort. PFs also contribute to positive changes in the urban environment strengthening biophilia, increasing the awareness of the need for GHG emissions reduction, and positively influencing the well-being of residents while enabling them to grow food in dense urban environments by themselves. Greening systems are a key element of the living architecture, whereas green façade technologies enable a wide range of options allowing designers to accomplish multiple objectives such as aesthetic value [52], cooling effects [53], and overall environmental benefits [54]. The developed PF prototypes rely on these characteristics to create synergies with good ergonomics and user-centered design, thereby providing high-quality, sustainable, and affordable architectural solutions. A total of 2135 design variants were created and analyzed, modified, and assessed in order to obtain optimal design solutions. Grasshopper parametric simulation tool with necessary plug-ins [55] was used to calculate relevant environmental performance criteria for the analyzed PF design variants.

VIKOR method [56,57], developed for solving MCDM problems involving conflicting and non-commensurable criteria, was applied to determine eight (8) optimal PF design variants corresponding to two façade categories—Window façade (WF) and balcony façade (BF)—for 4 orientations (South, North, West, East). Eight (8) PF prototypes will be installed, monitored, and tested at the Tropical Technologies Laboratory (T^2 Lab) located in a fairly open space at the precinct of the staff residences of the National University of Singapore (NUS).

2. Materials and Methods

Figure 1 presents the overall framework related to the development, assessment, and optimization of PF prototypes. In phase 1, the scope, design concept, and main strategies for the implementation of PFs were defined based on the available reference literature and discussions with local experts regarding both BIPV panels and BIA systems. In phase 2, the preliminary designs of two categories of façades for four orientations were explored considering two typical façade types in actual residential buildings in Singapore as well as the available budget and space in the T^2 Lab. As illustrated in Figure 2, eight test bed cells are located inside a 60 m^2 facility and are also used for other investigations of tropical technologies. In phase 3, the specifications of the BIPV and BIA systems on PFs were defined. The aspects considered, such as PV tilt angle and dimensions, planter location and configuration, etc., were crucial for the development of the PF design alternatives.

Once the design variants were listed, quantitative and qualitative assessment criteria—the so-called "criteria functions"—were defined in phase 4. The acceptable performance ranges considered were also defined according to Singapore's building codes and performance benchmarks recommended in the literature relative to the five criteria functions: Food production potential, electricity generation potential, indoor daylight, energy flow on façade, and view angles. A detailed description of each criterion function is presented in Section 2.2. Computational simulations were then conducted using Grasshopper's plug-ins in order to optimize the design of the eight PF prototypes. The description of the simulation algorithm and settings is presented in Section 2.3.

In phase 5, the simulation results were used as inputs in the VIKOR, MCDM method in order to select the eight optimal PF design variants. The VIKOR method selects a compromise solution based on the weighted-decision matrix considering a set of criteria functions, acceptable ranges and targeted performance values. The details of the VIKOR method are explained in Section 2.4. The final optimal PF prototypes were selected for hypothetical residential buildings. These however, may not coincide with the actual PF designs to be implemented in the T^2 Lab due to the availability of PV technology and the context. Phase 6 addressed this issue by adjusting the final prototypes design. The adjustment of the façade designs is explained in Section 4. Finally, in phases 7 and 8, the final PF prototypes are installed at the T^2 Lab (see Figure 2) and monitored, respectively.

2.1. The Model and Initial PF Arrangements

Figure 3 illustrates the main design strategies for the initial PF prototypes development. While the PF design variants are not targeting a specific residential building typology, a survey regarding the existing HDB public housing buildings in Singapore was conducted to determine common façade designs, materials, and arrangements. This helped to identify the two façade categories which were simplified and developed as a prototype to be optimized for four orientations in this study. Other relevant characteristics, including functional and constructive design parameters were also considered, although they are not the main focus of the study.

The PFs have conventional components (fenestration and opaque parapet) and non-conventional components (PV modules as external shading devices and planters outside the parapet or railing). Regarding the WF, two fenestration systems are used on the upper section and an opaque parapet of 1.1 m on the lower section of the façade. The fenestrations are designed to allow maximum wind porosity as well as easy access to planters. On the other hand, a floor-to-ceiling fenestration with operable glass louvers are used on the BF.

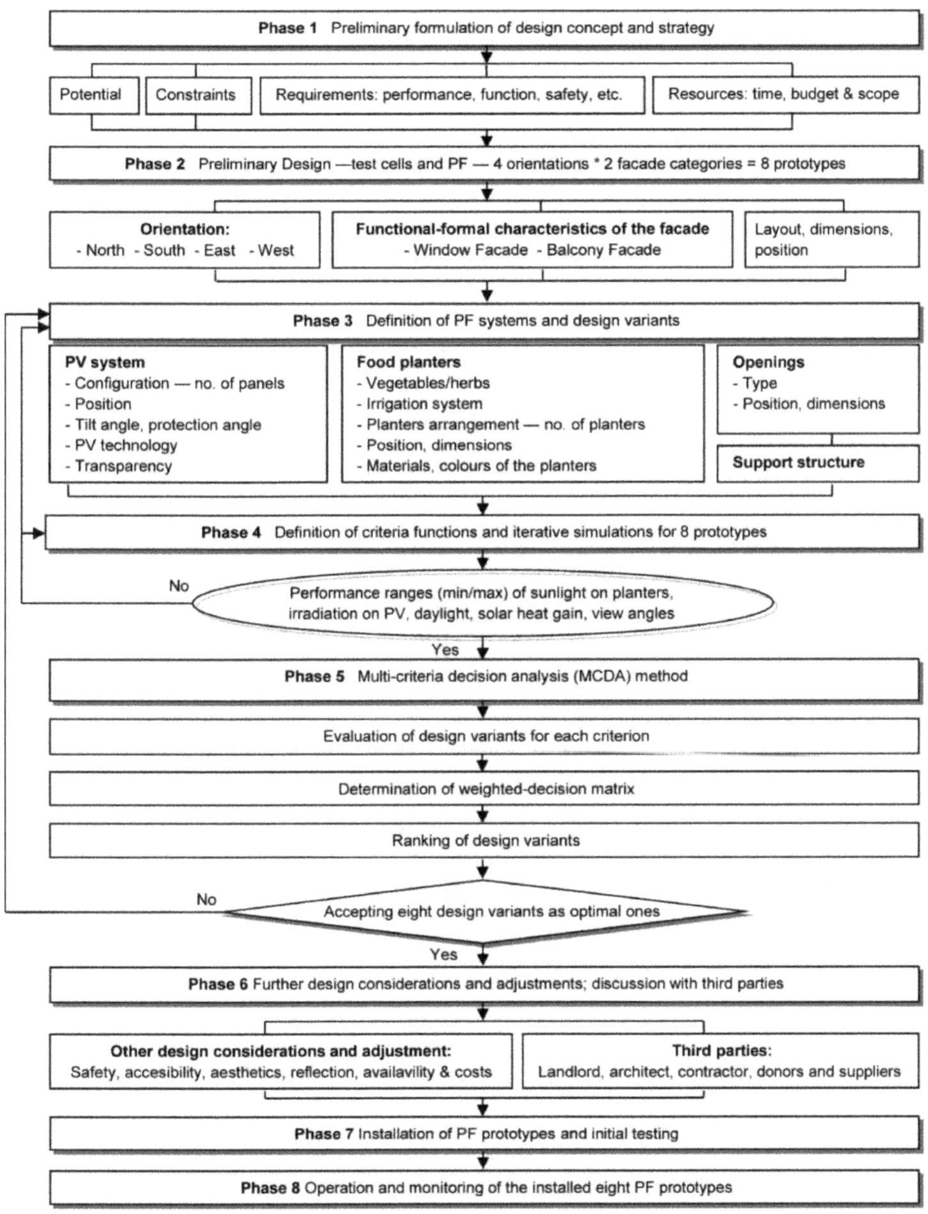

Figure 1. Graphical representation of the Productive Façade design development model.

Figure 2. Floor plan of the Tropical Technologies Laboratory (T^2 Lab) indicating the eight test bed cells and facades analyzed in this study. Adapted from: AWP Architects based on lead author's preliminary design.

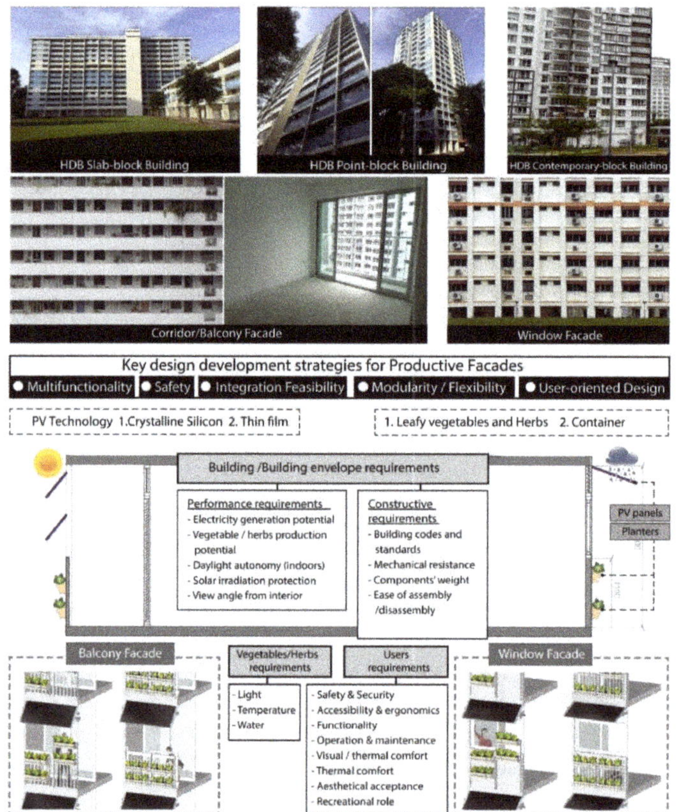

Figure 3. Key design development strategies for the PFs.

Figure 4 illustrates the following façade arrangements for WF and BF: (I-F) façade with a single PV panel attached on top of the façade, (II-F) façade with two PV panels, the top one is attached to the top of the façade, (I-P) façade with a single PV panel attached to the planter of the level above, (II-P) façade with two PV panels, the top one is attached to the planter of the level above, and (III-P) façade with three PV panels, the top one is attached to the planter of the level above.

Pursuant to the most common floor-to-floor height at HDB apartments, the test bed cells are 2.8 m high (2.6 m ceiling height). However, the cells are 1.8 m wide and 1.8 m deep. A number of constraints related to the position of PV modules and planters significantly reduces the number of potential design variants. The following restrictions may apply: PV modules that function as external shading devices are always placed on the upper third façade section or at the same height as the lowest planter from the upper floor; planters are always positioned in the bottom third façade section due to accessibility and safety reasons; in order to avoid obstructing the view, the central façade section is to be used as little as possible.

In order to provide more space for planters that are to be installed at the lower part of the façade, this study considers 1.1 m to be the window sill height. This allows planters to be positioned at 100 mm, 300 mm, 500 mm, 700 mm, and 900 mm, in two or three rows, while maintaining a 400 mm distance between planters. The sill height complies with BCA regulations [58].

Singapore is located very close to the Equator at $1°17'$ North latitude, meaning that solar irradiance is more evenly distributed among the four façade orientations in contrast to that of higher-latitude regions [19]. However, the design optimization of the BIPV external-shading devices on different façade orientations is nonetheless highly challenging and complex. The objectives that PF elements should achieve need to be comprehensively and carefully defined. The stated objectives of the shading PV modules are the following: Maximizing the electricity generation, reducing solar heat gain, allowing necessary illuminance indoors, and as unobstructed as a possible view to the outside, at least the minimum required amount of sunlight reaching the planters and PV modules on the same and lower stories, respectively.

The protection angle, defined as the angle between the vertical plane of the façade and the outside edge of the PV module, is the most important geometrical parameter that has to be considered in order to achieve the optimal design variant. The protection angle along with the tilt angle define the width of the BIPV shading element. For a single shading element, the angle is measured from the bottom of the window, whereas for double BIPV shading elements, it will also include the measurement from the top of the lower shading element (illustrated in Figure 4).

The number of PV modules should also be considered. The initial stage limited the number to two horizontal panels, while it subsequently considered three panels for East and West-oriented façades. Façade orientation also imposes limitations regarding PV panel tilt angles and protection angles as well as the number of hours during which a façade requires protection from direct solar radiation. The Residential Envelope Transmittance Value (RETV) calculation, based on the façade performance and orientation, was used to further define the minimum protection angle as required by the BCA of Singapore [59]. The RETV performance is dependent on the total thermal transmittance (U-value) of facade elements, the solar heat gain coefficient (SHGC) and orientation of a surface. Opaque surfaces only have static U-value, hence the only variable is the window-to-wall ratio (WWR) of the façade. For the WF, WWR is 0.54, whereas for the BF, it stands at 0.93. Additionally, a minimum threshold of direct sunlight hours was set to before 8:30 a.m. and after 4:30 p.m. for extreme incidence hours. Lastly, a minimum value of the protection angle range was also established so that the operative temperature would not be above 28 °C. In this way, it was possible to ideally intersect the protection and the tilt angle of the shading device. Protection angles that meet the specified requirements were determined to range from 28° to 37° for the North and South orientations, from 47° to 56° for the East orientation, and from 53° to 65° for the West orientation.

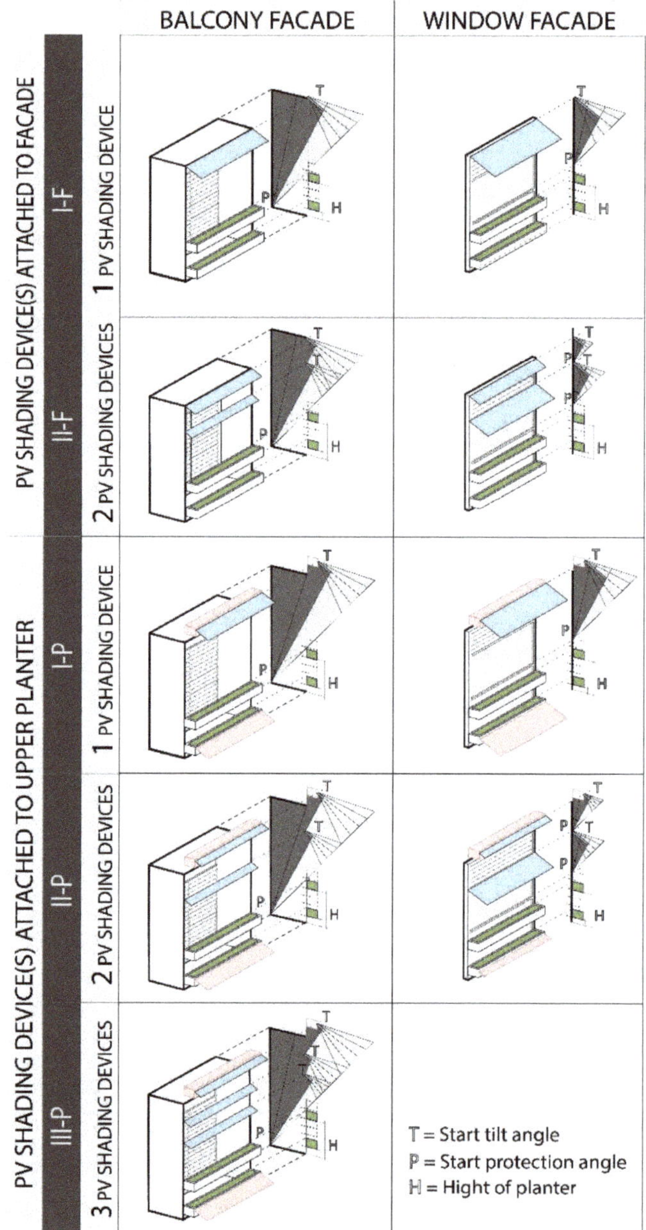

Figure 4. Types of PV configurations and the corresponding start protection and tilt angle.

2.2. Criteria Functions and Assumptions

2.2.1. Farming System and Food Production Potential

The factors affecting plant growth include light, temperature, water, rooting medium, and cultural practices. Due to the hot and humid weather conditions in Singapore, only a limited number of vegetable species are suitable for cultivation. Table 1 provides a list of commonly cultivated Singapore leafy vegetables and their evaluation in terms of the optimum daily light integral (DLI) [34]. They usually have shallow roots, which makes them space-efficient, a shorter growing period, and most of the final product, the aerial portion consisting of the stem and leaves, is edible.

Among other factors, light is a key factor affecting the growth of vegetables as it drives photosynthesis and plant development, morphology, and yield [60]. At the same time, it is the most crucial factor for integrating farming systems into building façades owing to the reduced amount of sunlight on vertical surfaces in comparison with horizontal unobstructed surfaces. Ideally, shade-tolerant crops with lower daylight demand should be selected for cultivation on building façades, such as lettuce (*Lactuca sativa*) and kangkong (*Ipomoea aquatica*) in Table 1.

Apart from light availability, excessive heat may inhibit root elongation, stimulate early bolting and accelerate plant development resulting in smaller-sized plants and a lower vegetable yield [61,62]. Vegetables like lettuce originated from cooler climate regions so their heat tolerance is low. However, there are varieties adapted to tropical climates and have been successfully grown in Singapore and other tropical regions. Therefore, due to the high light demand of kai lan, pak choy, red bayam, and cai xin, and the large plant size of kangkong, lettuce was selected for the purposes of this study after considering light availability, heat tolerance, and other factors.

The light intensity range required for growth of plants depends on the environment and time, the desired plant product, heat and CO_2 content of the air surrounding the plants [63]. The agricultural industry frequently uses DLI to determine the exact lighting condition for plants. It represents the total number of photosynthetically active photons that plants receive in 1 m^2 of growing space in one day. It is expressed in $mol/m^2/day$, and essentially reflects the combined results of the light intensity and duration of the photoperiod. Commercial farms will usually keep the minimum DLI of 10–12 $mol/m^2/day$ for optimum growth of plants [64]. For maximum production, leaf lettuce normally requires 14–17 $mol/m^2/day$ or more [65], although it can still be grown with as little as 4–10 $mol/m^2/day$. However, if the DLI is below 8 $mol/m^2/day$, the quality of the produce will be low [66,67], hence, the minimum DLI requirement for lettuce growth can be set at 8 $mol/m^2/day$, which is equivalent to 10,000 lux when DLI is converted to illuminance levels.

Table 1. List of common leafy vegetables and the required daily light integral (DLI) [34].

Name	Optimum DLI (mol/m^2·day)	DLI Category
Cai xin (*Brassica rapa* subsp. *chinensis* var. *parachinensis*)	24.51	High light
Kai lan (*Brassica oleracea* var. *alboglabra*)	47.22	Very high light
Kangkong (*Ipomoea aquatica*)	19.90	Moderate light
Lettuce (*Lactuca sativa*)	14.51	Moderate light
Pak choy (*Brassica rapa* subsp. *chinensis*)	39.96	Very high light
Red bayam (*Amaranthus tricolor*)	33.95	Very high light

2.2.2. BIPV Electricity Generation Potential

Electricity generation potential refers to the potential electricity generation from all PV modules for each façade. In order to focus on the geometrical impact of the PV panel on all design variants, the type of PV technology, performance ratio, and the effect of temperature were not considered in the simulations. Instead, the average of the lowest incident irradiation from all panels on each façade multiplied by the total PV area was obtained from the simulations and used as one of the criteria functions. The lowest incident irradiation were obtained from a row of test-points located at 100 mm from the façade and accounts for the partial shading produced by upper PV panels since PV cells on the same module are considered to be connected in series.

2.2.3. Indoor Daylight

Daylight Autonomy (DA) refers to the percentage of time during the year in which a certain, pre-defined illuminance value is achieved from 8 a.m. to 5 p.m. The bedroom WF and the living room BF were both subdivided into front and back subzones, whereas the DA of 50% was the targeted value. Taking into account the position of a desk in the bedroom, the front bedroom subzone requires 400 lux, whereas the back subzone designated for change of clothing requires only 100 lux. With regards to the living room, the front subzone requires 200 lux and the back subzone 100 lux. It should be noted that an equivalent DA was used rather than the actual values obtained in the simulations since the testbed cell size is smaller in comparison to the actual room dimensions in typical HDB buildings. Therefore, preliminary simulations referring to a living room and a bedroom with two façade arrangements were conducted on actual room dimensions to obtain the DA conversion coefficients for all cases using the reduced room dimensions. For example, the illuminance threshold in rooms with actual dimensions is 100 lux during 50% of the time, whereas in the reduced, smaller version, the said threshold is approximately 230–260 lux depending on the arrangement and dimensions of PV modules. The highest standard for maximum accuracy was achieved by using a larger number of reflections in order to carry out daylight simulation in rooms with actual dimensions, whereas simulations on the reduced scale model were conducted with low accuracy. This conversion process greatly reduces simulation time without compromising the prediction accuracy.

2.2.4. Energy Flow on Façade

The energy flow through the façade is defined as the heat gain minus heat loss (kWh) considering the rest of the walls as adiabatic. Net solar transmittance, which is the result of the façade design related to fenestration and shading devices, largely affects the energy flow of the façade. Therefore, instead of thermal comfort, energy flow values were used directly in order to isolate the impact of the façade design on thermal conditions and to simplify the simulations and analyses. For every case, constant thermal properties of the materials and no occupancy were applied. The energy flow calculations were conducted using the plug-in, Honeybee which connects with the EnergyPlus simulation engine for transient energy calculations. For this criteria function, no specific range was defined since the protection angle already assured compliance with RETV and the two other requirements, as explained in the above section. Therefore, all values from design variants are relative to each other.

2.2.5. View Angle

One of the criteria functions selected to assure the architectural quality of residential building façades is the outdoor view angle. It refers to the average view angle from two points inside the testbed cells towards the exterior. The location of the two points is 1.5 m from the façade at 1.17 m and 1.56 m above the floor. The said points correspond to the viewing height of a person 1.68 m tall, while sitting and standing. Any obstruction effects from the planters and PV panels are considered. For North and South façade orientations, as the most common and the recommended orientations in tropical residential buildings, the minimum view angle was set at 20°. For East and West façade orientations, the minimum view angle was set at 15°.

2.3. Grasshopper Algorithm

The aim of the developed algorithm is to provide a user-friendly platform that performs simulations corresponding to a large variety of façade variants, while complying with standard practices and recommendations found in literature—see Section 2.1—within a relatively short time-span, from 8 to 45 h simulation time, per façade type on a standard PC. A balance between simulation automation and user intervention is also achieved to support the design process. At first, the algorithm programmed the PF geometry to allow its transformation into multiple varying configurations according to pre-defined parameters and the range of component dimensions. This set-up enabled automatic cycles through all design variants for the same PF category and orientation.

Manual intervention was required in eight instances to input the parameter ranges for each of the eight PF subsets. Grasshopper plugins Ladybug [68] and Honeybee apply generic qualities to selected geometries and context to all cases. Regarding the building materials, lateral walls, and the ceiling are considered adiabatic while the back wall has a thermal transmittance (U-value) of 1.4 W/m^2K. Reflectance value of all interior surfaces is 0.65 corresponding to off-white color, except the floor whose value is 0.51. Single clear glass was used for all fenestrations with U-value of 6.5 W/m^2K and visible transmittance of 0.88. PV panels on North and South facades are opaque and those on East and West facades are semi-transparent with a transmittance coefficient of 0.44. All rooms have zero occupancy and no building or obstruction are considered on the opposite position of the facades. Regarding the ventilation settings, 'Window Natural Ventilation' was chosen as the ventilation type for all cells. The airflow rate is calculated based on the local weather file data and on the operable area and height of all exterior fenestrations. Half of the area of all fenestrations are set to be fully opened throughout the whole year.

The results are then organized and visual and mathematical analysis is performed in order to verify the results, and if required, further the design development. The results for individual façades are automatically represented in the design variants models as presented in Figure 5. The tabulated data were organized per variant in the form of a list with values separated by commas, storing variant identification, values for the criteria and support data for error checking and variant development. These data was then imported into Excel for the application of the VIKOR method. More details of the Grasshopper algorithm can be found in Tablada et al. [39].

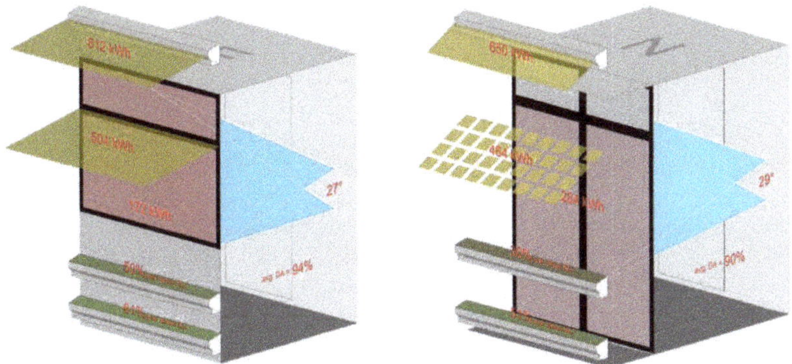

Figure 5. Examples of Rhino-Grasshopper models which are automatically updated according to geometrical inputs. Results are shown and stored after each simulation case.

2.4. VIKOR Method

The design of PFs is a highly complex and dynamic process. In order to solve the challenges pertaining to the process of finding the optimal PF design variant, VIKOR (in Serbian: VIšekriterijumska Optimizacija i Kompromisno Rešenje) method [56,57] is applied, a multi-criteria compromise method developed for the purpose of a multi-criteria optimization of complex systems. The VIKOR method determines a compromise solution from a set of alternatives often based on non-commensurable units (i.e., for optimisation specifically: %, W/m^2, kg) and conflicting requirements (i.e., PF criteria functions: food production potential, BIPV electricity generation potential, indoor daylight, energy flow on façade and view angle), while also allowing easy, highly flexible modelling of the decision maker's (DM) preference. The method has been successfully applied to resolve practical considerations in complex multidisciplinary fields related to sustainability and renewable energy planning [69,70], as well as in the building PV integration field [20,71]. It enables holistic evaluation of different design solutions and the optimization of the design variants and aids the sensitivity analysis which tests the robustness of the "optimality" of the selected design variant. The major advantages of the VIKOR method are that it compromises conflicting criteria, and a maximum "group utility" for the "majority" with a minimum of an individual regret for the "opponent" including the relative importance by the weights. The algorithmic steps of the VIKOR method and the Equations (A1)–(A11) related to it are presented in the Appendix A. The VIKOR compares each alternative by the distances to the ideal solution (point). Here, ideal consists of the maximum f_i^*, if the i-th function represents a benefit, and the minimum f_i^{-}, if the i-th function represents damages or costs, see Equation (A1). In terms of computational complexity, the method is not overly complicated or demanding. Both criteria functions and their weights can be varied easily during the design process and their impact on the optimal solution selection can be analyzed helping DMs and the design team obtain a better insight of the sensitivity of different solutions.

To add values of non-commensurable criteria, VIKOR converts them into the same units first. Normalization is used to eliminate the units of criterion functions, so that all the criteria are dimensionless. The VIKOR method uses linear normalization, whereas the normalized value does not depend on the evaluation unit of a criterion function unlike, for example, the TOPSIS, another widely applied MCDM method, which uses vector normalization, but the normalized values may depend on the evaluation unit [56].

The application of the VIKOR method helps evaluate PF design variants—i.e., alternatives—and arrive at a compromise solution, the one "closest" to an ideal solution, which is selected from a set of J alternatives, i.e.,: *A1*, *A2*, . . . , *AJ*, evaluated according to a set of n criteria functions. Value Q_j, Equation (A9) is used to select the optimal design alternative as it represents an approximation of the ideal balance, i.e., it compromises two decision-making strategies:

(1) "Maximum group benefit" defined by the value S_j, Equation (A3)—better alternatives are considered to be those deemed good according to the majority of criteria, and

(2) "Minimum of maximum deviation of ideal values" defined by the value R_j, Equation (A4)—those alternatives that are considered to be better must not be very bad according to any criteria.

Prior to the VIKOR method, the cases were filtered according to the established acceptable ranges for values of each parameter. For the evaluation and ranking of all selected types of PV configurations and all façade orientations through the VIKOR method, equal weights (0.2) were initially applied for all 5 criteria functions. However, they were not suitable for obtaining the optimal solution on a sub-set of cases, hence the criteria function weights for each sub-set were adjusted. $(1 - v) * QR_j$ part of the VIKOR ranking formula penalizes cases even when a single parameter value is close to the value of the worst performing solution within the set in order to avoid high ranking of a case with extremely poor performance within a single criterion. For example, in the all-cases set, DA parameter may vary between 54% and 98%, hence it is justifiable to strongly penalize the cases whose DA parameter value approaches 54%. For the North Window Façade (NWF) sub-set of cases, DA parameter may vary between 90% and 97% (i.e., all cases have satisfactory DA parameter), hence it is not justifiable to strongly penalize a case since even 1% difference could greatly influence its ranking. Therefore, a weighting system was created to account for the range between minimum and maximum values of the criteria parameters within the sub-set of cases with a lower weight being assigned to the criterion where the differences between the best performing and the worst performing cases are negligible.

For all sub-sets of cases, a total range of possible parameters for each criterion function is calculated. Then the average value of ranges among 8 sub-sets is derived for each of the 5 criteria functions and such average range is used as the basis for scaling criteria functions weights in the analysis of individual sub-sets of cases. The process of obtaining the weight per criteria function for each sub-set is presented in Appendix B. The final criteria function weights for all 8 PF prototype sub-sets are presented in Table 2. The lowest final weight of 0.08 was obtained for the "View-Angle" criterion in the WWF cell façade type since the View Angle of all cases in this subset varies between $15°$ and $18°$. The highest final weight of 0.37 was obtained for the irradiance on the PV surface in the same cell façade due to the relatively large difference between the maximum and minimum incident irradiance ($548\ W/m^2$) among that subset of cases.

Table 2. Weights ($W_{CF\ fac}$) and equivalent ranges ($\Delta_{AVG\ CF}$) per criteria function and cell façade type.

Cell Façade Type	Daylight Autonomy Interior	Energy Flow on Façade	Irradiance on PV	Vegetable Productivity Value	View Angle
$\Delta_{AVG\ CF}$	12.0%	$71\ W/m^2$	$568\ W/m^2$	19.6 kg	14.4°
NWF	0.13	0.32	0.17	0.11	0.27
SWF	0.17	0.30	0.15	0.09	0.29
EWF	0.22	0.27	0.25	0.11	0.15
WWF	0.16	0.19	0.37	0.21	0.08
NBF	0.18	0.20	0.21	0.12	0.29
SBF	0.15	0.22	0.23	0.08	0.32
EBF	0.25	0.12	0.21	0.29	0.14
WBF	0.24	0.13	0.14	0.34	0.15
AVG	**0.19**	**0.22**	**0.22**	**0.17**	**0.21**

3. Results

3.1. Selection of Optimal PF Arrangements per Façade Type

Table 3 presents the eight optimal cases, one per cell façade type following VIKOR method and the values obtained per criteria function. Figure 6 illustrates the configuration of the eight optimal PF prototypes.

Double-PV shading devices positioned at the height of the upper level planter is the most preferable variant for all façade types except for EBF and WBF. A double-PV panel allows increasing the total area of PV cells, while reducing the obstruction of sunlight towards the planters and maintaining the targeted DA inside the room. Unlike the WF type, the window and door facing the balcony is at floor level, therefore requiring more protection from direct solar radiation on the east and west façades. This implies that the optimal arrangement for EBF and WBF is a system of three PV panels.

As expected, the protection angles are lower for North and South orientations (around $30°$ from bottom of window/door) than for East and West orientations ($47–53°$). For this parameter, there are no substantial differences between the WFs and BFs of same orientation. Regarding the tilt angle, $50°$ is found to be preferable for all façade types except for SBF ($30°$).

The estimated electricity generation of the optimal cases is presented in Table 4 considering the conventional 15% efficiency crystalline silicon PV module. For a typical 4-room HDB unit (3 bedrooms) with total façade dimension (two opposite facades) of 20 m in length, the total electricity generated by the shading PV modules could be as high as 1860 kWh. This represents, approximately, 40% of the average annual electricity consumption for this type of unit (4474 kWh/year) [72]. The electricity generated by the PV shading device represents between 40% (SBF) and 46% (SWF and EBF) of the electricity generated by the same PV module type and surface located on a rooftop without obstruction. However, if each PV row is connected independently instead of in series, the electricity generation will not depend only on the lowest irradiance incident on the PV module. In that case, the total electricity generation will be 5–7% higher for PV panels on North and South facades and 13–42% higher for PV panels on east and west facades since it will also account for the higher irradiation incident on the most exposed rows of cells located closer to the exterior extreme of the module.

Regarding the number and position of planters, all cases accept only two planters per façade. Having three planters may compromise the amount of sunlight due to smaller spacing between and self-shading between the planters. The preferable height of the upper planter is 0.5 m except for the window façades facing North and South. When the PV system is double or triple-panelled, a lower position of the top planter reduces the obstruction from the lower PV module.

As explained in Section 2.4, the value of Q reflects the appropriateness of the compromise between criteria functions. The value that is the closest to zero, represents the most ideal balance. In this sense, the most ideal case is the BF facing west (Q = 0.0) followed by NWF and EBF (Q = 0.03). The least ideal among the optimal PF prototypes is the BF facing South, however, still with a very low Q of 0.06. In this instance, a combination of relatively low vegetable production and a relatively high incoming energy flow may have influenced the less balanced result.

Table 3. Optimal cases per façade type and values per criteria functions.

Façade Type	-	NWF	SWF	EWF	WWF	NBF	SBF	EBF	WBF
Variant Type	-	II-P	II-P	II-P	II-P	II-P	II-P	III-P	III-P
Protection angle PV-shading	°	31	31	47	53	28	31	50	53
Tilt angle PV-shading	°	50	50	50	50	50	30	50	50
Top planter height	-/m	700	700	500	500	500	500	500	500
Daylight Autonomy Interior	%	95.0	95.2	88.4	83.3	91.6	90.9	89.2	86.8
Energy Flow Façade	kWh	149	149	130	112	194	193	133	123
Lowest irradiance × Area	kWh	1189	1205	1736	1789	1028	1052	1837	1768
Vegetable production	kg/year	39.2	35.3	42.8	46.5	34.6	28.8	42.6	47.5
View Angle	°	28	28	18	15	37	39	25	24
Q	-	0.03	0.04	0.05	0.05	0.04	0.06	0.03	0.00

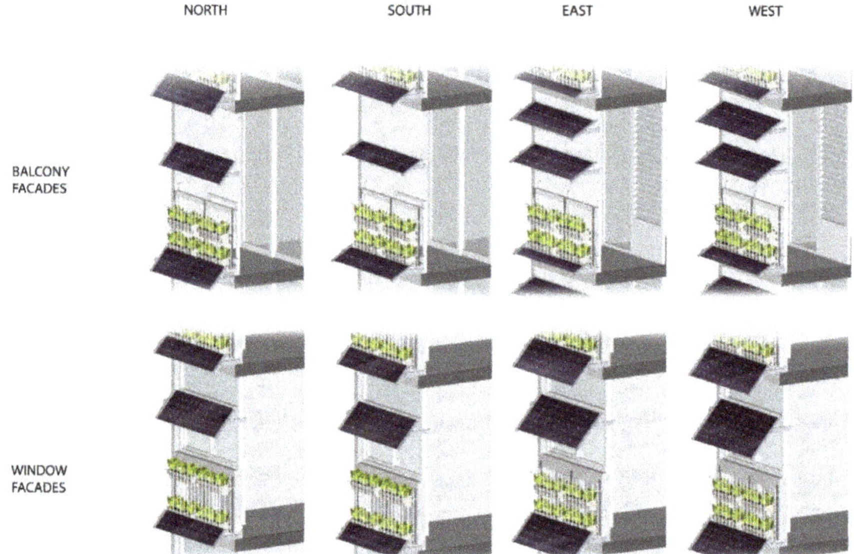

Figure 6. Artistic impression of the eight optimal PF prototypes. Most design variants are II-P except east and west BF which are III-P.

Table 4. Expected electricity generation from 15%-efficiency Si-monocrystalline modules on façades and rooftop.

Façade Type	-	NWF	SWF	EWF	WWF	NBF	SBF	EBF	WBF	Roof
PV Area	m²	1.6	1.6	2.3	2.6	1.4	1.6	2.4	2.6	1.0
Lowest irradiance/m²	kWh	743	753	748	696	714	658	756	691	1636
Electricity generation (15% effic.)	kWh	178.4	180.8	260.4	268.4	154.2	157.8	275.6	265.2	245.4
Percent from same area rooftop PV	%	45.4	46.0	45.7	42.5	43.6	40.2	46.2	42.2	100.0

3.2. Results for the Five Best Cases Per Façade Type

Little variation can be observed regarding the incidence of the configuration in the best five cases per façade prototype sub-sets. Table 5 presents the average values of protection and tilt angles and planter height for the best 5 cases per façade prototype.

Table 5. Average (Avg) values of 5 best cases compared to the optimal façade prototype per sub-set.

Façade Type	Variant Type		Protection Angle PV-Shading		Tilt Angle PV-Shading		Planter Qty/Height	
	Optimal	Times among 5 Best	Optimal	Avg 5 Best	Optimal	Avg 5 Best	Optimal	Avg 5 Best
NWF	II-P	3	31	29.8	50	50	2/700	2/620
SWF	II-P	4	31	29.2	50	50	2/700	2/620
EWF	II-P	4	47	48.2	50	50	2/500	2/580
WWF	II-P	5	53	53.6	50	44	2/500	2/580
NBF	II-P	4	28	29.8	50	44	2/500	2/500
SBF	II-P	5	31	29.8	30	42	2/500	2/500
EBF	III-P	5	50	49.4	50	48	2/500	2/500
WBF	III-P	5	53	53.6	50	46	2/500	2/540

The incidence of the façade type in the best five cases is very high. Only in NWF there are two cases different from II-P configuration. The average protection angle of the best five cases is also very close to that of the optimal cases. The average difference is 1.1°. However, regarding the tilt angle, the average difference is slightly larger and stands at 3.7°. This is caused by the lesser impact of the tilt angle on the energy flow of the façade and on DA for similar protection angle. The highest ranked cases apply high tilt angle on the PV shading since this assures a larger PV surface for electricity generation while it does not significantly obstruct the sunlight from reaching the planters below. It also offers good protection from direct solar radiation. However, lower tilt angles are preferable for SBF because the view angle and DA have higher weights in relation to the other criteria functions. Regarding the planter configuration, all PF prototypes accept only two planters. The height of the upper planter is higher for the North and South window façade types while lower planter heights are preferable for other façade types.

4. Final PF Arrangements for T^2 Lab

The main objective of the MCDM method in obtaining eight optimal façade arrangements was to implement them at the T^2 Lab. However, several contextual, practical and research-related reasons made the actual façade arrangement to be installed at the T^2 Lab façades somewhat different.

Regarding the context, it should be noted that the T^2 Lab is located in an open lawn, in between three high-rise residential towers accommodating university staff on the South-East side and public housing (HDB) buildings on the North-West side. One of the conditions stipulated by the Urban Redevelopment Authority (URA) in Singapore to allow using the terrain for research purposes was to ensure that no reflection of sunlight is produced from the PV modules onto façades of the surrounding residential buildings. In order to meet this condition, a raytracing study was conducted to verify if the optimal tilt angles from each façade do not cause reflection towards neighbouring façades. With the exception of EBF, all other façades had to be adjusted. This means that most PV tilt angles had to be changed to 20°.

With regard to practical considerations, some PV modules were donated by a collaborating institution. The dimensions of PV modules and quantity were fixed, therefore, the overall shading dimensions had to be adjusted and unified for the two façades on the north and the two façades on the South orientations. Additionally, the pre-designed window with two fixed sections on the East and West façades does not agree with the optimal position of the 3 PV panels. Therefore, the best designs with 2 PV panels were applied.

In addition to the limited amount of available PV modules, an online survey involving 100 PV experts and architects was conducted, as well as a survey among HDB residents, in order to obtain their feedback on several designs and other important issues such as accessibility and maintenance. One of the more consistent results referred to the preference, both of experts and residents, for the single PV panel over the double or triple panels.

Other changes, such as using three rows of planters on the West façade, were also applied to test the limits of the façade arrangements on the potential food and energy production. Considering those elements, Table 6 and Figure 7 present the final PFs arrangement to be implemented at the T² Lab.

Table 6. Façade arrangement to be installed at the T² Lab.

Façade Type	Variant Type	Protection Angle (Top/Middle/Bottom)	Tilt Angle PV-Shading	Planter Qty/Height (mm)
NWF	I-F	28	20	2/700
SWF	I-F	31	20	2/700
EWF	II-P	53.7/-/54.7	40	2/500
WWF	II-P	53.7/-/54.7	40	2/500
NBF	I-F	28.4	20	2/700
SBF	I-F	30	20	2/700
EBF	III-P	44/50/50	50	3/900
WBF	II-P	41.4/-/40.7	10	2/500

Figure 7. *Cont.*

Figure 7. Top: Artistic impression of final façade arrangement for the south and east façade orientations at the T^2 Lab. **Bottom**: Photograph of the T^2 Lab in August 2018 showing east and north facades.

5. Discussion

The results demonstrated that incident solar radiation according to façade orientation is the most crucial environmental variable defining the optimal PF arrangements. As expected, the East and West façades require more extended solar protection at the expense of the average lower outdoor view angle. However, pursuant to passive design strategies in residential public housing, most facades face orientations at angles not larger than 30° from North and South, which drastically reduces the protection angle and therefore the dimensions and tilt angle of the shading PV panel. The smaller area of facades facing East and West orientations are often blank walls that are more appropriate for the installation of vertical BIPV in substitution of conventional cladding [20]. Those facades with windows correspond with the private areas of the housing unit, which do not require ample view angles.

On the other hand, unlike previous studies which determined the preferable PV sun-shading tilt angles to be between 20° and 30° [24,29,30], the optimal tilt angles in this study are 50° for most cases considering other important criteria such as solar protection and the sunlight required by the planters. However, higher tilt angles are not recommended in dense urban environments due to the reflection from PV panels to neighboring buildings. Therefore, for East and West facades, larger number of PV panels with smaller tilt angle and dimensions is recommended.

Regarding the potential electricity generation, although the share of solar generation may be small at nation scale due to the extremely high energy demand in the Singapore's industrial sector (43% of electricity) [72], the share of solar generation in the residential sector could be considerably higher, taking into account that the household electricity consumption is 15% of the total electricity consumption. Therefore achieving 40% of the total electricity demand by the PV shading devices on north and south facades, is remarkable considering the shading effect of upper levels. However, this output should be lower for the facades located at lower levels considering the effect of the neighboring buildings in high density areas. The overshadowing effect and reduction of the sky view factor may be larger for east and west facades. The specific dimensions of PV modules according to façade type and orientation requires design customization. While this may lead to initial higher investment, BIPV customization can respond to some of the barriers on the application of conventional BIPV in terms of aesthetics, efficiency, and flexibility [73]. The benefits of customized BIPV applications are becoming more evident with the fast acceptance and adaptation of industry for their production. Therefore, the fabrication of affordable customized shading PV panels and the market availability of a wide diversity of design and dimensions is expected to be the norm in the future.

Regarding the arrangement of farming planters, the results are quite similar for all PF variants that accept only two rows of planters at relatively low height to reduce the impact of the PV panels above them. This optimal position, however, requires the use of lower operable fenestrations to access the planters. Safety elements should be incorporated in the façade system, as shown in Figure 6, especially for the WF. The production of vegetables on the facades is not meant to address the overall vegetables demand. Instead, it should complement other urban farming activities ranging from individual farming installations on each façade and community gardens in each neighborhood to public intensive farming facilities at district-scale. While the PV shading devices would involve a top-down implementation from HDB authorities to complement the energy supply from the grid, the implementation of the PF farming system may be left at the discretion of a household according to their needs and preferences. Despite the limited space for installation of the planters on facades, a façade of 20 m in length (two opposite 10 m facades on a typical HDB apartment) could enable each household to produce an estimated 35 to 66 kg of leafy vegetables. The amount varies according to the façade orientation and the type of crops. This amount represents 55–103% of the average leafy vegetables consumption of a 4-member household in Singapore (ca. 16 kg per year) [2]. The system also aims to reduce the cost and frequency of maintenance. Laborious activities are unnecessary due to the ergonomic considerations. For example, manual watering is required only once a day or once every two days since the planters include water reservoirs. Fertilizer should be applied once every two to three weeks and pesticide spray, once every two months. Harvesting can be done occasionally and leaves can be plucked when needed or when reaching the desired size.

Both systems can be implemented in higher latitudes providing a similar optimization framework in order to maximize energy and food production, while also complying with indoor thermal and visual performances.

6. Conclusions

Responding to the particular situation of Singapore in relation to the lack of agricultural areas for food production, the lack of energy resources and therefore, the high dependency on energy and food imports, energy, and food, this paper proposes the concept of productive façades (PF) for residential buildings that integrate PV and farming systems.

A design optimization methodology is described utilizing the Grasshopper parametric tool and VIKOR optimization method. The method used in this paper, which was applied to public housing residential building façades in Singapore, proved to be effective for optimizing complex façade designs in which multiple conflicting criteria were assessed. The optimization method allowed automated and manual procedures that both increased the computational efficiency within a limited time, and user control over the design variants and results.

Optimal PF designs were adjusted according to context and available resources. The PF systems have been installed at the Tropical Technologies Lab and their performance in terms of energy and food production will be monitored for at least a year (to capture possible seasonal effects) in conjunction with their impact on indoor thermal and visual conditions. Further improvement of the PF prototypes will be conducted after collecting and analyzing the measurement data for the first six months. Design parameters such as the number of planter rows, the amount of PV panels per façade, their dimensions, tilt angle, and position will be re-evaluated, especially for the East and West facades.

The PF concept in which food and energy-harvesting installations substitute other building envelope elements like sun shading, walls, and railings is a promising design direction that promotes resilience in residential buildings, especially in high-density urban areas in tropical Asia. It is also in line with the broader concept of continuous productive urban landscapes [74] aiming at integrating energy and food production in urban areas. The design development method and applicability of the results can be of great value for planners, urban designers, architects, engineers, and other environmental experts working towards carbon neutral and resilient urban areas at the low latitudes. Further refinement of the optimization method will be done by explicitly incorporating other criteria functions and extending its applicability to other climatic regions and building typologies.

Author Contributions: Project's principal investigator participated in the project administration and supervision, conceptualization of the project, definition of methodology, data analysis, writing, editing and review, A.T.; optimization framework, writing, visualisation and review, V.K.; investigation, farming methodology, simulations, visualization and writing; H.H.; Grasshopper algorithm development, simulations, validation, visualisation and writing, I.K.C.; conceptualization, project administration, review and editing, S.-K.L.; conceptualization, project administration, review and editing, C.Y.; and supervision, conceptualization, project administration, review and editing, S.S.-Y.L.

Funding: This research was funded by City Developments Limited (CDL), Singapore. CDL SGH Grant Call/Project 1.

Acknowledgments: The farming and BIPV systems support were partially financed by UNISEAL and Wiredbox (WBG (SG) Pte LTD), respectively. The PV modules were donated and fabricated by the Solar Energy Research Institute of Singapore (SERIS). The authors would also like to acknowledge the collaboration of Hugh Tan from Department of Biological Sciences (DBS), NUS and Veronika Shabunko and Thomas Reindl from Solar Energy Research Institute of Singapore (SERIS). Also our gratitude to Khoo Yong Sheng, Zhao Lu, Kyaw Zin Win, Lun Leung Alwin Jason, Srinath Nalluri, Jai Prakash and Liu Haohui, from SERIS and Song Shuang from DBS, NUS for their assistance and to Fadhlina Suhaimi from the Agri-Food and Veterinary Authority (AVA) of Singapore for her advice. The authors would also like to thank Serafim Opricović (Faculty of Civil Engineering, University of Belgrade) and Branislav Todorović, honorary Editor of Energy and Buildings (Faculty of Mechanical Engineering, University of Belgrade, Belgrade) for their professional assistance and support.

Conflicts of Interest: The authors declare no conflict of interest. The funders had no role in the design of the study; in the collection, analyses, or interpretation of data; in the writing of the manuscript, or in the decision to publish the results.

Appendix A

The VIKOR method [1–3] applies the following algorithmic steps:

1. Determining an ideal point. The ideal point is determined as the value of criteria functions according to the following formula:

$$f_i^* = ext_j \; f_{ij}, \, i = 1, \ldots, n \tag{A1}$$

   where *ext* represents the maximum if *i*-th criteria function stands for benefit or gain, or the minimum if it refers to costs or damages. f_i^* indicates the best value of all criteria functions; f_i^- indicates the worst value of all criteria functions.

2. Transformation of non-commensurable criteria functions. The following formula is applied to carry out the transformation into non-dimensional functions:

$$d_{ij} = (f_i^* - f_{ij})/D_i, \, D_i = f_i^* - f_i^-, \, i = 1, \ldots, n, \, j = 1, \ldots, J \tag{A2}$$

   where D_i is the length of scope of *i* criteria function; f_i^* corresponds to the best, and f_i^- to the worst value of the criteria function.

3. Calculation of S_j, R_j, Q_j values according to the following formulae:

$$S_j = \sum_{i=1}^{n} w_i \, d_{ij}, \, i = 1, \ldots, \, = 1, \ldots, J \tag{A3}$$

$$R_j = \max_i \left[w_i \, d_{ij} \right], \, = 1, \ldots, n, \, j = 1, \ldots, \tag{A4}$$

$$S^* = \min_j S_j, \; S^- = \max_j S_j, \, = 1, \ldots, \tag{A5}$$

$$R^* = \min_j R_j, \; R^- = \max_j R_j, \, = 1, \ldots, \tag{A6}$$

$$QS_j = (S_j - S^*)/(S^- - S^*), \, j = 1, \ldots, \tag{A7}$$

$$QR_j = (R_j - R^*)/(R^- - R^*), \ j = 1, \ldots, \tag{A8}$$

$$Q_j = v \times QS_j + (1 - v) \times QR_j, \ j = 1, \ldots, \tag{A9}$$

where w_i represents criteria weight and expresses the DM's preference as the relative importance of the criteria: $v = (n + 1)/2n$ is the weight of the "the maximum group utility" strategy and $(1 - v)$ represents the weight of the individual regret.

4. Ranking the alternatives. Alternatives are sorted according to values of measures: QS_j, QR_j and Q_j, and 3 ranking lists are obtained. The best alternatives are considered to be those that have the lowest measure values.

5. Proposing the compromise solution. Alternative *A1* as the first ranked alternative on the compromise ranking list Q_j is proposed as the best multi-criteria alternative, but only if it also meets the following conditions:

C1. "Sufficient advantage":

$$Q(A2) - Q(A1) \geq DQ \tag{A10}$$

where *A2* is the second ranked alternative on the Q ranking list, and *DQ* is the "advantage threshold":

$$DQ = \min(0.25; 1/(J - 1)) \tag{A11}$$

C2. "Acceptable stability": The best alternative *A1* has to be best ranked according to *S* and/or *R* measures.

These two conditions ensure that DMs are presented with all alternatives that can be seen as "close" to each other in terms of the multi-criteria applied.

If one of these two conditions is not met, then a set of compromise solutions is proposed as follows:

- If only the condition C2 is not met, alternatives *A(1)* and *A(2)* are proposed, and
- If the condition C1 is not met, alternatives *A(1)*, *A(2)*, ... , *A(M)* are proposed, where *A(M)* is determined by $Q(A(M)) - Q(A(1)) < DQ$ for maximal *M*.

Appendix B

Process of obtaining the weight per criteria function for each prototype sub-set.

$$\Delta_{\text{AVG CF}}(k) = \sum_{s=1}^{8} \frac{|\max(k, s) - \min(k, s)|}{8} \tag{A12}$$

$$R_{\text{CF}}(k, s) = \frac{\Delta_{\text{CF}}(k, s)}{\Delta_{\text{AVG CF}}(k)} \tag{A13}$$

$$W_{\text{CF}}(k, s) = \frac{R_{\text{CF}}(k, s)}{\sum_{k=1}^{5} R_{\text{CF}}(k, s)} \tag{A14}$$

where:

"k" is the number of the criterion, from 1 to 5;

"s" is the number of the sub-set, from 1 to 8;

Max (k,s) is the maximum value of the criterion parameter "c" within the subset "s";

Min (k,s) is the minimum value of the criterion parameter "c" within the subset "s";

$\Delta_{\text{AVG CF}}(k)$ is the average value of the range for the criterion "k" among all sub-sets;

$\Delta_{\text{CF}}(k,s)$ is the range of the criterion "k" in the subset "s";

$R_{\text{CF}}(k,s)$ is the relative size of the range of the criterion "k" in the subset "s" compared to the average range for the criterion "k" of all sub-sets of cases and

$W_{\text{CF}}(k,s)$ is the final weight of the criterion "k" within the subset "s".

References

1. Singapore Energy Statistics—2017. Energy Market Authority (EMA), 2017. Available online: https://www.ema.gov.sg/cmsmedia/Publications_and_Statistics/Publications/SES17/Publication_Singapore_Energy_Statistics_2017.pdf (accessed on 15 December 2017).
2. Annual Report 2016/2017. Agri-Food & Authority of Singapore (AVA), 2017. Available online: https://www.ava.gov.sg/docs/default-source/publication/annual-report/ava-ar-2016-17 (accessed on 11 May 2018).
3. Singapore's Intended Nationally Determined Contribution (INDC) and Accompanying Information. Available online: http://www4.unfccc.int/ndcregistry/PublishedDocuments/Singapore%20First/Singapore%20INDC.pdf (accessed on 23 March 2018).
4. Department of Statistics Singapore. Latest Data. Available online: https://www.singstat.gov.sg/whats-new/latest-data (accessed on 17 October 2018).
5. Which Country Is the World's Most Densely Populated? Available online: https://www.worldatlas.com/articles/most-densely-populated-countries-in-the-world.html (accessed on 23 March 2018).
6. A Sustainable Population for a Dynamic Singapore: Population White Paper. 2013. Available online: https://www.strategygroup.gov.sg/docs/default-source/Population/population-white-paper.pdf (accessed on 6 April 2018).
7. Singapore Budget 2018: Singapore's Population Expected to Be below 6.9 Million by 2030, The Business Time. Available online: http://www.businesstimes.com.sg/government-economy/singapore-budget-2018/singapore-budget-2018-singapores-population-expected-to-be (accessed on 6 April 2018).
8. Department of Statistics Singapore. SingStat Table Builder. Available online: http://www.tablebuilder.singstat.gov.sg/ (accessed on 6 April 2018).
9. Department of Statistics Singapore. Population and Population Structure—2017. Available online: https://www.singstat.gov.sg/-/media/files/publications/population/population2018.pdf (accessed on 17 October 2018).
10. Department of Statistics Singapore. Singapore in Figures 2017. Available online: https://www.singstat.gov.sg/find-data/search-by-theme/households/households/latest-data (accessed on 17 October 2018).
11. Meteorological Service Singapore. Annual Climatological Report. 2016. Available online: http://www.weather.gov.sg/wp-content/uploads/2017/02/Annual-Climatological-Report-2016.pdf (accessed on 13 October 2018).
12. EMA. Singapore Electricity Market Outlook (SEMO)—2016. Energy Market Authority (EMA), 2016. Available online: https://www.ema.gov.sg/cmsmedia/Singapore%20Electricity%20Market%20Outlook%20Final.pdf (accessed on 23 March 2018).
13. National Solar Repository of Singapore. Available online: http://www.solar-repository.sg/singapore-pv-market (accessed on June 2018).
14. Mohajeri, N.; Upadhyay, G.; Gudmundsson, A.; Assouline, D.; Kämpf, J.; Scartezzini, J.-L. Effects of urban compactness on solar energy potential. *Renew. Energy* **2016**, *93*, 469–482. [CrossRef]
15. Khoo, Y.S.; Nobre, A.; Malhotra, R.; Yang, D.Z.; Ruther, R.; Reindl, T.; Aberle, A.G. Optimal orientation and tilt angle for maximizing in-plane solar irradiation for PV applications in Singapore. *IEEE J. Photovolt.* **2014**, *4*, 647–653. [CrossRef]
16. Xu, L.; Reed, M.; Reindl, T. CityGML and Solar Potential Analysis for Buildings in Singapore. In Proceedings of the 6th Workshop CityGML Energy ADE, Ferrara, Italy, 23–25 November 2016.
17. Hachem, C.; Athienitis, A.; Fazio, P. Energy performance enhancement in multistory residential buildings. *Appl. Energy* **2014**, *116*, 9–19. [CrossRef]
18. Brito, M.C.; Freitas, S.; Guimaraes, S.; Catita, C.; Redweik, P. The importance of facades for the solar PV potential of a mediterranean city using lidar data. *Renew. Energy* **2017**, *111*, 85–94. [CrossRef]
19. Tablada, A.; Zhao, X. Sunlight availability and potential food and energy self-sufficiency in tropical generic residential districts. *Sol. Energy* **2016**, *139*, 757–769. [CrossRef]
20. Kosoric, V.; Lau, S.K.; Tablada, A.; Lau, S.S.Y. General model of photovoltaic (PV) integration into existing public high-rise residential buildings in Singapore—Challenges and benefits. *Renew. Sustain. Energy Rev.* **2018**, *91*, 70–89. [CrossRef]
21. International Energy Agency (IEA). Renewables 2017: Analysis and Forecast to 2022. 2017. Available online: www.iea.org/t&c/ (accessed on 6 April 2018).

22. International Energy Agency (IEA). Technology Roadmap, Solar Photovoltaic Energy. 2014. Available online: https://www.iea.org/publications/freepublications/publication/TechnologyRoadmapSolarPhotovoltaicEnergy_2014edition.pdf (accessed on 6 April 2018).
23. Zhang, W.; Lu, L.; Peng, J. Evaluation of potential benefits of solar photovoltaic shadings in Hong Kong. *Energy* **2017**, *137*, 1152–1158. [CrossRef]
24. Mandalaki, M.; Tsoutsos, T.; Papamanolis, N. Integrated PV in shading systems for Mediterranean countries: Balance between energy production and visual comfort. *Energy Build.* **2014**, *77*, 445–456. [CrossRef]
25. Stamatakis, A.; Mandalaki, M.; Tsoutsos, T. Multi-criteria analysis for pv integrated in shading devices for mediterranean region. *Energy Build.* **2016**, *117*, 128–137. [CrossRef]
26. Jayathissa, P.; Luzzatto, M.; Schmidli, J.; Hofer, J.; Nagy, Z.; Schlueter, A. Optimising building net energy demand with dynamic BIPV shading. *Appl. Energy* **2017**, *202*, 726–735. [CrossRef]
27. Luther, J.; Reindl, T. Solar Photovoltaic (PV) Roadmap for Singapore (A Summary), Solar Energy Research Institute of Singapore, Singapore. 2014. Available online: https://www.nccs.gov.sg/docs/default-source/default-document-library/solar-photovoltaic-roadmap-for-singapore-a-summary.pdf (accessed on 6 April 2018).
28. Wittkopf, S.; Seng, A.K.; Poh, P.; Pandey, A. BIPV Design for Singapore Zero-Energy Buildings. In Proceedings of the 25th Conference on Passive and Low Energy Architecture (PLEA), Dublin, Ireland, 22–24 October 2008.
29. Saber, E.M.; Lee, S.E.; Manthapuri, S.; Yi, W.; Deb, C. Pv (photovoltaics) performance evaluation and simulation-based energy yield prediction for tropical buildings. *Energy* **2014**, *71*, 588–595. [CrossRef]
30. Ong, B.Q.; Tablada, A. Investigating optimal BIPV energy yield in consideration of daylight and thermal performance in residential buildings. In Proceedings of the 33rd Conference on Passive and Low Energy Architecture (PLEA), Edinburgh, UK, 2–5 July 2017.
31. Sky Greens. Available online: www.skygreens.com (accessed on 23 March 2018).
32. Al-Kodmany, K. The vertical farm: A review of developments and implications for the vertical city. *Buildings* **2018**, *8*, 24. [CrossRef]
33. Astee, L.Y.; Kishnani, N.T. Building integrated agriculture utilising rooftops for sustainable food crop cultivation in Singapore. *J. Green Build.* **2010**, *5*, 105–113. [CrossRef]
34. Bay, J.H.P.; Owen, L.C.W.; Singh, S. Food production and density: The design of a high-rise housing development in Singapore. In *Growing Compact: Urban form, Density and Sustainability*; Bay, J.H.P., Lehmann, S., Eds.; Routledge: Abingdon, UK, 2017.
35. Song, X.P.; Tan, H.T.W.; Tan, P.Y. Assessment of light adequacy for vertical farming in a tropical city. *Urban For. Urban Green.* **2018**, *29*, 49–57. [CrossRef]
36. Suparwoko Taufani, B. Urban farming construction model on the vertical building envelope to support the green buildings development in Sleman, Indonesia. *Procedia Eng.* **2017**, *171*, 258–264. [CrossRef]
37. Tablada, A.; Kosoric, V.; Lau, S.K.; Yuan, C.; Lau, S. Productive facade systems for energy and food harvesting: Prototype optimisation framework. In Proceedings of the 33rd Passive Low Energy Architecture Conference (PLEA), Edinburgh, UK, 2–5 July 2017.
38. Tablada, A.; Chaplin, I.; Huajing, H.; Kosoric, V.; Kit, L.S.; Chao, Y.; Lau, S. Assessment of solar and farming systems integration into tropical building facades. In Proceedings of the Solar World Congress, Abu Dhabi, UAE, 29 October–2 November 2017.
39. Tablada, A.; Chaplin, I.; Huang, H.; Lau, S.; Yuan, C.; Lau, S.S.-Y. Simulation algorithm for the integration of solar and farming systems on tropical façades. In Proceedings of the 23rd International Conference of the Association for Computer-Aided Architectural Design Research in Asia, (CAADRIA), Beijing, China, 17–19 May 2018.
40. Ordenes, M.; Marinoski, D.L.; Braun, P.; Rüther, R. The impact of building-integrated photovoltaics on the energy demand of multi-family dwellings in Brazil. *Energy Build.* **2007**, *39*, 629–642. [CrossRef]
41. Farkas, K.; Frontini, F.; Maturi, L.; Munari Probst, M.C.; Roecker, C.; Scognamiglio, A. *IEA SHC Task 41: Solar Energy Systems in Architecture-Integration Criteria and Guidelines*; Probst, M.C.M., Roecker, C., Eds.; International Energy Agency: Paris, France, 2013.
42. Horvat, M.; Wall, M. (Eds.) *Solar Design of Buildings for Architects: Review of Solar Design Tools. Report T.41.B.3. IEA SHC Task 41, Subtask B*; International Energy Agency: Paris, France, 2012.
43. Ng, P.K.; Mithraratne, N.; Kua, H.W. Energy analysis of semi-transparent BIPV in Singapore buildings. *Energy Build.* **2013**, *66*, 274–281. [CrossRef]

44. Fath, K.; Stengel, J.; Sprenger, W.; Wilson, H.R.; Schultmann, F.; Kuhn, T.E. A method for predicting the economic potential of (building-integrated) photovoltaics in urban areas based on hourly radiance simulations. *Sol. Energy* **2015**, *116*, 357–370. [CrossRef]

45. Freitas, S.; Catita, C.; Redweik, P.; Brito, M.C. Modelling solar potential in the urban environment: State-of-the-art review. *Renewable and Sustainable Energy Rev.* **2015**, *41*, 915–931. [CrossRef]

46. Gautam, B.R.; Li, F.; Ru, G. Assessment of urban roof top solar photovoltaic potential to solve power shortage problem in Nepal. *Energy Build.* **2015**, *86*, 735–744. [CrossRef]

47. Draper, C.; Freedman, D. Review and analysis of the benefits, purposes, and motivations associated with community gardening in the United States. *J. Community Pract.* **2010**, *18*, 458–492. [CrossRef]

48. Marcus, C.C.; Barnes, M. *Healing Gardens: Therapeutic Benefits and Design Recommendations*; John Wiley & Sons: Hoboken, NJ, USA, 1999.

49. Bendt, P.; Barthel, S.; Colding, J. Civic greening and environmental learning in public-access community gardens in berlin. *Landsc. Urban Plann.* **2013**, *109*, 18–30. [CrossRef]

50. Despommier, D. Farming up the city: The rise of urban vertical farms. *Trends Biotechnol.* **2013**, *31*, 388–389. [CrossRef] [PubMed]

51. Kulak, M.; Graves, A.; Chatterton, J. Reducing greenhouse gas emissions with urban agriculture: A life cycle assessment perspective. *Landsc. Urban Plann.* **2013**, *111*, 68–78. [CrossRef]

52. Wong, N.H.; Tan, A.; Yok Tan, P.; Sia, A.; Chung, A.; Wong, N. Perception studies of vertical greenery systems in Singapore. *J. Urban Plan. Dev.* **2010**, *136*, 330–338. [CrossRef]

53. Wong, N.H.; Kwang Tan, A.Y.; Chen, Y.; Sekar, K.; Tan, P.Y.; Chan, D.; Chiang, K.; Wong, N.C. Thermal evaluation of vertical greenery systems for building walls. *Build. Environ.* **2010**, *45*, 663–672. [CrossRef]

54. Elgizawy, E.M. The effect of green facades in landscape ecology. *Procedia Environ. Sci.* **2016**, *34*, 119–130. [CrossRef]

55. McNeel, R. *Grasshopper—Generative Modeling with Rhino*; McNeel North America: Seattle, WA, USA, 2010; Available online: http://www.grasshopper3d.com/ (accessed on 10 November 2017).

56. Opricovic, S.; Tzeng, G.-H. Compromise solution by mcdm methods: A comparative analysis of vikor and topsis. *Eur. J. Oper. Res.* **2004**, *156*, 445–455. [CrossRef]

57. Opricovic, S.; Tzeng, G.-H. Extended vikor method in comparison with outranking methods. *Eur. J. Oper. Res.* **2007**, *178*, 514–529. [CrossRef]

58. Building and Construction Authority. Approved Document: Acceptable Solution—Issued by Commissioner of Building Control under Regulation 27 of the Building Control Regulations. Version 6.3; April 2017. Available online: https://www.bca.gov.sg/BuildingControlAct/others/Approveddoc.pdf (accessed on 6 April 2018).

59. Building and Construction Authority. Code on Envelope Thermal Performance for Buildings. Version 3Rb; January 2008. Available online: https://www.bca.gov.sg/PerformanceBased/others/RETV.pdf (accessed on 6 April 2018).

60. Inada, K.; Yabumoto, Y. Effects of light quality, daylength and periodic temperature variation on the growth of lettuce and radish plants. *Jpn. J. Crop Sci.* **1989**, *58*, 689–694. [CrossRef]

61. He, J. Farming of vegetables in space-limited environments. *Cosmos* **2015**, *11*, 21–36. [CrossRef]

62. Neto, H.S.L.; de Guimarães, M.A.; de Tello, J.P.J.; Mesquita, R.O.; do Vale, J.C.; Neto, B.P.L. Productive and physiological performance of lettuce cultivars at different planting densities in the Brazilian Semi-arid region. *Afr. J. Agric. Res.* **2017**, *12*, 771–779. [CrossRef]

63. Kalantari, F.; Mohd Tahir, O.; Mahmoudi Lahijani, A.; Kalantari, S. A review of vertical farming technology: A guide for implementation of building integrated agriculture in cities. *Adv. Eng. Forum* **2017**, *24*, 76–91. [CrossRef]

64. Morgan, L. Hydroponic Illumination & the Daily Light Integral. Available online: https://www.maximumyield.com/hydroponic-illumination-the-daily-light-integral/2/1450 (accessed on 20 September 2017).

65. Dorais, M. The use of supplemental lighting for vegetable crop production: Light intensity, crop response, nutrition, crop management, cultural practices. In Proceedings of the Canadian Greenhouse Conference, Toronto, ON, Canada, 9–10 October 2003.

66. Schiller, L. Is My Plant Getting Enough Light? Available online: http://www.ceresgs.com/is-my-plant-getting-enough-light/ (accessed on 20 September 2017).

67. Glenn, E.P.; Cardran, P.; Thompson, T.L. Seasonal effects of shading on growth of greenhouse lettuce and spinach. *Sci. Hortic.* **1984**, *24*, 231–239. [CrossRef]
68. Roudsari, M.S.; Pak, M. Ladybug: A parametric environmental plugin for grasshopper to help designers create an environmentally-conscious design. In Proceedings of the 13th International IBPSA Conference, Chambery, France, 25–28 August 2013.
69. Pohekar, S.D.; Ramachandran, M. Application of multi-criteria decision making to sustainable energy planning—A review. *Renew. Sustain. Energy Rev.* **2004**, *8*, 365–381. [CrossRef]
70. Mardani, A.; Zavadskas, K.E.; Govindan, K.; Amat Senin, A.; Jusoh, A. Vikor technique: A systematic review of the state of the art literature on methodologies and applications. *Sustainability* **2016**, *8*, 37. [CrossRef]
71. Kosoric, V.; Wittkopf, S.; Huang, Y. Testing a design methodology for building integration of photovoltaics (PV) using a PV demonstration site in Singapore. *Archit. Sci. Rev.* **2011**, *54*, 192–205. [CrossRef]
72. Energy Market Authority (EMA). Singapore Energy Statistics. 2018. Available online: https://www.ema.gov.sg/cmsmedia/Publications_and_Statistics/Publications/SES18/Publication_Singapore_Energy_Statistics_2018.pdf (accessed on 13 October 2018).
73. Attoye, D.E.; Tabet Aoul, K.A.; Hassan, A. A Review on Building Integrated Photovoltaic Façade Customization Potentials. *Sustainability* **2017**, *9*, 2287. [CrossRef]
74. Viljoen, A.; Bohn, K.; Howe, J. *Continuous Productive Urban Landscapes: Designing Urban Agriculture for Sustainable Cities*; Architectural Press: Oxford, UK, 2005.

 *sustainability*

Article

Modeling Nature-Based and Cultural Recreation Preferences in Mediterranean Regions as Opportunities for Smart Tourism and Diversification

André Samora-Arvela [1], Jorge Ferreira [1], Eric Vaz [2] and Thomas Panagopoulos [3,*]

[1] Interdisciplinary Centre of Social Sciences, Faculty of Social Sciences and Humanities, New University of Lisbon, 1069-061 Lisbon, Portugal; anesamora@gmail.com (A.S.-A.); jr.ferreira@fcsh.unl.pt (J.F.)

[2] Laboratory for Geocomputation, Faculty of Arts, Ryerson University, Toronto, ON M5B 1W7, Canada; evaz@ryerson.ca

[3] Research Centre for Tourism, Sustainability and Well-being, University of Algarve, Campus de Gambelas, 8005-139 Faro, Portugal

* Correspondence: tpanago@ualg.pt; Tel.: +351-289-800-900; Fax: +351-289-818-419

Received: 1 December 2019; Accepted: 23 December 2019; Published: 6 January 2020

Abstract: The tourism and recreational offer of Mediterranean destinations involves, essentially, the promotion of mass tourism, based on the appeal of the sun and beach, and the quality of its coastal assets. Alongside the impacts of climate change, poor tourism diversification represents a threat to the resilience of the territory. Thus, heterogenization of noncoastal tourism products presents an opportunity to strengthen regional resilience to present and future challenges, hence the need to study, comparatively, the complementary preferences of tourists and residents of these regions in order to unveil their willingness to diversify their recreational experience, not only in coastal spaces, but also—and especially—in interior territories with low urban density. Consequently, this strategic option may represent a way of strengthening resilience and sustainability through diversification. In this context, a survey was conducted among 400 beach tourists and 400 residents of a case study—namely, three municipalities of the Algarve region in southern Portugal—in order to analyze their degree of preference for activities besides the sun and beach, such as nature-based and cultural tourism activities, and to probe the enhancement potential of each tourism and recreational activity through the various landscape units considered by experts, stakeholders, and tour operators. The respective degree of preference and enhancement potential were indexed to the area of each landscape unit. Subsequently, respecting the existing recreational structure and constraints, a suitability map for territory enhancement and the implementation of smart tourism practices for each tourism activity and landscape unit is presented. Results show a significant preference for noncoastal outdoor recreational activities.

Keywords: sustainable tourism; smart tourism; mobile applications; nature recreation

---

1. Introduction

All Mediterranean regions focus on sun-and-beach tourism as the core of their socioeconomic development. This Mediterranean commonality corresponds to a rarely promoted landscape diversity when compared to the coastal beach sand, the fulcrum of the economic base of these territories. Many sun-and-sea destinations are stagnant, without innovation and without contributing to territorial cohesion, as most of them have low levels of employment and education [1]. This strong focus on beach tourism, coupled with the present and future impacts expected from climate change, especially sea-level rise, could weaken regional potential development by reducing the beach bathing area—a major resource for tourist enjoyment in Mediterranean destinations [2].

Due to their landscape diversity, a diversification of alternative products that differentiates Mediterranean destinations—strengthening their attractiveness and regional resilience—can be prolific through the presentation of inimitable products that can only be achieved through the sustainable use of the region's endogenous resources. This can be a way to guarantee the long-term sustainability of tourism destinations [3].

Tourism is seen as a potential factor in mitigating the disparities between rural and urban space [4] when low-density territories are declining in terms of landscape preservation or enhancement. It is essential to find alternative tourism and recreational activities to the standard sun-and-sea tourism in order to sustainably rejuvenate and reinvent Mediterranean destinations. It is paramount, in the first instance, to research the preference of mainstream bathing tourists for other options, such as nature-based and cultural tourism and related recreational activities.

The rejuvenation of destinations by finding and enhancing alternative tourism products should be done through smart tourism policies and practices, assumed as key points for sustainable development [1]. As such, this research presents the territorial model of a case study, namely the municipalities of Silves, Albufeira, and Loulé in the Algarve region, which attracts more than a third of all tourists visiting Portugal.

The aim of this study was to develop a multicriteria evaluation method that diversifies from sun-and-sea coastal tourism by identifying the areas to be enhanced by the public sector and to provide the necessary tools for developing smart tourism, such as websites and geolocation applications focused on tourism promotion, based on effective, efficient, and sustainable management of a wide range of Mediterranean landscape units and products that exist or may emerge in those tourist destinations. The method was based on tourists' and residents' preferences; evaluation of the enhancement potential as considered by experts, stakeholders, and tour operators; and suitability maps for outdoor recreation.

2. Theoretical Framework

2.1. Southwest European Coast Tourism Area Life Cycle

The evolution of tourist destinations is currently studied through the Tourism Life Cycle Model [5], which is based on five steps: Involvement, exploration, development, consolidation, and stagnation. The first stage, involvement, concerns a tourist's initial interaction with local communities, with an increasing impact on local economies [3]. The second stage, exploration, refers to the intensification of tourism consumption, which can create pressures on the environment and on residents' quality of life.

For Plog [6], in these early stages, tourists try to experience some adventures and some risks, not requiring a high level of institutionalized tourism organization. The development stage, on the other hand, is characterized by an increase in the number of tourists, accompanied by an increase in tourism activities, the emergence of new services (such as organized travel), a marked impact on the daily life of the resident community, and investment by international companies that start operating in developing destinations via accommodation and transportation, among others. Cohen [7] mentions that, at this stage, individual mass tourism gives way to organized mass tourism, that is, explorer tourists are progressively replaced by follower tourists [3].

In this line of development, many destinations face the challenge of the consolidation phase and further stagnation that refers to increasingly anemic tourist activity growth rates and a hegemony of tourist numbers to the detriment of the local population—A characteristic of sun-and-sea destinations, whose main attraction is the seasonal climatic pleasantness. This narrowing attractive base in these destinations often implies an increase in investment in the persistent promotion of sun-and-sea tourism assets that, worldwide, appear to be the least differentiating [3].

After these stages, tourist destinations can follow one of two opposing paths: Stagnation decline, which worsens the deterioration of tourism productivity in presenting tourism products, inherently less differentiating, or destination rejuvenation, associated with innovation, product diversification, and its specializing differentiation based on the plural landscape identity of each destination territory.

Through this theoretical framework, Romão [3] classifies the various destinations of the southwest coast of Europe according to the current situation stage of their life cycle, such as:

- Exploration: Basilicata, Campania, Lombardy, Molise, Sicilia (Italy), Murcia (Spain);
- Development: Calabria, Lazio, Piedmont, Puglia, Sardegna (Italy), Andalusia (Spain), and Azores (Portugal);
- Stagnation: Abruzzo, Marche, Emilia Romagna, Friuli-Venezia Giulia, Tuscany, Veneto (Italy), Provence-Alps-Cote d'Azur, Corse (France), Canary Islands, Catalonia, Valencian Community, Illes Balears (Spain), Algarve and Madeira (Portugal).

Ricard Butler [8], quoting Russo [9], acknowledges the importance of intervening in tourism management in order to prevent development that exceeds the destination capacity (capacity defined in terms of limits of economic, social, environmental, and physical parameters). If exceeded, the quality of visitor and resident experiences can decline, degrading the inimitable environmental and cultural resources of each destination, and losing visitors, tourist expenditure and, thus, investment capacity in the destination [9]. Agarwal [10] and Baum [11] describe how this tourism area life cycle evolved in regard to the number of stages and shape of the curve, and a comprehensive review of the model applications is contained in Lagiewski [12].

2.2. Diversification and Differentiation in Tourism

Among the various tourism product management strategies, Benur and Bramwell [13] point out two key processes: Concentration and diversification. These strategies can be planned at different scales, either by mass intensification or niche creation. The concepts of tourist destination and tourist experience are inextricably linked, as tourists perceive destinations as an integrated experience, derived from a large set of products and services provided in each destination that seeks to meet the motivations and expectations of each tourist [3].

Unlike concentrating on a single product, product diversification is based on offering a diverse set of activities and experiences to attract a wide and flexible multiplicity of tourist segments, which is ultimately the ultimate guarantee of competitiveness, resilience, and sustainability of tourist destinations and, as such, of their supporting territories and local communities [14].

It should be noted that tourism diversification is achieved by encouraging alternative products through their promotion and appreciation. Tourism diversification can only be effective if it leads to product differentiation among the tourist segments it aims to captivate, which is the affirmation of activities and services based on the preserved biophysical and cultural identity of a territory with the goal of ensuring a long-term and sustainable attractiveness, thus generating tourist fidelity, supported not by the low prices' leadership, but by the supply's nonimitable characteristics [3,14]. Thereby, the appreciation of destination uniqueness through diversification can be made efficient through innovation, that is, the result of technological evolution of tourism companies, institutions, tourism research and development, and the interactivity between clients and companies through information technologies.

The programming of inimitable resources of each destination into tourism products by synergy between communities and operators is still lacking and latent, as the recognition of their authenticity and uniqueness, while adding value, provides an opportunity for the rejuvenation of stagnant destinations. Only innovation can bring about the promotion of differentiating tourism diversification and thus assist the sustainable management of each destination, strengthening competitiveness by promoting its real authenticity, not its commodification [14].

2.3. Sustainable Tourism, Competitiveness, and Sustainability of Tourist Destinations

At this point, it is important to examine the concepts, interrelations, and interdependencies of sustainable tourism, competitiveness, and sustainability of tourism destinations. David Weaver [15], in *Sustainable Tourism: Theory and Practice*, defines sustainable tourism development as tourism that meets the needs of the present without eroding the capacity for future satisfaction, associating it with Budowski's definition [16], namely, tourism that uses and conserves resources without compromising its productive capacity so as to ensure its long-term viability, that is, tourism that minimizes negative impacts and maximizes positive aspects [15].

Sustainable tourism allows for the preservation, enhancement, and promotion of unique and distinctive features that serve as a basis for the creation of differentiating products that, presented in a diversified manner, affirm the competitiveness of each tourism destination. As such, for Buhalis [17], a destination is competitive if it attracts and satisfies potential tourists, as competitiveness is the ability of one destination to achieve greater success than other competing destinations, which derives from leadership of low prices for mass production with a view to minimizing production costs, differentiating products through tourism promotion of something unique, or an integrated strategy to reach various market segments through a combination of cost leadership and product differentiation within a framework of flexible specialization and permanent innovation [18].

Therefore, the determinants of competitiveness are comparative advantage, demand orientation, organization of the tourism industry, and preservation of local natural and cultural specificities [3]. Thus, the sustainability of a tourist destination comes from the adoption of sustainable tourism patterns that imply the respect of carrying capacity of a territory, precisely through environmental preservation, integration of local communities, and economic viability. A sustainable destination is supported not by the degradation or destruction of natural and cultural authenticity through homogenization of destinations, as is recurrent in sun-and-sea locations with their globalized palm tropicalisms, leading to their stagnation, but by the diversification and differentiation of products with the recognition of distinctive characteristics that support the attractiveness of tourism, that is, a rejuvenation strategy that gradually makes destinations sustainable and thus effectively competitive.

Along these lines, an effective and long-term competitive advantage should come from sustainable tourism through the preservation and enhancement of the destination's environmental and cultural differentiating uniqueness. In this sense, Hassan [19] states that destination development and marketing must pursue the following objectives:

- Promoting awareness and understanding among key stakeholders for sustaining the environment during tourism development;
- Promoting equity in development opportunities;
- Maximizing tourist satisfaction;
- Developing and sustaining the quality of life for the local communities and broadening their support through citizen/non-governmental organizations involvement;
- Maintaining balance among economic, social, and environmental needs; defining physical and social carrying capacity of the destination;
- Developing environmental impact assessments for any tourism development;
- Maintaining the local culture and promoting the image of its values, heritage, traditional way of life, and behaviour;
- Enhancing education and training in management for sustainable tourism development.

2.4. The Innovation and Rejuvenation Opportunity: Smart Tourism

In this context, it is important to define innovation: "innovation refers to the process of bringing any new, problem solving idea into use. Ideas for reorganizing, cutting cost, putting in new budgetary systems, improving communication or assembling products in teams are also innovations. Innovation is the generation, acceptance and implementation of new ideas, processes, products or services. Acceptance and implementation is central to this definition; it involves the capacity to change and adapt" [20]. It can be grounded not only in product and service innovations, which concern the changes considered as new, and process innovations, where Information and Communication Technology

(ICT) plays a major role, but also in managerial and management innovations and institutional innovations [21].

Here, it becomes paramount to establish that the concept of smart tourism integrates the potential role of digital technologies in developing collaborative processes between rival service providers at the destination level and in co-creating destinations and experiences through the increasing interaction between producers and consumers, generating benefits from some particular characteristics of tourism services (interoperability, spatiality, and temporality, implying direct interaction between suppliers and consumers, in the same place and at the same time) [1,22]. However, development should not be considered smart if it is not also sustainable. In addition, rural areas may benefit most from the slow tourism opportunities offered by trails for hiking and gravel biking, because slow tourism offers a more meaningful and deliberate way of connecting with the visited region [23].

In this sense, smart tourism, smart destination and smart region concepts derive from the alignment with the concept of a smart city [24]. A city destination or region is smart when "investments in human and social capital and traditional (transport) and modern (ICT) communication infrastructure fuel sustainable economic growth and a high quality of life, with a wise management of natural resources, through participatory governance." Thus, smart tourism adjuvates sustainable spatial planning in destinations/regions through the development of spaces, infrastructure, and services through knowledge-based innovation strategies, supported by the knowledge and information of specific resources of each region, which governance should be done in concert with public participation at the destination level [25,26].

Nowadays, websites, social networks, geolocation, and mobile applications (apps) are the most widely used technology tools in digital marketing, information, and communication, but their potential in Mediterranean tourism is not still fully understood by institutions and enterprises [27]. In this context, major factors have to be considered when designing and presenting an app. According to Lu et al. [28], the acceptance of technology depends on its perceived ease of use and usefulness, the diffusion of innovation by presenting relative advantage over previous ways of performing the same task, the degree to which an innovation is consistent with the values, beliefs, experiences, and needs of the users, visibility to others.

Thereby, it is recognized the need for delineating rural tourism promotion strategies based on the use of mobile technology. Meanwhile, there is high dispersion and disorderly developed digital information and uneven wireless internet infrastructures. Lu et al. [28] suggest a need to establish "centralized tourism databases and appropriate operational systems to collect, store, exchange, and release tourism data; launch smart tourism service platforms (websites, apps, and digital screens) to enhance tourist experience and destination image; and install vending machines and QR code readers in stations, terminals, hotels, restaurants, and tourism sites to facilitate the adoption of mobile technology in tourism process".

3. Materials and Methods

3.1. Study Area

The study area corresponds to the municipalities of Silves, Albufeira, and Loulé (green area of Figure 1), Algarve region, southern Portugal, marked by its Mediterranean climate and permeated by three fundamental landscape units: Interior Mountains (A), Barrocal Midlands (B), and Coastal Landscape (C) [29].

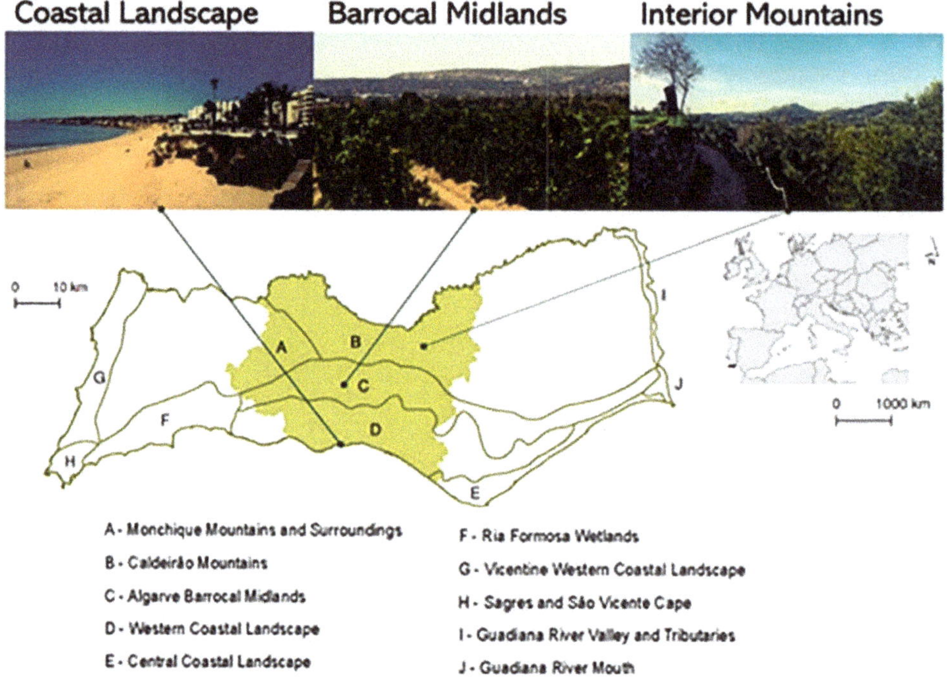

Figure 1. The study area municipalities of Silves, Albufeira, and Loulé within Algarve region, southern Portugal.

The coastline of those municipal territories is consigned to a disorderly urban-tourist occupation, derived from the strict promotion of sun and beach product, whose use is intensive and seasonal. Between the coast and the mountains, the Barrocal Midlands landscape unit is characterized by limestone soils, its agricultural matrix of fig, almond, carob, and olive groves, and the presence of irrigated orange orchards, which is being penetrated by the proliferation of second homes.

It must be acknowledged that agricultural abandonment is relatively transverse to all landscape units of this territory as a consequence of the rural exodus, the commutative transition from agricultural labour to tourism and tertiarization. These problems determine the poor municipal and regional agricultural competitiveness as a result of the fragmented agricultural land structure and the small average size of rustic property, which constituted and constitutes a constraint to agriculture modernization (50% of farmers in the region have a mean 2 ha farmland) [30].

As such, the integrated landscape management capacity appears to be precarious. The valados walls (dry stone walls to support from terraces and waterfront margins) are not being conserved and are gradually crumbling. Old agricultural plots have been substituted with shrubland, while the Algarve destination simply and simplistically promotes the image of palm tropicalisms imported to its coastline [31]. This situation is even more noticeable in the interior mountainous landscape, marked by the presence of declining cork tree and xerophytic forests, inherent abandonment, desertification and fires.

Without conditions of competitive and sustainable use of its agricultural matrix, one strategy for reinforcing the competitiveness and sustainability of the rural space of these municipalities can and should be the enhancement of their endogenous resources in favour of diversification, differentiation, and innovation in the tourist use of their territories [32]. A diversification aimed at nature-based and cultural tourism activities could be a strategic opportunity, as 60% of tourists state that Algarve's heritage and cultural offerings are an important attraction in the choice to visit the region [33], and 87%

of the tourists who visit the region try to have other experiences beyond the sun and sea. Also, 78% of tourists point out gastronomy as one of the main attractions [34,35], while the degree of preferential taste for regional Algarve products and sweets is high [36].

It should be noted that these opportunities are crucial in the search for the reinforcement of regional resilience and respective strengthening of the capacity for cultural adaptation of these communities to climate change. It is urgent for these municipalities and others in Mediterranean destinations to adopt a transformative strategy of diversification and promotion of their tourism products through the enhancement potential of diversified endogenous resources, beyond sun and beach, as a means of adaptation, especially with regard to sea level rise, which could reduce the coastal beach area and, consequently, disrupt the socioeconomic base of these regions, which depend so much on it. The attenuation of this dependence through tourism diversification adoption can be a transition towards the reinforcement of the resilience and sustainability of Mediterranean regions.

3.2. Methods

The methodology of this article aims at the realization and construction of a territorial model of tourism and recreational enhancement through the following methods.

1. Analysis of questionnaire survey results that assessed the degree of preference of tourists and residents and enhancement potential considered by experts, stakeholders, and tour operators for the promotion of alternative tourism and recreation activities, besides sun and beach, namely nature-based and cultural tourism. The sample size was 400 tourists and 400 residents, and the sample selection was performed using the stratified random sampling method of number of guests in tourist accommodation establishments by municipality in 2015, country of usual residence [37], and number of resident individuals living in the study area [38]. Tourists and residents were surveyed in 2017. Subsequently, the enhancement potential of each tourism and recreational activity was studied through a survey of experts, stakeholders, and tour operators, which, in proportion to the degree of preference of tourists and residents, allowed for assessment of the total potential for enhancement of each activity in each landscape unit.

The preference degree of taste was determined using the numerical scale 1—I don't like it; 2—I like it a little; 3—I like it relatively; 4—I like it reasonably; 5—I like it; 6—I like it a lot; 0—Does not know/does not answer. The enhancement potential was studied using the following scale: 1—Not to enhance; 2—To enhance a little; 3—To enhance relatively; 4—To enhance reasonably; 5—To enhance; 6—To enhance a lot; 0—Does not know/does not answer.

2. Investigation into the existing territorial structure of recreational activities most appreciated in preference and identified as having the greatest potential for enhancement. Digitalization of this information and spatialization in geographic information system maps.

3. Building a suitability map of total enhancement potential of each recreation activity assessed in each landscape unit through the multicriteria evaluation of preference degree of tourists and residents and enhancement potential considered by experts, stakeholders, and tour operators, the distance from existing recreational structures and the zoning of spatial planning policies: Suitability and factors for multicriteria evaluation [39,40].

4. Results

Given the mean degree of preference of tourists and residents, and the mean enhancement potential considered by experts, stakeholders, and tour operators for each tourism and recreational activity within the Interior Mountains, it is noted that some are at a level higher than 5 (I like it/To enhance). Walks, orienteering, and fitness circuits being the preferred activity (5.53), followed by health and wellness (5.36), rural accommodation (5.27), gastronomy and wines (5.11), and landscape touring and picnics (5.09), represent the main activities to be enhanced within the Interior Mountains landscape (Table 1). The exceptions with the lowest scores were golf (3.21) and hunting (3.12).

Table 1. Mean degree of tourists' and residents' preference and enhancement potential considered by experts, stakeholders, and tour operators for each activity within the Interior Mountains.

Interior Mountain Area	A	B	C	D	E	F	G	H	Total
Walks, orienteering and keep-fit circuits	4.85	5.51	4.95	5.27	6.00	6.00	5.80	5.00	5.53
Cycling	4.27	5.51	4.22	5.27	4.33	5.71	5.13	5.00	4.84
Mountain biking	3.10	5.51	3.51	5.27	4.33	5.57	5.20	5.00	4.54
Birdwatching	2.25	5.51	3.10	5.27	6.00	5.86	5.50	5.00	4.79
Golf	2.43	5.51	2.61	5.27	1.33	4.00	4.00	5.00	3.21
Bathing, canoeing and fishing	4.26	5.51	4.10	5.27	4.67	4.71	4.40	5.00	4.58
Climbing, abseiling and zip-lining	3.12	5.51	3.12	5.27	4.67	5.29	3.60	5.00	4.26
Geocaching	1.90	5.51	2.52	5.27	5.33	5.00	4.57	5.00	4.20
Hunting	1.60	5.51	2.06	5.27	3.33	3.71	2.47	5.00	3.12
Horse riding	2.74	5.51	3.15	5.27	4.67	5.00	4.00	5.00	4.21
Other outdoor activities in sports centre	3.77	5.51	4.86	5.27	3.00	5.57	4.80	5.00	4.52
Rural accommodation	4.78	5.51	4.86	5.27	5.33	5.86	5.33	5.00	5.27
Camping	3.69	5.51	4.28	5.27	3.67	5.29	4.00	5.00	4.37
Campervanning	3.34	5.51	3.71	5.27	3.67	5.57	4.07	5.00	4.30
Temporary integration in eco-farms	3.20	5.51	3.82	5.27	4.67	5.71	4.21	5.00	4.55
Health and wellness	4.89	5.51	5.21	5.27	5.67	5.86	5.00	5.00	5.36
Gastronomy and wines	4.72	5.51	4.98	5.27	4.33	5.71	5.73	5.00	5.11
Touring architectural and archaeological	4.40	5.51	4.46	5.27	5.33	5.86	5.67	5.00	5.21
Landscape touring and picnics	4.63	5.51	4.50	5.27	5.00	5.71	5.27	5.00	5.09
Outdoor fairs and markets	4.61	5.51	4.81	5.27	4.00	5.00	5.00	5.00	4.75
Outdoor shows and performances	4.63	5.51	5.18	5.27	3.50	5.29	5.13	5.00	4.79
Outdoor religious processions	2.54	5.51	3.78	5.27	3.33	5.14	4.13	5.00	4.05

A: 15% * Mean of tourists' preference for each recreation activity (N = 400); B: 5% * Mean of tourists' preference for each landscape unit (N = 400); C: 15% * Mean of residents' preference for each recreation activity (N = 400); D: 5% * Mean of residents' preference for each landscape unit (N = 400); E: 20% * Mean of enhancement potential considered by experts for each recreation activity and landscape unit (N = 3); F: 20% * Mean of enhancement potential considered by stakeholders for each recreation activity and landscape unit (N = 7); G: 15% * Mean of enhancement potential considered by tour operators for each recreation activity (N = 15); H: 5% * Mean of enhancement potential considered by tour operators for each landscape unit (N = 15).

Within the Barrocal Midlands, some activities also exceeded level 5 for surveyed tourists, residents, experts, stakeholders, and tour operators. Rural accommodation (5.42), walks, orienteering, and keep-fit circuits were also the most appreciated and recognized here (5.41). Gastronomy and wines (5.26), touring architectural and archaeological sites (5.16), health and wellness (5.04), and landscape touring and picnics (5.04) were the most preferred and suitable for enhancement in the Barrocal Midlands landscape. The exception was also reiterated for golf (3.36) and hunting (2.87) (Table 2).

Table 2. Mean degree of tourists' and residents' preference and enhancement potential considered by experts, stakeholders, and tour operators for each activity within the Barrocal Midlands.

Barrocal Midlands	A	B	C	D	E	F	G	H	Total
Walks, orienteering and keep-fit circuits	4.85	5.30	4.95	5.26	5.33	6.00	5.80	5.53	5.41
Cycling	4.27	5.30	4.22	5.26	4.67	5.71	5.13	5.53	4.92
Mountain biking	3.10	5.30	3.51	5.26	5.00	5.57	5.20	5.53	4.69
Birdwatching	2.25	5.30	3.10	5.26	5.00	5.86	5.50	5.53	4.60
Golf	2.43	5.30	2.61	5.26	2.00	4.00	4.00	5.53	3.36
Bathing, canoeing and fishing	4.26	5.30	4.10	5.26	4.33	4.71	4.40	5.53	4.53
Climbing, abseiling and zip-lining	3.12	5.30	3.12	5.26	3.33	5.29	3.60	5.53	4.00
Geocaching	1.90	5.30	2.52	5.26	5.00	5.00	4.57	5.53	4.15
Hunting	1.60	5.30	2.06	5.26	2.00	3.71	2.47	5.53	2.87
Horse riding	2.74	5.30	3.15	5.26	4.33	5.00	4.00	5.53	4.15
Other outdoor activities in sports centre	3.77	5.30	4.86	5.26	3.67	5.57	4.80	5.53	4.67
Rural accommodation	4.78	5.30	4.86	5.26	6.00	5.86	5.33	5.53	5.42
Camping	3.69	5.30	4.28	5.26	5.00	5.29	4.00	5.53	4.66
Campervanning	3.34	5.30	3.71	5.26	3.67	5.57	4.07	5.53	4.32
Temporary integration in eco-farms	3.20	5.30	3.82	5.26	4.33	5.71	4.21	5.53	4.50
Health and wellness	4.89	5.30	5.21	5.26	4.00	5.86	5.00	5.53	5.04
Gastronomy and wines	4.72	5.30	4.98	5.26	5.00	5.71	5.73	5.53	5.26
Touring architectural and archaeological	4.40	5.30	4.46	5.26	5.00	5.86	5.67	5.53	5.16
Landscape touring and picnics	4.63	5.30	4.50	5.26	4.67	5.71	5.27	5.53	5.04
Outdoor fairs and markets	4.61	5.30	4.81	5.26	4.00	5.00	5.00	5.53	4.77
Outdoor shows and performances	4.63	5.30	5.18	5.26	4.00	5.29	5.13	5.53	4.90
Outdoor religious processions	2.54	5.30	3.78	5.26	3.33	5.14	4.13	5.53	4.07

A: 15% * Mean of tourists' preference for each recreation activity (N = 400); B: 5% * Mean of tourists' preference for each landscape unit (N = 400); C: 15% * Mean of residents' preference for each recreation activity (N = 400); D: 5% * Mean of residents' preference for each landscape unit (N = 400); E: 20% * Mean of enhancement potential considered by experts for each recreation activity and landscape unit (N = 3); F: 20% * Mean of enhancement potential considered by stakeholders for each recreation activity and landscape unit (N = 7); G: 15% * Mean of enhancement potential considered by tour operators for each recreation activity (N = 15); H: 5% * Mean of enhancement potential considered by tour operators for each landscape unit (N = 15).

For the Coastal Landscape, walks, orienteering, and keep-fit circuits were still the most preferred activity and the one with the most potential of enhancement for this landscape (5.23), along with health and wellness (5.20), gastronomy and wines (5.15), outdoor shows and performances (5.06), and touring architectural and archaeological sites. Exception were geocaching (3.84), outdoor religious processions (3.82), climbing, abseiling and zip-lining (3.69), golf (3.38), and hunting (2.75) (Table 3).

In view of the above, there may be a redirection of tourism promotion in the region, which should favour activities beyond the sun-and-sea product, especially the following: Walking in trails and boardwalks, hiking, orienteering, and keep-fit circuits; cycling, gravel biking; birdwatching; bathing, canoeing and fishing in rivers or lakes; horse riding; other outdoor activities in sports centres (soccer, basketball, tennis, etc.); rural accommodation; camping; campervanning; temporary integration in ecovillages, ecocommunities or ecofarms; health and wellness; gastronomy and wines; touring architectural and archaeological heritage, and museums; landscape touring and picnics, and outdoor fairs and markets.

Regarding the existing recreational structure, the research outcomes have identified the major paths, trails, and suitable spaces for the various recreational activities studied, through the surveys, in order of considered preference and enhancement potential. An example of this is the pre-existing paths for walks, orienteering, and keep-fit circuits (Figure 2). These layers can and should be used in digital marketing of Algarve destinations through the delineation of geolocation algorithm app for the tourism brand Algarve, characterized by integration of these existing layers and by the app's acceptance, diffusion, and social outcomes.

Table 3. Mean degree of tourists' and residents' preference and enhancement potential considered by experts, stakeholders, and tour operators for each activity within the Coastal Landscape.

Barrocal Midlands	A	B	C	D	E	F	G	H	Total
Walks, orienteering and keep-fit circuits	4.85	5.30	4.95	5.26	5.33	6.00	5.80	5.53	5.41
Cycling	4.27	5.30	4.22	5.26	4.67	5.71	5.13	5.53	4.92
Mountain biking	3.10	5.30	3.51	5.26	5.00	5.57	5.20	5.53	4.69
Birdwatching	2.25	5.30	3.10	5.26	5.00	5.86	5.50	5.53	4.60
Golf	2.43	5.30	2.61	5.26	2.00	4.00	4.00	5.53	3.36
Bathing, canoeing and fishing	4.26	5.30	4.10	5.26	4.33	4.71	4.40	5.53	4.53
Climbing, abseiling and zip-lining	3.12	5.30	3.12	5.26	3.33	5.29	3.60	5.53	4.00
Geocaching	1.90	5.30	2.52	5.26	5.00	5.00	4.57	5.53	4.15
Hunting	1.60	5.30	2.06	5.26	2.00	3.71	2.47	5.53	2.87
Horse riding	2.74	5.30	3.15	5.26	4.33	5.00	4.00	5.53	4.15
Other outdoor activities in sports centre	3.77	5.30	4.86	5.26	3.67	5.57	4.80	5.53	4.67
Rural accommodation	4.78	5.30	4.86	5.26	6.00	5.86	5.33	5.53	5.42
Camping	3.69	5.30	4.28	5.26	5.00	5.29	4.00	5.53	4.66
Campervanning	3.34	5.30	3.71	5.26	3.67	5.57	4.07	5.53	4.32
Temporary integration in eco-farms	3.20	5.30	3.82	5.26	4.33	5.71	4.21	5.53	4.50
Health and wellness	4.89	5.30	5.21	5.26	4.00	5.86	5.00	5.53	5.04
Gastronomy and wines	4.72	5.30	4.98	5.26	5.00	5.71	5.73	5.53	5.26
Touring architectural and archaeological	4.40	5.30	4.46	5.26	5.00	5.86	5.67	5.53	5.16
Landscape touring and picnics	4.63	5.30	4.50	5.26	4.67	5.71	5.27	5.53	5.04
Outdoor fairs and markets	4.61	5.30	4.81	5.26	4.00	5.00	5.00	5.53	4.77
Outdoor shows and performances	4.63	5.30	5.18	5.26	4.00	5.29	5.13	5.53	4.90
Outdoor religious processions	2.54	5.30	3.78	5.26	3.33	5.14	4.13	5.53	4.07

A: 15% * Mean of tourists' preference for each recreation activity (N = 400); B: 5% * Mean of tourists' preference for each landscape unit (N = 400); C: 15% * Mean of residents' preference for each recreation activity (N = 400); D: 5% * Mean of residents' preference for each landscape unit (N = 400); E: 20% * Mean of enhancement potential considered by experts for each recreation activity and landscape unit (N = 3); F: 20% * Mean of enhancement potential considered by stakeholders for each recreation activity and landscape unit (N = 7); G: 15% * Mean of enhancement potential considered by tour operators for each recreation activity (N = 15); H: 5% * Mean of enhancement potential considered by tour operators for each landscape unit (N = 15).

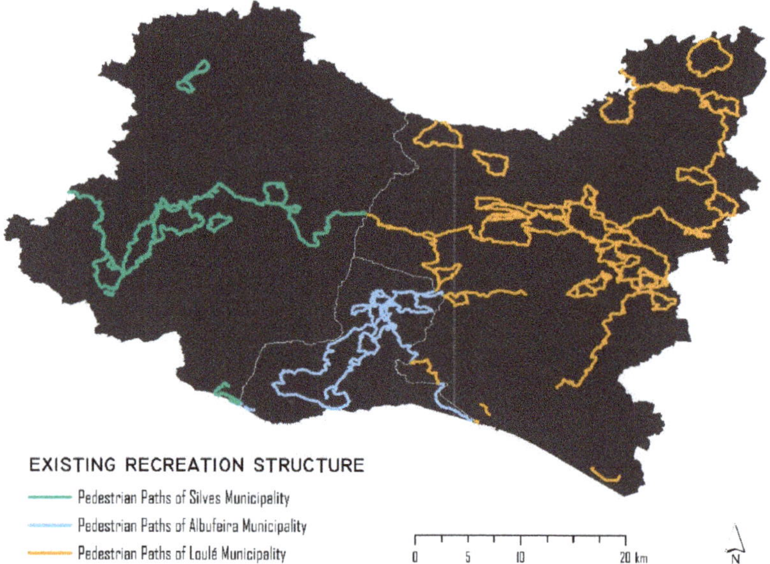

Figure 2. Trails and boardwalks of Silves, Albufeira, and Loulé municipalities in the Algarve region of Southern Portugal.

In addition to the existing recreation structure, there are spheres in the study area that can still be incremented and interconnected, namely those that do not have paths, trails, or recreational spaces. However, this strategy must respect the landscape's ecological sensitivity in order to represent a strategy for territorial cohesion and sustainable tourism development.

In this context, a suitability map (Figure 3 with the example for walks, orienteering, and keep-fit circuits), grounded on a score scale from 0—No suitability to 1—High suitability, was created to help the public and private sectors in defining strategies to increment the recreation structure for each nature-based and cultural activity, studied through the surveys, based on the multicriteria evaluation of the factors listed in Table 4 (for the same example).

Figure 3. Suitability map of enhancement potential for the creation of new pedestrian paths of Silves, Albufeira, and Loulé municipalities, Algarve region, southern Portugal.

Table 4. Multicriteria evaluation factors, respective fuzzy functions and factor weight.

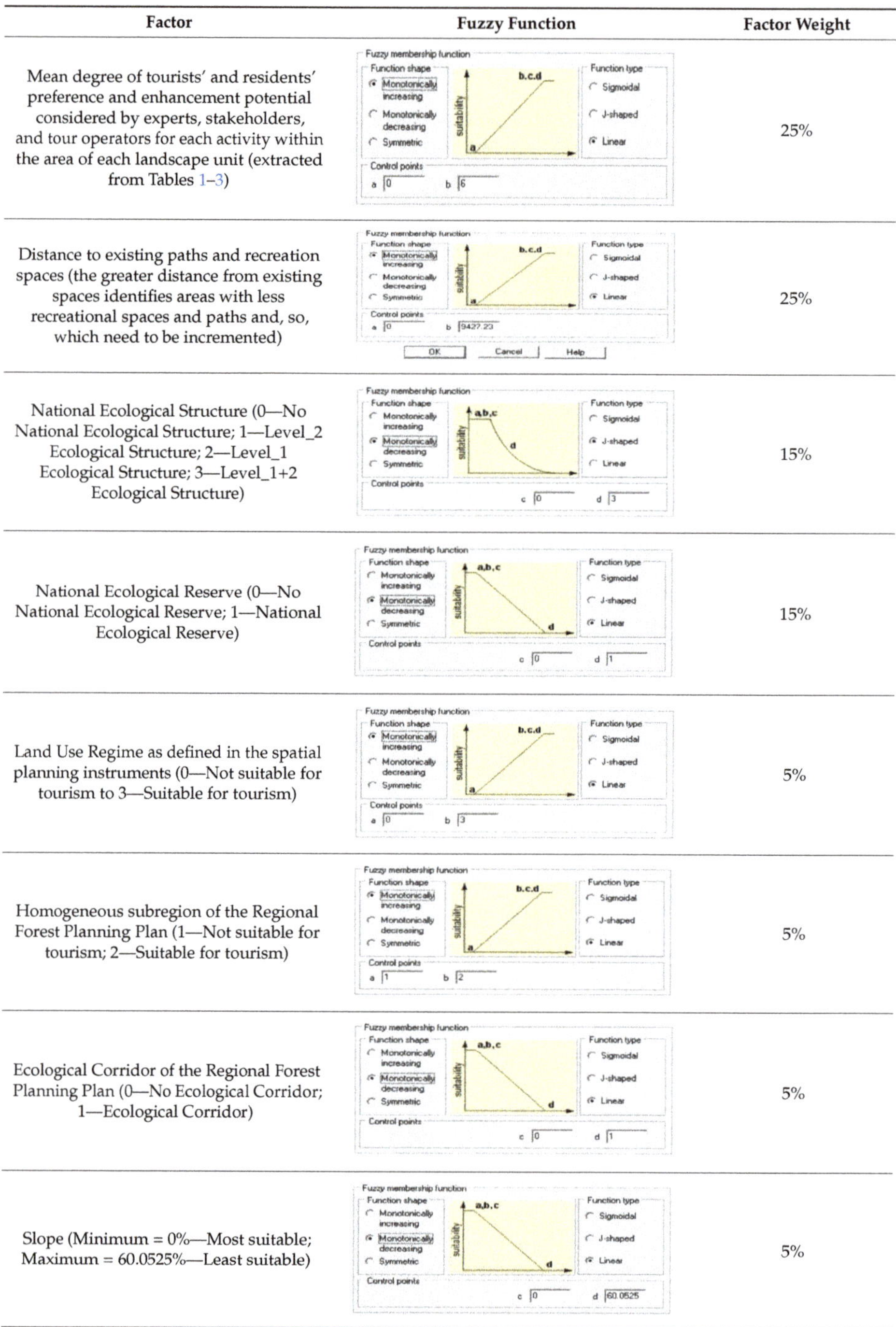

Factor	Fuzzy Function	Factor Weight
Mean degree of tourists' and residents' preference and enhancement potential considered by experts, stakeholders, and tour operators for each activity within the area of each landscape unit (extracted from Tables 1–3)		25%
Distance to existing paths and recreation spaces (the greater distance from existing spaces identifies areas with less recreational spaces and paths and, so, which need to be incremented)		25%
National Ecological Structure (0—No National Ecological Structure; 1—Level_2 Ecological Structure; 2—Level_1 Ecological Structure; 3—Level_1+2 Ecological Structure)		15%
National Ecological Reserve (0—No National Ecological Reserve; 1—National Ecological Reserve)		15%
Land Use Regime as defined in the spatial planning instruments (0—Not suitable for tourism to 3—Suitable for tourism)		5%
Homogeneous subregion of the Regional Forest Planning Plan (1—Not suitable for tourism; 2—Suitable for tourism)		5%
Ecological Corridor of the Regional Forest Planning Plan (0—No Ecological Corridor; 1—Ecological Corridor)		5%
Slope (Minimum = 0%—Most suitable; Maximum = 60.0525%—Least suitable)		5%

5. Conclusions

The present study highlights the activities that should be promoted, besides the sun-and-sea product. It makes it possible to analyze the structure of existing paths, trails, and recreational spaces for each of the activities studied in the surveys applied to tourists, residents, experts, stakeholders, and tour operators, and to identify areas that do not have recreational facilities, so as to be scaled up and interconnected with existing ones.

This research identifies the high potential for enhancement of the most preferred and considered of the alternative recreational activities, differentiated by each landscape unit, namely walks, orienteering, and keep-fit circuits, health and wellness, rural accommodation, gastronomy and wines and landscape touring and picnics for the Algarve Interior Mountains landscape; rural accommodation, walks, orienteering, and keep-fit circuits, gastronomy and wines, touring architectural and archaeological sites, health and wellness, and landscape touring and picnics for the Barrocal Midlands landscape; and walks, orienteering, and keep-fit circuits, health and wellness, gastronomy and wines, outdoor shows and performances, and touring architectural and archaeological sites for the Coastal landscape.

This territorial model should serve as the basis for a strategy of diversifying tourism products with a view for strengthening the regional resilience, where digital marketing, supported by the need for creation of geolocation apps, can be the main method of promotion [41]. The survey study identified the activities that can be enhanced and promoted in addition to the sun-and-sea product, in order to guide a management strategy and promotion of this destination, based not only on the enhancement potential considered by experts, stakeholders, and tour operators, but also in the diversification of products that are differentiated by the corresponding adherence to preferences of tourists and residents. Tourism diversification and differentiation should come from the sustainable management of these municipal territories and their tourism assets, providing advantages in presenting a rejuvenated and long-term viable destination, ensuring its sustainability [42].

The limitations of the study, which constitute important opportunities for future research, correspond to the need to study a governance structure that encourages synergies between resident communities, landowners, and tour operators, which may enable this diversification transition. Measuring the environmental and socioeconomic impact of this proposed recreational enhancement is paramount as an ongoing research need, since monitorization is imperious to the proposal implementation in order to avoid environmental degradation, gentrification, and tourism massification in today's low-density territories of the Algarve's interior mountains and Barrocal midlands, which still preserve more of their landscapes' ecological sensitivity compared to the coastal unit.

Author Contributions: Conceptualization, A.S.-A., J.F., E.V., and T.P.; methodology, A.S.-A. and J.F.; formal analysis, A.S.-A.; investigation, A.S.-A.; writing—Original draft preparation, A.S.-A. and T.P.; writing—Review and editing, T.P.; supervision, E.V., T.P. and J.F. All authors have read and agreed to the published version of the manuscript.

Funding: This research was funded by the Foundation for Science and Technology (FCT), grant number PTDC/GES-URB/31928/2017 and Interreg Atlantic Area project TrailGazerBid: "An analytical & technical framework to measure returns from trail investment."

Acknowledgments: The first author would like to thank dearly the expert's participation, here enunciated by alphabetic order: Professor Ana Paula Barreira, Professor Pedro Prista and Professor Rosário Oliveira. In the same way, the authors thank the stakeholders' and tour operators' participation, as well as the participation of all anonymous tourists and residents who were willing to contribute to this study. This paper was financed by the FCT- Fundação para a Ciência e Tecnologia through the PhD grant SFRH/BD/102328/2014.

Conflicts of Interest: The authors declare no conflict of interest.

References

1. Romão, J.; Neuts, B. Territorial Capital, Smart Tourism Specialization and Sustainable Regional Development: Experiences from Europe. *Habitat Int.* **2017**, *68*, 64–74.
2. Samora-Arvela, A.; Ferreira, J.; Panagopoulos, T.; Vaz, E. Tourists, Residents and Experts Rethink the Future of Mediterranean Regions: Can Tourism Diversification Contribute to Climate Change Adaptation? The Case of the Algarve Region, Southern Portugal. In *Regional Intelligence: Spatial Analysis and Anthropogenic Regional Challenges in the Digital Age*; Vaz, E., Ed.; Springer: Cham, Switzerland, 2020.
3. Romão, J. *Turismo e Lugar–Diferenciação Territorial, Competitividade e Sustentabilidade em Turismo*; Escolar Editora: Lisbon, Portugal, 2013.
4. Panagopoulos, T.; Antunes, M.D.C.; Karanikola, P.; Lakoova, L. Rural Renaissance through Synergies of Land and Sea Communities. In Proceedings of the 6th International Conference on Environmental Management, Engineering, Planning & Economics, Thessaloniki, Greece, 25–30 June 2017; pp. 1217–1222.
5. Butler, R. The Concept of a Tourism Area Life Cycle of Evolution: Implications for Management of Resources. *Can. Geogr.* **1980**, *24*, 5–12. [CrossRef]
6. Plog, S. Why Destination Areas Rise and Fall in Popularity. *Hotel. Restaur. Adm. Q.* **1972**, *14*, 55–58. [CrossRef]
7. Cohen, E. Towards a Sociology of International Tourism. *Soc. Res.* **1972**, *39*, 164–182.
8. Butler, R. Tourism in the Future: Cycles, Waves and Wheels? *Futures* **2008**, *41*, 346–352. [CrossRef]
9. Russo, P. The Vicious Circle of Tourism Development in Heritage Cities. *Ann. Tour. Res.* **2002**, *29*, 165–182. [CrossRef]
10. Agarwal, S. Coastal Resort Restructuring and the TALC. In *The Tourism Area Life Cycle: Conceptual Theoretical Issues*; Butler, R.W., Ed.; Channelview Publications: Clevedon, UK, 2006; pp. 201–218.
11. Baum, T. Revisiting the TALC: Is there an Off-Ramp? In *The Tourism Area Life Cycle: Conceptual and Theoretical Issues*; Butler, R.W., Ed.; Channelview Publications: Clevedon, UK, 2006; pp. 219–230.
12. Lagiewski, R.M. The application of the TALC: A literature Survey. In *The Tourism Area Life Cycle: Conceptual and Theoretical Issues*; Butler, R.W., Ed.; Channelview Publications: Clevedon, UK, 2002; pp. 27–50.
13. Benur, A.; Bramwell, B. Tourism Product Development and Product Diversification in Destinations. *Tour. Manag.* **2015**, *50*, 213–224. [CrossRef]
14. Samora-Arvela, A.; Vaz, E.; Ferrão, J.; Ferreira, J.; Panagopoulos, T. Diversifying Mediterranean Tourism as a Strategy for Regional Resilience Enhancement. In *Resilience and Regional Dynamics*; Pinto, H., Noronha, T., Vaz, E., Eds.; Springer: Cham, Switzerland, 2018; pp. 105–127. [CrossRef]
15. Weaver, D. *Sustainable Tourism–Theory and Practice*; Elsevier: Oxford, UK, 2006.
16. Budowski, G. Tourism and Environmental Conservation: Conflict, Coexistence, or Symbiosis? *Environ. Conserv.* **1976**, *3*, 27–31. [CrossRef]
17. Buhalis, D. Marketing the Competitive Destination of the Future. *Tour. Manag.* **2000**, *21*, 97–116. [CrossRef]
18. Porter, M. Industry Scenarios under Competitive Strategy under Uncertainty. In *Competitive Advantage–Creating and Sustaining Superior Performance*; The Free Press: New York, NY, USA, 1985; pp. 445–481.
19. Hassan, S. Determinants of Market Competitiveness in an Environmentally Sustainable Tourism Industry. *J. Travel Res.* **2000**, *38*, 239–245. [CrossRef]
20. Hall, C.M.; Williams, A.M. *Tourism and Innovation*; Routledge: London, UK, 2008.
21. Hjalager, A. A Review of Innovation Research in Tourism. *Tour. Manag.* **2010**, *31*, 1–12. [CrossRef]
22. Boes, K.; Buhalis, D.; Inversini, A. Smart Tourism Destinations: Ecosystems for Tourism Destination Competitiveness. *Int. J. Tour. Cities* **2016**, *2*, 108–124. [CrossRef]
23. Karanikola, P.; Panagopoulos, T.; Tampakis, S.; Tsantopoulos, G. Cycling as a Smart and Green Mode of Transport in Small Touristic Cities. *Sustainability* **2018**, *10*, 268. [CrossRef]
24. Caragliu, A.; Del Bo, C.; Nijkamp, P. Smart Cities in Europe. *J. Urban Technol.* **2011**, *18*, 65–82. [CrossRef]
25. Binkhorst, E.; Den Dekker, T. Agenda for Co-Creation Tourism Experience Research. *J. Hosp. Mark. Manag.* **2009**, *18*, 311–327. [CrossRef]
26. Kim, J.; Fesenmaier, D.R. Measuring Emotions in Real Time: Implications, for Tourism Experience Design. *J. Travel Res.* **2015**, *54*, 419–429. [CrossRef]
27. Palos-Sanchez, P.; Saura, J.R.; Reyes-Menendez, A.; Esquivel, I.V. Users Acceptance of Location-Based Marketing Apps in Tourism Sector: An Exploratory Analysis. *J. Spat. Organ. Dyn.* **2018**, *6*, 258–270.

28. Lu, J.; Mao, Z.; Wang, M.; Hu, L. Goodbye Maps, Hello Apps? Exploring the Influential Determinants of Travel App Adoption. *Curr. Issues Tour.* **2015**, *18*, 1059–1079. [CrossRef]
29. Cancela d'Abreu, A.; Pinto Correia, T.; Oliveira, R. *Contributos para a Identificação e Caracterização da Paisagem em Portugal Continental*; Direcção-Geral do Ordenamento do Território e Desenvolvimento Urbano: Lisbon, Portugal, 2004; Volume V, pp. 171–218.
30. Brito, S.P. *Território e Turismo no Algarve*; Edições Colibri: Faro, Portugal, 2009.
31. Santos, F.S. *Intervir na Paisagem*; Argumentum: Lisbon, Portugal, 2017.
32. Panagopoulos, T.; Karanikola, P.; Tampakis, S.; Gounari, N.; Tampakis, A. Rural Renaissance–Fostering Innovation and Business Opportunities in the Quarry Sector of Paggaio Municipality. In Proceedings of the 8th International Conference on Information and Communication Technologies in Agriculture, Food and Environment, Chania, Greece, 21–24 September 2017; pp. 415–421.
33. Silva, J.A.; Mendes, J.; Valle, P.; Guerreiro, M. *Consumos culturais dos turistas do Algarve*; Direcção Regional da Cultura do Algarve: Faro, Portugal, 2007.
34. Valle, P.; Guerreiro, M.; Mendes, J.; Silva, J.A. The Cultural Offer as a Tourist Product in Coastal Destinations: The Case of Algarve, Portugal. *Tour. Hosp. Res.* **2011**, *11*, 233–247. [CrossRef]
35. Mendes, J.; Henriques, C.; Guerreiro, M. Dos recursos às temáticas culturais na gestão do turismo cultural no Algarve. *Int. J. Sci. Manag. Tour.* **2015**, *4*, 31–48.
36. Samora-Arvela, A.; Vaz, E.; Ferreira, J.; Panagopoulos, P. Turismo e Património Gastronómico: A valorização turística de um cabaz de doçaria algarvia. *Public Policy Port. J.* **2017**, *2*, 55–69.
37. *INE, Inquérito à Permanência de Hóspedes na Hotelaria e Outros Alojamentos em 2015*; Instituto Nacional de Estatística: Lisbon, Portugal, 2016.
38. *INE, Censos 2011–Resultados Definitivos*; Instituto Nacional de Estatística: Lisbon, Portugal, 2013.
39. Jiang, H.; Eastman, J.R. Application of Fuzzy Measures in Multi-Criteria Evaluation in GIS. *Int. J. Geogr. Inf. Sci.* **2000**, *14*, 173–184. [CrossRef]
40. Eastman, J.R. *TerrSet Geospatial Monitoring and Modeling System*; Clark University: Worcester, MA, USA, 2016; pp. 345–389.
41. Simões, J.M.; Ferreira, C.C. The Tourism-Territory Nexus: Challenges for Planning. In Proceedings of the International Conference Sustainable Tourism: Issues, Debates & Challenges, Crete & Santorini, Greece, 22–25 April 2010; Wickens, E., Soteriades, M., Eds.; pp. 1060–1068.
42. Ritchie, J.; Crouch, G. *The Competitive Destination: A Sustainable Tourism Perspective*; CABI International Publishing: Wallinford, UK, 2003.

Article

Social Factors Key to Landscape-Scale Coastal Restoration: Lessons Learned from Three U.S. Case Studies

Bryan M. DeAngelis [1,*], Ariana E. Sutton-Grier [2], Allison Colden [3], Katie K. Arkema [4,5], Christopher J. Baillie [6], Richard O. Bennett [7], Jeff Benoit [8], Seth Blitch [9], Anthony Chatwin [10], Alyssa Dausman [11], Rachel K. Gittman [6], Holly S. Greening [12], Jessica R. Henkel [13], Rachel Houge [14], Ron Howard [15], A. Randall Hughes [16], Jeremy Lowe [17], Steven B. Scyphers [16], Edward T. Sherwood [18], Stephanie Westby [19] and Jonathan H. Grabowski [16]

[1] The Nature Conservancy, URI Bay Campus, Narragansett, RI 02882, USA
[2] Earth System Science Interdisciplinary Center, University of Maryland, College Park, MD 20740, USA; ariana.suttongrier@gmail.com
[3] Chesapeake Bay Foundation, Annapolis, MD 21403, USA; AColden@cbf.org
[4] Natural Capital Project, Woods Institute for the Environment, Stanford University, Stanford, CA 94305, USA; karkema@stanford.edu
[5] School of Environmental and Forest Sciences, University of Washington, Seattle, WA 98195, USA
[6] Department of Biology and Coastal Studies Institute, East Carolina University, Greenville, NC 27858, USA; bailliec19@ecu.edu (C.J.B.); gittmanr17@ecu.edu (R.K.G.)
[7] United States Fish and Wildlife Service, Hadley, MA 01035, USA; rick_bennett@fws.gov
[8] Restore America's Estuaries, Arlington, VA 22201, USA; jbenoit@estuaries.org
[9] The Nature Conservancy, Baton Rouge, LA 70802, USA; sblitch@tnc.org
[10] National Fish and Wildlife Foundation, Washington, DC 20005, USA; Anthony.Chatwin@nfwf.org
[11] The Water Institute of the Gulf, Baton Rouge, LA 70802, USA; adausman@thewaterinstitute.org
[12] Coastwise Partners, St. Petersburg, FL 34219, USA; hgreening@coastwisepartners.org
[13] Gulf Coast Ecosystem Restoration Council, New Orleans, LA 70130, USA; jessica.henkel@restorethegulf.gov
[14] United States Environmental Protection Agency Gulf of Mexico Program, Gulfport, MS 39501, USA; Houge.Rachel@epa.gov
[15] United States Department of Agriculture, Natural Resource Conservation Service, Gulf Coast Ecosystem Restoration Team, Madison, MS 39110, USA; Ron.Howard@ms.usda.gov
[16] Department of Marine and Environmental Sciences, Northeastern University, Marine Science Center, Nahant, MA 01908, USA; rhughes@northeastern.edu (A.R.H.); s.scyphers@northeastern.edu (S.B.S.); j.grabowski@northeastern.edu (J.H.G.)
[17] San Francisco Estuary Institute, Richmond, CA 94804, USA; jeremyl@sfei.org
[18] Tampa Bay Estuary Program, St. Petersburg, FL 33701, USA; esherwood@tbep.org
[19] National Oceanic and Atmospheric Administration, Restoration Center, Annapolis, MD 21401, USA; stephanie.westby@noaa.gov
* Correspondence: bdeangelis@tnc.org

Received: 21 November 2019; Accepted: 16 January 2020; Published: 23 January 2020

Abstract: In the United States, extensive investments have been made to restore the ecological function and services of coastal marine habitats. Despite a growing body of science supporting coastal restoration, few studies have addressed the suite of societally enabling conditions that helped facilitate successful restoration and recovery efforts that occurred at meaningful ecological (i.e., ecosystem) scales, and where restoration efforts were sustained for longer (i.e., several years to decades) periods. Here, we examined three case studies involving large-scale and long-term restoration efforts including the seagrass restoration effort in Tampa Bay, Florida, the oyster restoration effort in the Chesapeake Bay in Maryland and Virginia, and the tidal marsh restoration effort in San Francisco Bay, California. The ecological systems and the specifics of the ecological restoration were not the focus of our study. Rather, we focused on the underlying social and political contexts of each case study and found common themes of the factors of restoration which appear to be important for maintaining support for large-scale restoration efforts. Four critical elements for sustaining public and/or political support for large-scale restoration include: (1) resources should be invested in building public support prior to significant investments into ecological restoration; (2) building political support provides a level of significance to the recovery planning efforts and creates motivation to set and achieve meaningful recovery goals; (3) recovery plans need to be science-based with clear, measurable goals that resonate with the public; and (4) the accountability of progress toward reaching goals needs to be communicated frequently and in a way that the general public comprehends. These conclusions may help other communities move away from repetitive, single, and seemingly unconnected restoration projects towards more large-scale, bigger impact, and coordinated restoration efforts.

Keywords: coastal restoration; oyster; marsh; seagrass; restoration success; coastal habitat

---

1. Introduction

Throughout the United States, extensive investments have been made to restore lost ecological functions and services resulting from habitat loss and degradation. The restoration of coastal marine habitats, such as salt marshes, submerged aquatic vegetation, oyster reefs, mangroves, and corals, has occurred in every coastal state and U.S. territory. Coastal restoration has increased in terms of both number and scale of projects over the past decade, yet many restoration projects are still small relative to the degree of habitat loss that has occurred over the past two centuries [1,2]. This restoration lag is likely due to many factors including the lack of suitable area for projects, the cost of habitat restoration, and the availability of funding [3,4]. Furthermore, many restoration projects are implemented with minimal acknowledgement or understanding of how an individual restoration project contributes to ecosystem-scale (e.g., bay or estuary-wide) functioning or regional management goals [5]. The lack of funding for long-term monitoring of restoration projects further reduces the ability to disentangle the degree to which these activities help recover ecosystem functioning.

There have been several excellent academic reviews that have addressed and emphasized the ecological theory that must be considered when developing recovery plans (e.g., ecological baselines, stable and unstable ecological states, setting quantitative restoration objectives). These contributions to the literature have been paramount in providing restoration practitioners with a better understanding of the science underpinning ecological restoration and recovery, and the importance of advancing that science (e.g., [6–18] and others). There have been historically fewer reviews, however, that have addressed the suite of societally enabling conditions that existed in ecosystem-scale projects where coastal restoration efforts were sustained for longer periods. This may be in part because large-scale restoration efforts are relatively rare. However, it may also be because most of the initial focus of coastal ecosystem restoration research has been on understanding the ecological processes and outcomes of restoration, while there has been less focus on the social factors important to coastal restoration. Specifically, there has been little research examining what societal factors are important to maintain public and/or political support for large-scale restoration, even though this is a major potential barrier to ecosystem recovery.

To better understand the human and societal conditions that lead to successful coastal restoration and ecosystem recovery, we reviewed three case studies involving large-scale coastal restoration efforts and determined whether there are common principles for sustaining support for these large efforts that can guide future efforts. The case studies are the seagrass restoration effort in Tampa Bay (TB), Florida, the oyster restoration effort in the Chesapeake Bay (CB) in Maryland and Virginia, and the tidal marsh restoration effort in San Francisco Bay (SFB). While each case is geographically and ecologically different, we focused on the societal commonalities across the three case studies that point to important social factors that are needed to facilitate coastal ecosystem restoration and recovery. Furthermore, we explored the important roles of different stakeholder groups, including citizens, governments and politicians, and scientists.

All three case studies demonstrate the potential of coordinated, large-scale restoration efforts to achieve landscape-scale conservation goals. Based on lessons from these case studies, we draw conclusions that may help other communities move away from repetitive, single, and seemingly unconnected restoration projects towards more large-scale, bigger impact, societally-supported and coordinated restoration efforts.

2. Materials and Methods

2.1. Selecting Case Studies

We developed an initial list of potential landscape-scale restoration case studies around the U.S. using the following criteria: (1) the restoration had to be either completed or with enough active project implementation completed to assess the degree of restoration; (2) the case had to be at a geographic scale that was larger than the singular project level, and involve substantial regional and local coordination to implement it; (3) there had to be enough information available on the restoration efforts to develop a comprehensive case study; and (4) the list needed to represent multiple regions around the U.S. and a diversity of restored coastal habitat types to avoid developing generalities that could potentially be specific to one region or habitat type. To create an initial list of candidate cases that met the criteria above, we first consulted an expert coastal restoration working group of more than a dozen federal, academic, and non-governmental organization (NGO) professionals in coastal restoration. Using the initial list created by the working group, we selected 9 potential cases to query for additional information (See Table S1 in Supplemental Materials). To collect information on those 9 cases in a standardized manner, we created a questionnaire with eight questions which we sent directly to specific local experts who were familiar with each case (See Questionnaire in Supplementary Materials). The questionnaire resulted in the collection of qualitative information on each of the candidate cases. The questionnaire included questions on the goals of the restoration efforts (e.g., output or outcome based); whether the restoration was singular or multi-habitat based; the geographic scope of the restoration efforts; the level of participation from partners and other stakeholders in the restoration planning phase; the status of the past and current restoration; information on funding; and the level of public awareness of the restoration efforts. Based on the questionnaire responses, we selected the seagrass restoration effort in Tampa Bay (TB), Florida, the oyster restoration effort in the Chesapeake Bay (CB) in Maryland and Virginia, and the tidal marsh restoration effort in San Francisco Bay (SFB), California (Figure 1).

2.2. Reviewing Cases

To review each case, we mined the peer-reviewed and gray literature for information and reviewed any management plans developed for the case. We also conducted interviews with local experts, particularly those who were involved with the development of the restoration plans for each case. We gathered information specifically about four topics: (1) the background, history and ecological context of the geographic area; (2) a history of the restoration plan (i.e., how and why it was developed, and the restoration goals); (3) the status, results, and impacts of the restoration; and (4) the role of stakeholder involvement, including resource management and funding, in the restoration.

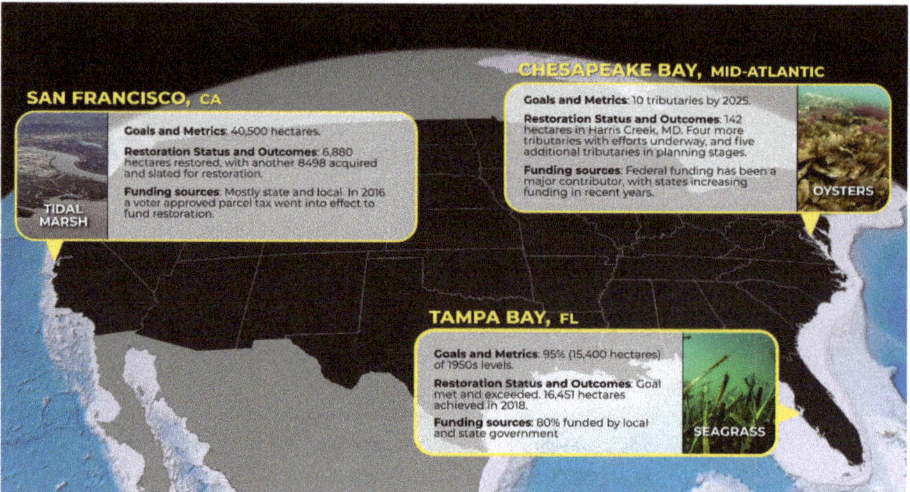

Figure 1. Infographic of summary Goals and Metrics, Restoration Status and Outcomes, and Funding Sources for three case study locations. Tampa Bay: Photo Credit, James R. White. Restoration focused on rehabilitation of seagrasses via improvements in water quality, but also to restore four other key habitats to the proportion they were in the 1950s relative to seagrasses. Other aquatic habitats like mangroves are at or near this goal, and some are increasing in extent. Funding has averaged USD 250M per year. Chesapeake Bay: Photo Credit, Oyster Recovery Partnership. Goals were based on "Oyster Success Metrics" defining reef- and landscape-level criteria necessary for a tributary to be considered "restored". The 142 hectares restored in Harris Creek is presently the largest oyster reef restoration project in the world. Since 2011, more than USD 51M of federal dollars has been spent on oyster restoration in MD alone. San Francisco Bay: Photo Credit, Dicklyon. The 40,500 hectares recommended by the Goals Project was based around improved habitat quality and quantity to support key species and presented at various geographic scales. In 2002, voters approved USD 200M to implement projects recommended in the Goals Project report. The 2016 voter-approved parcel tax is expected to raise USD 25M annually for restoration.

3. Results

3.1. Tampa Bay, Florida

3.1.1. Background and Ecological Context

Tampa Bay (TB), Florida is arguably one of the United States' greatest success stories regarding ecosystem restoration, and it is recognized internationally for its remarkable progress towards recovery [19–27]. TB is a relatively large (water surface area of 1031 km^2) embayment on the west coast of Florida with a watershed of approximately 5700 km^2 [24,28]. The subtropical estuary primarily includes seagrass meadows, emergent tidal wetlands (mangroves, salt marshes, salt barrens), tidal flats, and oyster reefs/bars [29]. Population growth has put pressure on these coastal ecosystems since the 1880s. By 1980, urban development activities (e.g., poorly treated wastewater, port channel dredging, and shoreline dredge and fill) had negatively impacted coastal wetlands and seagrass beds [28,30,31]. By the early 1980's 44% of emergent wetlands and 81% of seagrass areal extent were lost [32]. Circulation and salinity patterns were changed, and nutrient pollution had so degraded water quality by 1980 that many considered the bay to be "dead" [30].

3.1.2. Restoration Plan at Scale: History, Development, and Goals

Citizens of TB demanded action [22,25] in the 1980's, and as a result, legislation was enacted requiring more stringent treatment standards for wastewater plants discharging to TB. Recognizing the need for a comprehensive bay restoration and protection plan, the Tampa Bay National Estuary Program (TBNEP) was established in 1991 to address the harmful effects of population growth and coastal development on the water quality and coastal wetlands of TB. National Estuary Programs are place-based Environmental Protection Agency (EPA)-funded programs that use federal dollars to leverage additional funding and partner support (EPA National Estuary Program website). The TBNEP was responsible for the development and implementation of a science-based management and restoration plan for the TB estuary and leveraged an interlocal funding agreement to become the Tampa Bay Estuary Program (TBEP) in 1998 [31]. The TBEP helped coordinate and oversee organizing technical efforts to develop goals for restoring the estuary, but the impetus to implement projects fostering ecological change was from the community via considerable citizen input and pressure from both public and private entities and stakeholders [25]. The TBEP developed a Comprehensive Conservation and Management Plan in December 1996—subsequently updated in 2006 and 2017 [29]—that included measurable goals for the achievement of the Bay's designated uses and to support full aquatic life protection by identifying a diverse set of actions and strategies to improve environmental quality [22].

For TB, seagrasses are the "canary in the coal mine", as much of the focus of the recovery efforts revolved around meeting water quality goals that promote seagrass recovery. Seagrass recovery goals were established from aerial photography of the 1950s (a period prior to major development impacts), and a TBEP Policy Board decision to restore the Bay to 95% of its 1950s seagrass acreage. To achieve this goal, empirical analyses were used to derive nitrogen-loading targets sufficient to maintain water quality requirements of *Thalassia testudinum* [21]. For four other key habitats (mangroves, salt marsh, freshwater wetlands and salt barrens), quantifiable restoration and protection targets were set by calculating the relative proportion of each of these habitats in comparison to their original amounts in the 1950s [31]. As such, the recovery of TB is often not referred to as "restoration", but rather "rehabilitation", given the acknowledgment that returning to a state prior to significant anthropogenic impact is neither feasible nor attainable [33]. This concept, termed "Restoring the Balance," had broad appeal to both the TB public and resource managers [22].

3.1.3. Restoration Plan Status and Outcomes

TB is considered a worldwide model for estuary recovery. As of the 2018 assessment [34], the bay-wide seagrass recovery goal of 15,378 hectares (38,000 acres) was surpassed with an estimated 16,451 hectares (40,652 acres). Likewise, other important estuarine habitats, like mangroves, are increasing in extent [35].

Several reviews of the TB recovery efforts have identified the development of quantitative restoration and recovery goals as being a critical component of the overall recovery movement because they allowed collective agreement on a clear path forward to achieve a 'healthier' Tampa Bay, thereby bringing everyone together around those common goals [21–23,25,36]. It also enabled the TBEP to relay positive progress towards clear benchmarks of water quality and ecosystem recovery, which further fostered community buy-in and momentum for continuing the investments and commitments to nutrient-load reduction projects that would help toward the goal.

3.1.4. Role of Stakeholder Involvement

The TBEP concluded that establishing quantitative goals early in the process resulted in meaningful participation by local stakeholders, as evidenced by their voluntary participation in the comprehensive nutrient management strategy for TB [25]. Citizen and stakeholder involvement have been a critical component to meeting seagrass recovery goals in TB. Initial state regulations implemented in the 1980's requiring wastewater treatment facilities to significantly reduce nutrient discharges were a direct result of citizens' call for action. Again, in the early 1990's as part of the TBEP's development of a comprehensive restoration plan for Tampa Bay, citizens identified improving water quality, fishing, and swimming conditions as primary recovery goals. This support ultimately led to the development of specific, numeric water quality targets and seagrass restoration goals for the Bay. Furthermore, implementing the actions set forth in the recovery plans required broad partnerships and collaborative projects among scientists, resource managers, citizens, and public agencies to collectively achieve the environmental and economic benefits currently realized from a 'healthy' Tampa Bay [29].

On-the-ground habitat restoration has only been one component of the suite of ecological restoration activities conducted in TB. Diverse habitat protection and management activities have been pursued by local and regional entities throughout the estuary's watershed. Other work implemented to meet the TB's recovery goals revolved around infrastructure modifications and improvements, or best management practice implementation, primarily focused on directly reducing atmospheric or stormwater sources of nitrogen inputs to the Bay. From 1990–2017, more than 450 nutrient load reduction projects have been completed, ranging from municipal wastewater treatment facility upgrades to residential, agricultural, and urban storm water runoff reduction projects, improvements in fertilizer manufacturing and shipping activities, and pet waste reduction campaigns in neighborhoods and parks [37,38].

3.1.5. Funding

According to Russel and Greening [26], public agencies contributed approximately USD 250 million per year across nine different program areas corresponding to TB resource management priorities, including pollution control, wastewater and storm-water management, living resources, habitat preservation and restoration, land acquisition, dredged material management, regulation and enforcement, public awareness, and administration planning and coordination. The TBEP estimates that approximately 80% was funded from local or state sources, and while there were some federal grants, they summed to a relatively small percentage in comparison to regional investments by the Southwest Florida Water Management District and other local governments [33]. The role of the TBEP, however, cannot be over-stated. While the TBEP is not a large, direct contributor of funding for infrastructure and restoration activities contributing to bay wide water quality improvements, their scientific, advisory, and coordination efforts underpinned and helped garner the necessary community support needed to rally around a shared recovery goal for TB. For example, TBEP is a neutral facilitator and convener of the public/private Tampa Bay Nitrogen Management Consortium (TBNMC), an alliance of more than 45 local governments, regulatory agencies, key industries, and utilities formed to work collaboratively to meet nitrogen management targets supportive of seagrass recovery goals. The TBNMC has contributed more than USD 0.7 billion since the mid 1990's on various nitrogen load reduction projects [29]. The degree of organization, coordination and collaboration necessary to initiate and maintain the many restoration activities being conducted in the TB estuary and its watershed would have been extremely difficult without federal, state and local government commitment, and funding to support TBEP's role.

3.2. Chesapeake Bay, Maryland and Virginia

3.2.1. Background and Ecological Context

The Chesapeake Bay (CB) is the largest estuary in the United States, with a watershed of 165,800 km^2 that spans six states. There are more than 150 major rivers in the watershed, but roughly 80% of the freshwater input to the CB comes from the Susquehanna, Potomac, and James Rivers. The estuary is relatively shallow, averaging 6.5 m deep, with a deeper channel (20–30 m) running through the main stem. CB consists of many habitats such as tidal marshes, seagrass beds, oyster reefs, hard bottom, and mud flats [39].

With a watershed area roughly fourteen times the surface area of the estuary [39], land use has had a profound influence on the productivity and structure of the CB ecosystem, which has changed significantly in the past 200 years [40,41]. Human settlement in the early 17th century was followed by rapid deforestation that increased nutrient and sediment loading to the system, with particularly negative impacts to oyster reefs [40,42–44]. As nutrient loading increased, oysters initially benefited from greater primary productivity, but continued eutrophication led to persistent seasonal hypoxia and the silting over of the remnant oyster reefs [42,43]. Water quality issues were exacerbated by the overharvesting of oysters, which reduced the yield per recruit to 8.4% of the unfished population [44] and further worsened water quality by reducing filtration capacity [45,46]. By the 1950s, it was evident that the system had exceeded a water quality tipping point, leading to a rapid decline in several important coastal habitats (seagrasses, saltmarshes, oyster reefs [47,48]), including a 99.7% decline in oyster abundance in the Upper Chesapeake Bay since the early 1800's [49]. Cumulative economic losses of more than $4 billion over the past three decades have affected the coastal communities of Maryland and Virginia due to loss of oyster harvest revenue and impacts to associated industries [50]. Unquantified losses of ecosystem services other than extractive value and related industries are likely much higher [51].

3.2.2. Restoration Plan at Scale: History, Development, and Goals

In light of deteriorating water quality and ecosystem impacts, citizens appealed to elected officials to take action. A key development in the Bay clean-up process was when Senator Charles "Mac" Mathias from Maryland responded to citizens' appeals by commissioning a 5-year, USD 27 million study to pinpoint the causes of the Chesapeake's problems. This study led to the development of the EPA's Chesapeake Bay Program and the first Chesapeake Bay Agreement, signed in 1983 [52]. The agreement consisted of a simple one-page pledge by the Governors of Maryland, Pennsylvania, and Virginia, along with representatives from Washington, D.C., the EPA, and the Chesapeake Bay Commission, to work together to restore the health of CB. Following this initial effort, the Chesapeake 2000 Agreement was the first to set a quantitative oyster restoration goal—to increase the oyster population in the Bay ten-fold by 2010. Yet, even this ambitious goal failed to produce a significant improvement in oyster populations, as the agreement lacked a specific implementation plan [53] and surveys of the oyster population were inadequate to determine progress towards the ten-fold population goal [54].

In 2009, President Obama issued Executive Order 13508, which instructed federal agencies to develop a coordinated federal strategy for the restoration and protection of CB, including its oyster populations, within 180 days of its issuance [55]. This directive resulted in the Strategy for Protecting and Restoring the CB Watershed [56], which established the goal of restoring the oyster populations of 20 CB tributaries by 2025. This was the first quantifiable goal that focused on large-scale restoration. Through this directive and the resulting goal, the region was able to quickly galvanize the technical expertise, funding, and coordination of federal efforts to begin addressing this large-scale coordinated effort [57].

In 2011, restoration partners came together to define a priori metrics, through consultation with external oyster scientists, that would define restoration success. The "Oyster Success Metrics," developed in 2011, defined reef- and landscape-level criteria necessary for a tributary to be considered "restored" [58]. From these metrics, restoration partners and scientists worked backward to determine the restoration effort in each area that would most likely achieve target oyster densities, biomass, and reef acreage as well as the necessary monitoring protocols for assessing if targets were met.

The 2014 CB Watershed Agreement solidified state and federal partners' commitments to large-scale oyster restoration in ten tributaries by 2025, a revised goal that more accurately reflected the feasibility of the project. The "10 tributaries by 2025" goal is the primary driver for current oyster restoration efforts in CB.

3.2.3. Restoration Plan Status and Outcomes

On-the-ground work to implement tributary-scale oyster restoration began in 2011. By 2016, construction on the first restoration tributary, Harris Creek, in Maryland, and the largest oyster restoration project in the world to date, was complete, resulting in the restoration or enhancement of 142 hectares (350 acres) of oyster reef habitat [59]. Restoration activities are currently underway in four other CB tributaries, and all 10 tributaries have been at least tentatively selected and are in the survey and planning phase [60].

The ability to track and report progress toward the 10-tributary restoration goal has helped to enhance public support for the project. A bipartisan opinion poll conducted in February 2018 indicated that 83% of Maryland voters support tributary-scale oyster restoration in the state [61].

Although the Oyster Success Metrics focus on quantitative outputs (e.g., area restored, oyster density), they are linked to ecosystem outcomes through additional criteria, including multiple oyster age classes and reef footprint and accretion. These metrics are intended as a quantitative proxy for ecosystem services (e.g., fish and macrofauna habitat provisioning, water quality improvements) not directly measured through the monitoring program [58]. Additional research programs spurred by the large-scale restoration goal are working to directly assess the ecosystem service benefits of large-scale oyster restoration. Thus far, results of these studies have indicated that large-scale oyster restoration will significantly increase blue crab (*Callinectes sapdius*) biomass, thereby benefitting associated blue crab fisheries [62]. Significant advancements in quantifying denitrification on restored oyster reefs have also led to the approval of oyster aquaculture as an in-water best management practice for nitrogen and phosphorus removal by the EPA [63].

3.2.4. Role of Stakeholder Involvement

The restoration efforts in CB are unique, as the governance structure of the Chesapeake Bay Program leads to a primarily top-down approach where most of the coordination and funding occurs at the federal level [64]. This approach is appropriate for CB, where efforts are multi-jurisdictional and require cooperation amongst multiple states to achieve a common objective [65]. While federal agencies are responsible for coordination, oyster restoration requires full support from the states as restoration work is occurring in waters under their jurisdiction. Thus, states, along with local governments, watershed groups, and other relevant stakeholder groups, are full partners in these efforts, both financially and logistically [65]. Additionally, each of the outcomes of the Chesapeake Bay Watershed Agreement is assigned to a Goal Implementation Team, which consists of federal and state agency partners along with consulting scientists and local stakeholders, such as local watershed associations [65]. Through these Teams, local and regional interests are given a forum through which to contribute to restoration planning and policy.

3.2.5. Funding

Executive Order 13,508 provided a clear, common goal around which federal and state agencies could target restoration work. Though several agencies, particularly the Army Corps of Engineers and NOAA, had already been engaged in oyster restoration in Chesapeake Bay, setting large-scale targets for restoration necessitated the cooperation and coordination of state and federal agencies to achieve the funding levels required to achieve these goals. Through the mechanism of federal-state cost-share agreements, federal dollars were leveraged with state funding, usually at a ratio of 75% federal and 25% state, though the funding arrangements differed by agency and some did not require state matching funds. Additionally, tributaries in which large-scale oyster restoration is conducted are protected from harvest through statute or regulation. This assurance of protection has resulted in positive feedbacks that have bolstered larger restoration efforts. For instance, it has catalyzed public–private partnerships such as the Chesapeake Bay Stewardship Fund [66] that brought corporate philanthropy to oyster restoration, and invited further investment from watershed organizations and community groups interested in contributing to areas closed to harvest.

3.3. San Francisco Bay, California

3.3.1. Background and Ecological Context

The San Francisco Bay (SFB), together with the upstream inland Delta, comprises the largest estuary (~4000 km^2) on the U.S. Pacific Coast, and remains one of California's most important ecosystems. The evolution of the SFB involves a complicated history of natural and human-induced factors including sediment ebbs and flows, sea level changes, diking, and development [67–69]. Prior to the mid-19th century, the SFB and the inland Delta were comprised of approximately 1300 km^2 of open water and another 2200 km^2 of fresh-, brackish- and salt-water marsh [70,71]. The region was heavily modified by humans to support a rapidly growing population with the gold rush of the 1800's, including diking wetlands for agricultural land [71]. Simultaneously, gold-seekers were perfecting hydraulic mining where high-pressure streams of water led to destruction of the hills and flushing of a great deal of sediment into the rivers and creeks, delivering nearly a billion cubic meters of sediments between 1849–1914 [72]. By 1930, most of the of freshwater marshes were diked and farmed, and 80% of the Bay's salt marshes and intertidal mudflats were turned into salt ponds, cow pastures, or residential and commercial real estate [71], and the Bay was continually being filled to provide more space for ports, industry, garbage dumps and other development well into the 1960s. The result of the anthropogenic pressures on SFB was the loss of wildlife habitats and a reduction in tide-related flushing, which in turn has led to progressive deterioration of the Bay's water quality [67–69,71].

3.3.2. Restoration Plan at Scale: History, Development, and Goals

There was a growing public concern for the health of the Bay, and in 1961 three women—Silvia McLaughlin, Catherine "Kay" Kerr, and Esther Gulick—took action against the filling of the Bay to create the association that is now known as Save the Bay [73]. At Save the Bay's urging, the McAteer-Petris Act was enacted in 1965, serving as the key legal provision preventing the indiscriminate filling of the Bay, and establishing the San Francisco Bay Conservation Development Commission (BCDC)—the world's first coastal protection agency [74]. The BCDC was the first agency set up to look at the Bay as a whole system, a switch from the previous management, where municipalities only considered their own parts of the Bay. While the primary mission of the BCDC is to protect the Bay, in 1987 the EPA, as part of its National Estuary Program, established the San Francisco Estuary Project (SFEP), with the mission of restoring the health of the Bay's ecosystem. Bringing together the environmental community, private sector and government, the SFEP was a collaborative effort that focused much-needed attention on the San Francisco Estuary [75]. In addition to identifying the Estuary's most critical problems, a major project of the SFEP was a Comprehensive Conservation and Management Plan (CCMP) [75], which was signed by the Governor of California and the Administrator of the U.S. EPA in 1993, and was then updated in 2007 and 2016. The CCMP identified 145 actions necessary to "restore and maintain the estuary's chemical, physical, and biological integrity", as well as specifying the creation of an estuary-wide plan to "protect, enhance, restore, and create wetlands in the Estuary", and that this plan will be based on habitat goals designed to protect wildlife [75].

By 1995, a large group of Bay scientists and resource managers, including nine state and federal agencies, came together to develop a "shared vision" for habitat change in the whole estuary. This effort was called the San Francisco Bay Area Wetlands Ecosystem Goals Project (covering Suisan Bay to the South Bay) [70]. The 1999 report was later updated in 2015 [76] to address the projected effects of climate change. While the acreage goals of the 1999 report remained the same, the 2015 update synthesized the latest science, and incorporated projected changes through 2100 to generate new recommendations for achieving a healthy ecosystem. The focus of the Goals Project is based around improved habitat quality and quantity to support key species. In addition to wildlife being specified in the CCMP, this decision was justified because concern about species and human health drives most federal and state environmental laws and policies. Furthermore, they surmised that protecting key species by improving their habitats would concurrently improve other important wetland functions [70].

The approach for developing the habitat goals involved several iterative steps that included more than 65 qualified experts. Five focus teams were developed for plants, fish, and wildlife. The focus teams developed lists of key species and identified their habitat requirements. Seven key habitats were identified within the baylands, and seven key habitats were identified outside of the baylands but within the baylands ecosystem. The project next mapped the historic and current habitat area of each. The focus teams blended the habitat recommendations into a conceptual vision that balanced the competing needs of the many baylands species. Ultimately, this two-year process allowed them to calculate area for each of the key habitats and compared the proposed future habitat area to the historic and modern amounts [70].

The outcome of these efforts resulted in specific habitat goal recommendations, presented in terms of area, that were required to support key species. The habitat goals were presented at various geographic scales, including recommendations for four main subregions, as well as for segments of each subregion. Notably, the regional area goals called for tidal marsh restoration on an unprecedented scale: 24,281 hectares (60,000 acres) to be restored, to reach a total of 40,466 hectares (100,000 acres). Setting goals to restore this degree of salt marsh required anticipated reductions of other associated habitats (e.g., salt ponds); thus, the report suggested offsetting the reductions by maximizing wildlife management effectiveness in those associated habitats, thereby still increasing the region's overall ability to support shorebirds, waterfowl, mammals, and other wildlife [70]. The 2015 science update to the original Goals report implemented adaptive management and improved upon the original 1999 goals. The update addressed issues arising since 1999 such as climate change and reduction in sediment supply. It also built on 15 years of landscape-scale restoration experience, ultimately adapting the Goals to reflect increased knowledge and science since the original report.

3.3.3. Restoration Plan Status and Outcomes

Prior to the publication of the Goals Project, tidal wetland restoration projects were few and relatively small in scale, with the largest around 350 acres [76,77]. By providing a consensus-based scientific vision of the kinds, amounts, and distribution of habitats needed to sustain healthy populations of fish and wildlife for the entire region, the Goals Project gave regulators, resource managers, and citizens the framework necessary to pursue large-scale restoration for bay habitats. For example, the South Bay Salt Pond Restoration Project is the largest tidal wetland project on the US West Coast, the footprint of which encompasses nearly the entirety of the southern end of the Bay, and will restore 6111 hectares (15,100 acres) when complete [78]. Nineteen years after the Goals Project report was published, 6880 hectares (17,000 acres) of wetland habitat have been restored, and another 8498 hectares (21,000 acres) of diked baylands has been acquired and slated for restoration to tidal marsh and associated habitats [79]. Beyond setting the quantitative goals for restoration, the Goals Project provides guidance to coordinate the restoration and acquisition investments, ensuring the projects and land acquisitions are best suited to achieve landscape-scale benefits for the entire Bay system.

3.3.4. Role of Stakeholder Involvement

The SFB is arguably one of the greatest stories of how stakeholder involvement, particularly from community members, played a pivotal role in ecosystem recovery. The story of three women, and the role they played in "saving the Bay", is practically folklore in the region. Their actions not only created one of the most well-known conservation organizations in the U.S., but it helped kick-start a series of actions that ultimately led to a significant change in how the ecology and ecosystem of the Bay were viewed and managed. According to experts, one of the most significant outcomes was the entire stakeholder community "getting on the same page" in terms of aligning and focusing efforts on a common set of goals [79].

The approach for developing the Goals Project involved several steps, following a designed organizational structure that included stakeholder involvement throughout the process. This included a steering committee of representatives from multiple resource management and science organizations. Focus teams were developed that consisted of more than 65 science contributors, selected to participate in collaborative workgroups. An independent science review panel was created to review the draft Goals. Throughout the development of the Report, public outreach was extensive. The public outreach meetings provided many benefits to the process, including developing a better sense of the issues of concern, improving technical products, and ideas on how to present the Goals in a way that would make them most useful [70]. The process for the 2015 science update included a steering committee of representatives from resource management and science organizations; collaborative and open participation by science contributors organized into workgroups; an independent science review panel; and a core administrative team, including the science coordinator [76].

3.3.5. Funding

The inclusion of specific, quantitative recommendations (i.e., reestablishing 100,000 acres of tidal wetlands) in the Goals Project has been integral to leveraging new funding sources for restoration. Indeed, after the Goals Project was released, funding for baylands restoration projects increased appreciably. Indicative of its importance, in 2002, the Goals Project was explicitly cited in Proposition 50, a proposition approved by voters that allocated the Wildlife Conservation Board up to USD 200 million for the implementation of restoration projects mentioned in the report. Importantly, the Goals Project has also benefitted many smaller bay restoration projects, as both state and federal agencies have increasingly used its science-based guidance to identify restoration and conservation projects that address grant program mandated habitat and water-quality enhancement objectives. In a recent historic vote, the people of the Bay Area leveed upon themselves the first regional parcel tax measure in California's history, which will raise USD 25 million annually, resulting in USD 500 million over twenty years (Measure AA) [80].

The report and its update have become a cornerstone of policy, planning, coordination, and advocacy for the acquisition, protection, and restoration of the SFB baylands. Many public agencies have incorporated the Goals Project into regional planning and policy documents. The Goals Project has also spurred regional entities in working with members of the US House and Senate to seek a federal funding program (e.g., the San Francisco Bay Improvement Act of 2010, the San Francisco Bay Restoration Act of 2015, the San Francisco Bay Restoration Act of 2019) comparable to other nationally significant bay-restoration programs to accelerate the restoration of the bay [76].

4. Discussion

The three case studies differ in their geographies and the species and ecosystems being restored, but we observed similar themes among them that point to important social factors of effective landscape-scale ecosystem restoration and recovery efforts. Here, we examine each of these themes in more detail and provide an insight into the significance of each in the ability of the three case studies to achieve sustained and coordinated landscape-scale ecosystem restoration.

4.1. Recognizable Ecological Crisis with Public Demand for Action

In each of the cases reviewed, ecosystem degradation was well-documented by the scientific community and recognized and considered to be at a point of crisis by the public. Identifying the processes leading to degradation or decline of a natural system has been proposed as the initial step of a restoration process [5,9,10]. While this is a key step, scientific understanding of declining ecological conditions may not be enough to motivate large-scale restoration efforts. In the three cases reviewed here, not only was the decline well-documented, but there was also a corresponding public demand for action that resulted from the communities' awareness of that decline. In each case, the strident public outcry led to political intervention which then resulted in actual restoration action. These examples highlight the importance of the public understanding the extent and consequences of the environmental crisis (e.g., ecological, societal, economic, cultural). In cases where the appropriate level of public support and demand for change does not yet exist, building public motivation may be an important first step [81], even prior to the dedicating of resources to active ecological restoration.

4.2. Political Response Catalyzing the Development of Estuary-Level Recovery Plans

In an excellent review of the role of ecological restoration in the turn of the millennium, Hobbs and Harris [5] suggested that political opportunism often is more critical in setting restoration priorities than any rational process. The cases we reviewed provide examples of where political support catalyzed the development of recovery plans and "set the tone" for recovery efforts.

The political support that arose from public outcry for action, and its catalytic role in developing estuary-level plans for ecological recovery, was an important commonality in each of the cases evaluated. Furthermore, in all three cases, financial and/or political support from the EPA was a fundamental component. The significance of politically-motivated calls for comprehensive recovery plans should not be underestimated. Furthermore, in all three cases there was federal expertise and coordination provided to support the recovery planning efforts. In TB, the TBEP, which was federally funded, provided the coordination and catalyst that facilitated both public and private investment in the restoration. In San Francisco, the EPA was instrumental in the development of the SFEP, and later many federal agencies were part of the effort to design the restoration goals and contributed a great deal of expertise that helped the project succeed. Further, in the CB, federal involvement was explicitly directed in Executive Order 13508, which called on seven federal agencies to work on what became the "Strategy for Protecting and Restoring the Chesapeake Bay Watershed". Federal partners bring considerable nationally gained expertise, knowledge, and a capacity that can be critical in helping guide the comprehensive recovery planning efforts.

4.3. Development of Science-Based, Estuary-Level, Comprehensive Plans for Action with Clear, Measurable Goals

The concept of setting restoration goals is not new and has been addressed multiple times in the academic literature (e.g., [5–7,9,16–18,82,83] and others). Ehrenfeld [7] declared that there is no one paradigm or context for setting restoration goals. The cases reviewed here support that statement, as each went through entirely different processes to develop the recovery plan that resulted in entirely different recovery goals. Arguably, however, developing the vision for recovery—that was both founded/grounded in science and supported by the community—was the most critical element of the recovery plans.

Several key similarities amongst the three recovery plans goals may have led to their sustained success: First, none of the cases set restoration and recovery goals solely based on returning to an historic benchmark. It has been well documented that over-dependency on historical baselines as restoration goals is often unrealistic or unachievable (e.g., [6,9,14–16]). In Tampa Bay, while the seagrass restoration goals were based around a historical extent, the restoration goals of several other key habitats focused on recovering the proportions of the habitat present during an earlier, less-disturbed, period. In San Francisco, a similar approach was based on evaluating the habitat needs of targeted species, with a guiding principle of increasing the quantity and quality of wetlands without trying to "reach" the past. While in Chesapeake Bay, the aspirational oyster goal of ten tributaries restored over a 10-year horizon reflected anticipated resources required to achieve those goals (See Figure 1).

Second, each of the recovery plans were translated into quantifiable management goals that were easily understood by the public, with specific targets that enabled the clear communication of progress on restoration goals. Establishing measurable goals is critical to maximizing the chances of obtaining and demonstrating restoration success [16,84]. Furthermore, the goals should be easily observable by the public [10,84]. Thus, while recovery goals can and should be based on a range of outcomes and trajectories (e.g., [6,7,15,16] and many others), they simultaneously need to be translated into terms that the public can understand and witness progress towards. The area to be restored (e.g., acres) is often used because it is a tangible metric that is easily communicated to the public. However, goals should reflect the primary motivation for the restoration, and the service the community is seeking to "get back" from restoration of that habitat. For example, in SFB, the "goals" are framed in acres restored; however, the area-based goal is a function of the ecological service to benefit key wildlife outcomes. Despite the differences in the restoration planning process, including the community participation in it, the planning process was a critical theme in each of these cases. For example, TBEP recognized that goals needed to be framed in a manner that could be easily and convincingly communicated to the public. Changes to habitat landscapes over time are a visible and intuitive aspect of estuaries that the public can easily see, understand, and relate to. In CB, the goal of ten tributaries over ten years is easy for the public to comprehend, even if the specific ecologic metrics that define "restored" were painstakingly developed [58].

Finally, restoration goals were established in all three regions at appropriate spatial and temporal scales and with realistic recovery time-scales in mind. Longer-term (decadal) restoration trajectories that are less predictable, but more representative of real system attributes, are more realistic to accommodate variability [14]. Spatially, recovery plans need to set a trajectory that can be accomplished through the implementation of several smaller projects. In other words, it is unrealistic to expect "large-scale" to always mean bigger individual projects, since projects are often limited by funding, the amount of land available, or other factors. The role of the recovery plan is to ensure that smaller-scale projects are connected ecologically. For example, in CB, recovery goals were set to achieve a restoration of 50%–100% of the restorable bottom in each identified tributary. Those goals will be accomplished via several smaller projects that all contribute to the overall goal. The role of the science is to ensure the planning, prioritization, selection and implementation of projects that allow for each of them to contribute to the landscape-scale ecological outcome (e.g., network of larval source and sink reefs, enhanced nitrogen removal through siting, etc.).

4.4. Funding Provided to Implement the Plan

The importance of adequately funding the projects cannot be understated. Gaining initial access to funding enabled the implementation of restoration techniques and allowed the efforts to begin to make progress towards their goals. However, the funding for the three cases studied did not come from the same sources. For the CB, the project was primarily federal- and state-funded, while in TB and San Francisco, the funding was a combination of local, regional and state funding, with federal contributions making up the smallest proportion of funding. It is rather remarkable that both the TB and the SFB projects were able to complete landscape-scale restoration with limited federal funds. This finding suggests that there are many ways to fund landscape-scale restoration, including combining state and federal funds (CB), having citizens vote to tax themselves to fund the work (as occurred in SFB), and relying primarily on funding from local and state public agencies (TB).

4.5. The Public Has Remained Engaged

Citizen involvement in these cases is also critical to recognize. In TB, for example, citizens worked to implement backyard interventions (i.e., rain gardens, reduced fertilization during summer wet seasons, etc.), and there was a dog waste pick-up campaign linked to supporting the Bay clean-up efforts. In the CB, the watershed organizations were participating in oyster restoration projects to help clean up the Bay. In the SFB, the majority of citizens voted to tax themselves. Each of these efforts gave citizens a way to directly contribute to the restoration and to "buy-in" to the effort via their own actions. This buy-in is likely a very important reason as to why there was such strong, direct citizen support for the projects, which is one of the most important factors in effective landscape-scale restoration.

5. Conclusions

Large-scale, long-term, ecological recovery requires a combination of public and political motivation to build momentum for change, funding and partnerships, and science-based specific restoration goals and metrics of success. Based on these three case studies, we conclude that the science of restoration and ecological recovery is paramount in guiding, setting goals, and communicating results—but without sustained public and political support and funding, significant change is unlikely to happen. Restoration guidance documents have noted the importance of effective communication and outreach to relevant stakeholders when building restoration projects [85]. However, our findings highlight the importance of a priori efforts to build the community and stakeholder support necessary to drive systemic restoration recovery of the ecosystem.

We found the following four critical themes for sustained large-scale restoration: First, where public support and demand for change does not yet exist, putting substantial resources into building public motivation may be an important first step, and could provide long-term benefits in garnering political support and help sustain community engagement. A number of mechanisms for building this public support could be used, including the use of social media, ad campaigns, etc. There is an important need for additional social science research, to better our understanding of what methods, mechanisms, and communication tools are most useful in garnering public and/or political support for ecological restoration, as well as to gain a better understanding of what degree of public/political support is needed to catalyze a movement toward ecological recovery. Second, while political support may not be a requirement for recovery, with it typically comes a level of resource investment to the recovery planning efforts and the motivation to set and achieve meaningful recovery goals. Furthermore, political support may translate to federal involvement, which can be useful when working across jurisdictional lines and brings considerable geographically diverse expertise and capacity to comprehensive recovery planning. Third, recovery plans need to be science-based with clear, measurable goals that resonate with the public. It is critical that the goals are based in science that considers realistic recovery end-points and ecological states, and there are a variety of tested approaches available for developing quantitative goals. Most importantly, the goals need to be communicable and transparent to the general public. Fourth, communication is critical for continued public support and enthusiasm. Therefore, the monitoring and accountability of progress toward reaching goals is essential, and the progress needs to be communicated to political leaders and the public frequently and in a comprehensible way. How to best run a communication campaign to share updates about restoration projects with the public and political leaders is a subject for future social science research. Such research could help determine preferred communication strategies for communicating project progress in order to ensure continued public support.

Achieving all four of these principles is not easy, and yet these case studies illustrate how important the principles were to the coordinated and sustained landscape-scale restoration efforts that we reviewed. From these cases, we can conclude that landscape-scale restoration was most effective when citizens, scientists, and governments worked together with a common goal of restoring the health, integrity, and function of an ecosystem. In other words, it takes a village.

Supplementary Materials: The following are available online at http://www.mdpi.com/2071-1050/12/3/869/s1, Table S1: Basic information on each of the 9 case study projects and Questionnaire Sent to Local Project Experts of 9 Projects.

Author Contributions: All authors (except H.S.G., E.T.S., J.L. and S.W.) conceived and outlined the framework for this perspective as part of a Science for Nature and People Partnership (SNAPP) Coastal Restoration Working Group, led by J.H.G., B.M.D., K.K.A. and R.K.G.; B.M.D., A.E.S.-G. and A.C. wrote the paper; H.S.G., E.T.S., J.L. and S.W. provided significant local expertise and co-drafted individual case studies; All authors contributed to editing and revising of the paper. All authors have read and agreed to the published version of the manuscript.

Funding: This research was funded by a grant from the Science for Nature and People Partnership (SNAPP) to J. Grabowski, K. Arkema, and B. DeAngelis.

Acknowledgments: This research was conducted by the Coastal Restoration expert working group supported by the Science for Nature and People Partnership (SNAPP), a collaboration of The Nature Conservancy, the Wildlife

Conservation Society, and the National Center for Ecological Analysis and Synthesis (NCEAS) at the University of California, Santa Barbara. SNAPP is a first-of-its-kind collaboration that delivers evidence-based, scalable solutions to global challenges at the intersection of nature conservation, sustainable development, and human well-being. The scientific results and conclusions, as well as any views or opinions expressed herein, are those of the author(s) and do not necessarily reflect the views of the RESTORE Council, USDA NRCS, USEPA, USFWS, NOAA, NFWF, the U.S. Department of Interior, or the U.S. Department of Commerce. We thank M. Ribera for her assistance in developingFigure 1.

Conflicts of Interest: The authors declare no conflict of interest.

References

1. Bayraktarov, E.; Saunders, M.I.; Abdullah, S.; Mills, M.; Beher, J.; Possingham, H.P.; Mumby, P.J.; Lovelock, C.E. The cost and feasibility of marine coastal restoration. *Ecol. Appl.* **2016**, *26*, 1055–1074. [CrossRef] [PubMed]

2. Hernández, A.B.; Brumbaugh, R.D.; Frederick, P.; Grizzle, R.; Luckenbach, M.W.; Peterson, C.H.; Angelini, C. Restoring the eastern oyster: How much progress has been made in 53 years? *Front. Ecol. Environ.* **2018**, *16*, 463–471. [CrossRef]

3. Miller, J.R.; Hobbs, R.J. Habitat Restoration—Do We Know What We're Doing? *Restor. Ecol.* **2007**, *15*, 382–390. [CrossRef]

4. Perring, M.P.; Standish, R.J.; Price, J.N.; Craig, M.D.; Erickson, T.E.; Ruthrof, K.X.; Whiteley, A.S.; Valentine, L.E.; Hobbs, R.J. Advances in restoration ecology: Rising to the challenges of the coming decades. *Ecosphere* **2015**, *6*, art131. [CrossRef]

5. Hobbs, R.J.; Harris, J.A. Restoration Ecology: Repairing the Earth's Ecosystems in the New Millennium. *Restor Ecol.* **2001**, *9*, 239–246. [CrossRef]

6. Duarte, C.M.; Conley, D.J.; Carstensen, J.; Sánchez-Camacho, M. Return to Neverland: Shifting Baselines Affect Eutrophication Restoration Targets. *Estuaries Coasts* **2009**, *32*, 29–36. [CrossRef]

7. Ehrenfeld, J.G. Defining the Limits of Restoration: The Need for Realistic Goals. *Restor. Ecol.* **2000**, *8*, 2–9. [CrossRef]

8. Hassett, B.; Palmer, M.; Bernhardt, E.; Smith, S.; Carr, J.; Hart, D. Restoring watersheds project by project: Trends in Chesapeake Bay tributary restoration. *Front. Ecol. Environ.* **2005**, *3*, 259–267. [CrossRef]

9. Hobbs, R.J. Setting Effective and Realistic Restoration Goals: Key Directions for Research. *Restor Ecol.* **2007**, *15*, 354–357. [CrossRef]

10. Hobbs, R.J.; Norton, D.A. Towards a Conceptual Framework for Restoration Ecology. *Restor. Ecol.* **1996**, *4*, 93–110. [CrossRef]

11. Hobbs, R.J.; Higgs, E.; Hall, C.M.; Bridgewater, P.; Chapin, F.S.; Ellis, E.C.; Ewel, J.J.; Hallett, L.M.; Harris, J.; Hulvey, K.B.; et al. Managing the whole landscape: Historical, hybrid, and novel ecosystems. *Front. Ecol. Environ.* **2014**, *12*, 557–564. [CrossRef]

12. Holl, K.D.; Howarth, R.B. Paying for Restoration. *Restor. Ecol.* **2000**, *8*, 260–267. [CrossRef]

13. Holl, K.D.; Crone, E.E.; Schultz, C.B. Landscape Restoration: Moving from Generalities to Methodologies. *BioScience* **2003**, *53*, 491. [CrossRef]

14. Hughes, F.M.R.; Colston, A.; Mountford, J.O. Restoring Riparian Ecosystems: The Challenge of Accommodating Variability and Designing Restoration Trajectories. *Ecol. Soc.* **2005**, *10*, art12. [CrossRef]

15. Palmer, M.A. Reforming Watershed Restoration: Science in Need of Application and Applications in Need of Science. *Estuaries Coasts* **2009**, *32*, 1–17. [CrossRef]

16. Rohr, J.R.; Bernhardt, E.S.; Cadotte, M.W.; Clements, W.H. The ecology and economics of restoration: When, what, where, and how to restore ecosystems. *Ecol. Soc.* **2018**, *23*, art15. [CrossRef]

17. Sanderson, E.W. How Many Animals Do We Want to Save? The Many Ways of Setting Population Target Levels for Conservation. *BioScience* **2006**, *56*, 911. [CrossRef]

18. Tear, T.H.; Kareiva, P.; Angermeier, P.L.; Comer, P.; Czech, B.; Kautz, R.; Landon, L.; Mehlman, D.; Murphy, K.; Ruckelshaus, M.; et al. How Much Is Enough? The Recurrent Problem of Setting Measurable Objectives in Conservation. *BioScience* **2005**, *55*, 835. [CrossRef]

19. Cloern, J. Our evolving conceptual model of the coastal eutrophication problem. *Mar. Ecol. Prog. Ser.* **2001**, *210*, 223–253. [CrossRef]

20. Tomasko, D.A.; Corbett, C.A.; Greening, H.S.; Raulerson, G.E. Spatial and temporal variation in seagrass coverage in Southwest Florida: Assessing the relative effects of anthropogenic nutrient load reductions and rainfall in four contiguous estuaries. *Mar. Pollut. Bull.* **2005**, *50*, 797–805. [CrossRef]

21. Greening, H.; Janicki, A. Toward Reversal of Eutrophic Conditions in a Subtropical Estuary: Water Quality and Seagrass Response to Nitrogen Loading Reductions in Tampa Bay, Florida, USA. *Environ. Manag.* **2006**, *38*, 163–178. [CrossRef] [PubMed]

22. Cicchetti, G.; Greening, H. Estuarine Biotope Mosaics and Habitat Management Goals: An Application in Tampa Bay, FL, USA. *Estuaries Coasts* **2011**, *34*, 1278–1292. [CrossRef]

23. Greening, H.S.; Cross, L.M.; Sherwood, E.T. A Multiscale Approach to Seagrass Recovery in Tampa Bay, Florida. *Ecol. Restor.* **2011**, *29*, 82–93. [CrossRef]

24. Yates, K.K.; Greening, H.; Morrison, G. *Integrating Science and Resource Management in Tampa Bay, Florida*; US Geological Survey: Reston, VA, USA, 2011; p. 298.

25. Greening, H.; Janicki, A.; Sherwood, E.T.; Pribble, R.; Johansson, J.O.R. Ecosystem responses to long-term nutrient management in an urban estuary: Tampa Bay, Florida, USA. *Estuar. Coast. Shelf Sci.* **2014**, *151*, A1–A16. [CrossRef]

26. Russell, M.; Greening, H. Estimating Benefits in a Recovering Estuary: Tampa Bay, Florida. *Estuaries Coasts* **2015**, *38*, 9–18. [CrossRef]

27. Sherwood, E.T.; Greening, H.S.; Johansson, J.O.R.; Kaufman, K.; Raulerson, G.E. Tampa Bay (Florida, USA): Documenting Seagrass Recovery since the 1980's and Reviewing the Benefits. *Southeast. Geogr.* **2017**, *57*, 294–319. [CrossRef]

28. Lewis, R.R.I.; Estevez, E.D. *The Ecology of Tampa Bay, Florida: An Estuarine Profile*. U.S. Fish and Wildlife Service Biological Report 85. Available online: http://palmm.digital.flvc.org/islandora/object/uf%3A71939#page/cover1/mode/1up (accessed on 22 January 2020).

29. Tampa Bay Estuary Program Charting the Course: The Comprehensive Conservation and Management Plan for Tampa Bay. 2017. Available online: http://www.tampabay.wateratlas.usf.edu/upload/documents/192_tbep_ccmp_2017-web.pdf (accessed on 5 June 2018).

30. Fehring, W.K. *Data Bases for Use in Fish and Wildlife Mitigation Planning in Tampa Bay, Florida: Project Summary*; Greiner Engineering Inc.: Tampa, FL, USA, 1986.

31. Lewis, R.R.; Robinson, D. Setting priorities for Tampa Bay habitat protection and restoration: Restoring the balance. *Tech. Publ.* **1995**, *9*.

32. Lewis, R.R., 3rd; Clark, P.A.; Fehring, W.K.; Greening, H.S.; Johansson, R.O.; Paul, R.T. The Rehabilitation of the Tampa Bay Estuary, Florida, USA, as an Example of Successful Integrated Coastal Management. *Mar. Pollut. Bull.* **1999**, *37*, 468–473. [CrossRef]

33. Sherwood, E.T.; Greening, H.S. Personal communication, 2018.

34. Burke, M.; Raulerson, G. *2018 Tampa Bay Water Quality Assessment*. Technical Report #01-19 of the Tampa Bay Estuary Program, Prepared by the Tampa Bay Estuary Program. 2019. Available online: https://www.tbeptech.org/TBEP_TECH_PUBS/2019/TBEP_01_19_2018_Decision_Matrix_Update_FINAL.pdf (accessed on 5 June 2018).

35. Radabaugh, K.R.; Powell, C.E.; Moyer, R.P. *Coastal Habitat Integrated Mapping and Monitoring Program Report for the State of Florida*; Fish and Wildlife Research Institute: Petersburg, FL, USA, 2017.

36. Sherwood, E.T.; Greening, H.S.; Janicki, A.J.; Karlen, D.J. Tampa Bay estuary: Monitoring long-term recovery through regional partnerships. *Reg. Stud. Mar. Sci.* **2016**, *4*, 1–11. [CrossRef]

37. Beck, M.W.; Sherwood, E.T.; Henkel, J.R.; Dorans, K.; Ireland, K.; Varela, P. Assessment of the Cumulative Effects of Restoration Activities on Water Quality in Tampa Bay, Florida. *Estuaries Coasts* **2019**, *42*, 1774–1791. [CrossRef]

38. Morrison, G.; Greening, H.S.; Sherwood, E.T.; Yates, K.K. Management Case Study: Tampa Bay, Florida. In *Reference Module in Earth Systems and Environmental Sciences*; Elsevier: Amsterdam, The Netherlands, 2014.

39. Chesapeake Bay Program. 2018 Facts and Figures. Available online: https://www.chesapeakebay.net/discover/facts (accessed on 5 June 2018).

40. Zimmerman, A.R.; Canuel, E.A. A geochemical record of eutrophication and anoxia in Chesapeake Bay sediments: Anthropogenic influence on organic matter composition. *Mar. Chem.* **2000**, *69*, 117–137. [CrossRef]

41. Kemp, W.M.; Boynton, W.R.; Adolf, J.E.; Boesch, D.F.; Boicourt, W.C.; Brush, G.; Cornwell, J.C.; Fisher, T.R.; Glibert, P.M.; Hagy, J.D.; et al. Eutrophication of Chesapeake Bay: Historical trends and ecological interactions. *Mar. Ecol. Prog. Ser.* **2005**, *303*, 1–29. [CrossRef]

42. Kirby, M.X.; Miller, H.M. Response of a benthic suspension feeder (Crassostrea virginica Gmelin) to three centuries of anthropogenic eutrophication in Chesapeake Bay. *Estuar. Coast. Shelf Sci.* **2005**, *62*, 679–689. [CrossRef]

43. Smith, G.F.; Bruce, D.G.; Roach, E.B.; Hansen, A.; Newell, R.I.E.; McManus, A.M. Assessment of Recent Habitat Conditions of Eastern Oyster Crassostrea virginica Bars in Mesohaline Chesapeake Bay. *N. Am. J. Fish. Manag.* **2005**, *25*, 1569–1590. [CrossRef]

44. Rothschild, B.J.; Ault, J.S.; Goulletquer, P.; Heral, M. Decline of the Chesapeake bay oyster population: A century of habitat destruction and overfishing. *Mar. Ecol. Prog. Ser.* **1994**, *111*, 29–39. [CrossRef]

45. Breitburg, D.L.; Craig, J.K.; Fulford, R.S.; Rose, K.A.; Boynton, W.R.; Brady, D.C.; Ciotti, B.J.; Diaz, R.J.; Friedland, K.D.; Hagy, J.D.; et al. Nutrient enrichment and fisheries exploitation: Interactive effects on estuarine living resources and their management. *Hydrobiologia* **2009**, *629*, 31–47. [CrossRef]

46. Zu Ermgassen, P.S.E.; Spalding, M.D.; Grizzle, R.E.; Brumbaugh, R.D. Quantifying the Loss of a Marine Ecosystem Service: Filtration by the Eastern Oyster in US Estuaries. *Estuaries Coasts* **2013**, *36*, 36–43. [CrossRef]

47. Boesch, D. Factors in the Decline of Coastal Ecosystems. *Science* **2001**, *293*, 1589c. [CrossRef]

48. Lotze, H.K.; Lenihan, H.S.; Bourque, B.J.; Bradbury, R.H.; Cooke, R.G.; Kay, M.C.; Kidwell, S.M.; Kirby, M.X.; Peterson, C.H.; Jackson, J.B.C. Depletion, Degradation, and Recovery Potential of Estuaries and Coastal Seas. *Science* **2006**, *312*, 1806–1809. [CrossRef]

49. Wilberg, M.J.; Livings, M.E.; Barkman, J.S.; Morris, B.T.; Robinson, J.M. Overfishing, disease, habitat loss, and potential extirpation of oysters in upper Chesapeake Bay. *Mar. Ecol. Prog. Ser.* **2011**, *436*, 131–144. [CrossRef]

50. Chesapeake Bay Foundation. On the Brink: Chesapeake's Native Oysters. 2010. Available online: http://www.cbf.org/document-library/cbf-reports/Oyster_Report_for_Release02a3.pdf (accessed on 5 June 2018).

51. Grabowski, J.H.; Brumbaugh, R.D.; Conrad, R.F.; Keeler, A.G.; Opaluch, J.J.; Peterson, C.H.; Piehler, M.F.; Powers, S.P.; Smyth, A.R. Economic Valuation of Ecosystem Services Provided by Oyster Reefs. *BioScience* **2012**, *62*, 900–909. [CrossRef]

52. Chesapeake Bay Program. 2018 Bay Program History. Available online: https://www.chesapeakebay.net/who/bay_program_history (accessed on 5 June 2018).

53. Chesapeake 2000 Agreement. 2000. Available online: https://www.chesapeakebay.net/documents/cbp_12081.pdf (accessed on 5 June 2018).

54. Kennedy, V.S.; Breitburg, D.L.; Christman, M.C.; Luckenbach, M.W.; Paynter, K.; Kramer, J.; Sellner, K.G.; Dew-Baxter, J.; Keller, C.; Mann, R. Lessons Learned from Efforts to Restore Oyster Populations in Maryland and Virginia, 1990 to 2007. *J. Shellfish Res.* **2011**, *30*, 719–731. [CrossRef]

55. Executive Order 13508. 74 FR 23099; 2009. Available online: https://www.federalregister.gov/documents/2009/05/15/E9-11547/chesapeake-bay-protection-and-restoration (accessed on 5 June 2018).

56. Federal Leadership Committee for the Chesapeake Bay. 2010 Strategy for Protecting and Restoring the Chesapeake Bay Watershed. Available online: https://federalleadership.chesapeakebay.net/file.axd?file=2010%2f5%2fChesapeake+EO+Strategy%20.pdf (accessed on 5 June 2018).

57. Federal Leadership Committee for the Chesapeake Bay. 2010 Fiscal Year 2011 Action Plan: Strategy for Protecting and Restoring the Chesapeake Bay Watershed. Available online: https://federalleadership.chesapeakebay.net/file.axd?file=2010%2f9%2fChesapeake+EO+Action+Plan+FY2011.pdf (accessed on 5 June 2018).

58. Sustainable Fisheries Goal Implementation Team. 2011 Restoration Goals, Quantitative Metrics, and Assessment Protocols for Evaluating Success on Restored Oyster Reef Sanctuaries, Final Report to the Chesapeake Bay Program Sustainable Fisheries Goal Implementation Team. Available online: https://www.chesapeakebay.net/channel_files/17932/oyster_restoration_success_metrics_final.pdf (accessed on 5 June 2018).

59. Blankenship, K. Harris Creek Reef Restoration at 350 Acres, is Largest Ever. Available online: https://www.bayjournal.com/article/harris_creek_reef_restoration_at_350_acres_is_largest_ever (accessed on 5 June 2018).

60. Felver, R. 2018 Final Two Tributaries Chosen for Oyster Restoration in Virginia. Available online: https://www.chesapeakebay.net/news/blog/final_four_tributaries_chosen_for_oyster_restoration (accessed on 5 June 2018).

61. Mason-Dixon. Polling & Strategy. *February 2018 Maryland Poll*; Mason-Dixon: Washington, DC, USA, 2018. Available online: http://www.trbas.com/media/media/acrobat/2018-03/70212851885400-01092209.pdf (accessed on 5 June 2018).

62. Knoche, S.; Ihde, T. *Estimating Ecological Benefits and Socio-Economic Impacts from Oyster Reef Restoration in the Choptank River Complex, Chesapeake Bay*; National Fish and Wildlife Federation and NOAA Chesapeake Bay Office: Annapolis, MD, USA, 2018; p. 68.

63. Cornwell, J.; Rose, J.; Kellogg, L.; Luckenbach, M.; Bricker, S.; Paynter, K.; Moore, C.; Parker, M.; Sanford, L.; Wolinski, B.; et al. Panel Recommendations on the Oyster BMP Nutrient and Suspended Sediment Reduction Effectiveness Determination Decision Framework and Nitrogen and Phosphorus Assimilation in Oyster Tissue Reduction Effectiveness for Oyster Aquaculture Practices. *Oyster BMP Expert Panel First Increm. Rep.* **2016**, 1.

64. Office of Management and Budget. 2017 Chesapeake Bay Restoration Spending Crosscut: Report to Congress. Available online: http://chesapeakeprogress.com/files/CBARA_Reporting_FY18_Chesapeake_Bay_Crosscut_FINAL.pdf (accessed on 5 June 2018).

65. Chesapeake Bay Watershed Agreement. 2014. Available online: https://www.chesapeakebay.net/channel_files/24334/2014_chesapeake_watershed_agreement.pdf (accessed on 5 June 2018).

66. Chesapeake Bay Stewardship Fund. Available online: http://Nfwf.org/Chesapeake/Pages/home.aspx (accessed on 20 November 2019).

67. Atwater, B.F. Ancient processes at the site of southern San Francisco Bay, movement of the crust and changes in sea level. In *San Francisco Bay–The Urbanized Estuary*; Conomos, T.J., Ed.; American Association for the Advancement of Science- Pacific Division: Ashland, OR, USA, 1979; pp. 31–45.

68. San Francisco Bay: The urbanized estuary: Investigations into the Natural History of San Francisco Bay and Delta with reference to the influence of man. In *Proceedings of the Fifty-Eighth Annual Meeting of the Pacific Division/American Association for the Advancement of Science held at San Francisco State University, San Francisco, CA, USA, 12–16 June, 1977*; Conomos, T.J.; American Association for the Advancement of Science; American Society of Limnology and Oceanography (Eds.) The Division: San Francisco, CA, USA, 1979; ISBN 978-0-934394-01-7.

69. Nichols, F.H.; Cloern, J.E.; Luoma, S.N.; Peterson, D.H. The Modification of an Estuary. *Sci. New Ser.* **1986**, *231*, 567–573. [CrossRef] [PubMed]

70. Goals Project. *Baylands Ecosystem Habitat Goals. A Report of Habitat Recommendations Prepared by the San Francisco Bay Area Wetlands Ecosystem Goals Project*; S.F. Bay Regional Water Quality Control Board: Oakland, CA, USA, 1999.

71. Cohen, A. *An Introduction to the San Francisco Estuary*; Save the Bay, San Francisco Estuary Project, and San Francisco Estuary Institute: Oakland, CA, USA, 2000.

72. Gilbert, G.K. *Hydraulic-Mining Débris in the Sierra Nevada*; U.S. Government Printing Office: Washington, DC, USA, 1917.

73. Save the Bay. A Brief History. Available online: https://savesfbay.org/who-we-are/a-brief-history (accessed on 20 December 2019).

74. BCDC—History of the San Francisco Bay Conservation and Development Commission. Available online: https://bcdc.ca.gov/history.html (accessed on 18 December 2019).

75. SFEP. Comprehensive Conservation and Management Plan, San Francisco Estuary Project, Oakland, CA. 1993. Available online: https://www.sfestuary.org/wp-content/uploads/2018/01/1993-CCMP.pdf (accessed on 22 January 2020).

76. Goals Project. The Baylands and Climate Change: What We Can Do. Baylands Ecosystem Habitat Goals Science Update 2015. Available online: https://baylandsgoals.org/wp-content/uploads/2019/06/Baylands-Complete-Report.pdf (accessed on 22 January 2020).

77. Zentner, J. Wetland projects of the California state coastal conservancy: An assessment. *Coast. Manag.* **1988**, *16*, 47–67. [CrossRef]

78. Home|South Bay Salt Ponds. Available online: https://www.southbayrestoration.org/ (accessed on 18 December 2019).

79. Grenier, J.L. Personal communication, 2019.
80. Gies, E. Fortress of Mud. *Nature* **2018**, *562*, 178–180. [CrossRef] [PubMed]
81. Leslie, H.M.; McLeod, K.L. Confronting the challenges of implementing marine ecosystem-based management. *Front. Ecol. Environ.* **2007**, *5*, 540–548. [CrossRef]
82. Palmer, M.A.; Bernhardt, E.S.; Allan, J.D.; Lake, P.S.; Alexander, G.; Brooks, S.; Carr, J.; Clayton, S.; Dahm, C.N.; Follstad Shah, J.; et al. Standards for ecologically successful river restoration: Ecological success in river restoration. *J. Appl. Ecol.* **2005**, *42*, 208–217. [CrossRef]
83. Carwardine, J.; Klein, C.J.; Wilson, K.A.; Pressey, R.L.; Possingham, H.P. Hitting the target and missing the point: Target-based conservation planning in context. *Conserv. Lett.* **2009**, *2*, 4–11. [CrossRef]
84. McDonald, T.; Jonson, J.; Dixon, K.W. National standards for the practice of ecological restoration in Australia. *Restor. Ecol.* **2016**, *24*, S4–S32. [CrossRef]
85. McDonald, T.; Gann, G.D.; Jonson, J.; Dixon, K.W. *International Standards for the Practice of Ecological Restoration—Including Principles and Key Concepts*; Society for Ecological Restoration: Washington, DC, USA, 2016; p. 48.

Article

Urban Forestry in Brazilian Amazonia

Thiago Almeida Vieira [1,*] **and Thomas Panagopoulos** [2,*]

1   Institute of Biodiversity and Forests, Federal University of Western of Pará, Santarém 68035–100, Brazil
2   Research Center for Tourism, Sustainability and Well-being, University of Algarve, 8000 Faro, Portugal
*   Correspondence: thiago.vieira@ufopa.edu.br (T.A.V.); tpanago@ualg.pt (T.P.)

Received: 2 March 2020; Accepted: 14 April 2020; Published: 16 April 2020

Abstract: Urban forests provide multiple benefits in improving people's lives and can be an important tool for achieving the goal of carbon neutral cities. In this study, we analyzed the diversity of plant species from urban forests in cities in the Brazilian Amazonia, based on data from scientific articles, through a systematic literature review. Our analysis revealed that 530 taxa, of which 479 were identified at the species level and 51 at the genus level, covering 38,882 individuals were distributed in 29 cities. The three most frequent species were *Ficus benjamina*, *Mangifera indica*, and *Licania tomentosa*. Exotic species were more frequent than native. The three most frequent species had almost 42% of the inventoried individuals. The choice of species has been made mainly by the local population, without monitoring by the public authorities. Recommendations for sustainable management of urban forests in Amazonia include investing in training of management bodies, periodic inventories, and awareness actions about the benefits of urban green infrastructure and on the advantages of native species. Policies for the sustainable management of urban green areas are necessary. The municipal governments must continuously monitor indicators of urban ecosystem services and provide financial resources for maintaining and increasing those area rates per person.

Keywords: exotic species; urban biodiversity; urban ecosystems; carbon neutral cities

---

1. Introduction

The Amazonia is known worldwide for its forests, mostly of humid ombrophilous type, and for its rivers with clear, dark, or white (muddy) waters, with emphasis on the Amazon River, the largest on the planet. However, their ecosystems are damaged by human action, such as the urbanization process. Until the 1960s, cities in the Brazilian Amazonia were small, often associated with fluvial circulation, with non-modern rural life and few forests explored yet, which gave them strong links to nature [1]. Today, the impacts of urbanization are also observed in this region.

The current form of growth in cities has resulted in losses and degradation of natural ecosystems in urban areas, causing the drastic loss of ecosystem services and low resilience to disturbances, such as those caused by climate change [2]. Among the pressures, there is the large consumption of water for residential and commercial use, damage caused by the generation of waste and its inadequate disposal, expansion over natural areas to meet housing demands, etc. [3], leading to negative impacts on water, air quality, and the maintenance of the habitat of different species [4]. The reality of cities in the Brazilian Amazonia still results from several models of land appropriation, land use, and the linear exploitation of natural capital in this biome [5].

Urban expansion planning must be carried out effectively. One of the aspects to be observed concerns the management of urban green areas. Urban forests, including trees that are not only in woodlands, but also on streets, along streams, and in parks, provide important ecosystem services for urban and peri-urban populations [6]. Trees in cities contribute to the stability of the urban ecosystem [7], provide food [8], and have aesthetic aspects that contribute to the generation of economic and social benefits [9]. Urban green spaces are also important for attraction of tourism in city destinations [10]. Urban green infrastructure conserves biodiversity, allows interaction between people, and the contemplation of nature, which helps to break with everyday stress [11]. Trees, for example, are also a way of reducing atmospheric carbon dioxide and for that they need to have their structure maintained [12]. A tree can absorb up to 150 kg of CO_2 per year [13], sequester carbon, and consequently mitigate climate change and contribute to the goal of carbon neutral cities.

In general, studies of urban trees and shrub diversity have largely focused on just one city [14]. Meanwhile, there is need to study this parameter at a regional scale. Knowledge of the patterns of urban species diversity and what can influence them contributes to better planning of conservation actions, especially for the population of trees on the streets of the city [15,16]. A study conducted in the city of Manaus found a positive relationship between socioeconomic variables and the valorization of vegetation in an urban area by the residents of the neighborhood [17]. Thus, we assume as a hypothesis that there is a relationship between the patterns of species diversity of urban forests and the socioeconomic characteristics of cities in the Amazon.

Thus, the aim of this study was to analyze the diversity of urban forest species in the Brazilian Amazonia, relate it to the socioeconomic characteristics of cities, and propose improvements in policy and management of urban green infrastructure.

2. Materials and Methods

2.1. Data Collection

Scientific articles were the source of data of the diversity of plant species grown in cities of the Brazilian Amazonia, which includes the states of Acre, Amapá, Amazonas, Mato Grosso, Pará, Rondônia, Roraima, and Tocantins and part of Maranhão (also called Legal Amazonia). We conducted a systematic review of the literature based on scientific articles available on international indexing databases (Google Scholar, Scielo, Scopus, and Web of Science). We did additional research in a specialized Brazilian journal—*Revista da Sociedade Brasileira de Arborização Urbana* (ISSN 1980–7694).

To search for the articles, combinations of keywords were used that indicated the object of study (Urban Forest; Urban Forestry) and the location (Amazonia; Amazon; Acre; Amapá; Amazonas; Maranhão; Mato Grosso; Pará;Rondônia; Roraima; Tocantins). The research was done using the above words in both Portuguese and English. Year of publication of the articles was not delimited. From this, for initial screening, the titles and abstracts were read, and we excluded publications that could not meet the object of this study. Thus, studied articles were selected that contained a number of inventoried individuals and with the respective scientific names. When two or more articles contained the same species list, only the first one was considered.

From public databases, we collected socioeconomic and territorial information from municipalities: population size (number of people) [18]; territorial area (km^2) [19]; Gross Domestic Product—GDP (provided in Real—R\$, and converted to Euro—€) [20]; Urban Forestry Rate of streets—UFor.R (%) [21]; Urbanization of Public Roads Rate—UPR (%) [22]; Human Development Index—HDI (ranges from 0 to 1) [23]; fleet of vehicles of 2018 [24]; temperatures, humidity, and rainfall [25]. The websites of the city halls and/or municipal chamber of each city were consulted to raise the value of finance resources allocated to the management of green areas (consulting the Annual Budget Law of 2019 or the closest available year).

2.2. Data Analysis

The scientific names of the species were tabulated with the names of the botanical families and number of individuals inventoried by the studies and verified at *Tropicos* (www.tropicos.org). All of this information was organized by Amazonian cities. The species were classified as native and exotic according to *Flora do Brasil 2020* (http://floradobrasil.jbrj.gov.br/), which provides the origin of native species by phytogeographic domain in Brazil.

To assess the diversity between cities and to make comparisons between them, the following indices were used. Shannon–Wiener [26], which varies from 0 to Ln of the number of species sampled and determined by the equation:

$$H' = - \sum pi.\text{Lnpi}$$

(1)

where (H ') Shannon–Wiener diversity index, (Ln) Napierian logarithm, (pi) ni/N, (ni) number of individuals sampled for each species, and (N) total number of individuals.

The Simpson index, which varies from 0 to 1, was also used, and the higher it is, the greater the probability that individuals are of the same species, that is, greater dominance and less diversity.

$$\lambda = \sum p_i{}^2$$

(2)

where pi = ni/N, "ni" represents the number of specimens of each species, and "N" the total number of specimens in the sample.

The relative frequency of each species was calculated by the ratio between the number of individuals of the species and the total number of specimens, multiplied by 100 [27]. The relative density was calculated by the relationship between the number of individuals of one species and the total number of individuals of all species, multiplied by 100 [28].

In order to understand possible influences between socioeconomic variables and ecological parameters of plant species inventoried by studies on urban forests in the Amazon, simple regression analyses were performed between pairs of variables. When necessary, we standardized variables using standard deviation. Principal component analysis was carried out in order to verify climatic similarities between the cities studied. A *t* test was performed in order to check for differences between the frequency and density of native and exotic species in cities. To verify that socioeconomic variables (population size, Gross Domestic Product, Human Development Index) influence ecological attributes (richness, diversity, and density), a multiple linear regression analysis was performed, as shown in Appendix A. In a second moment, we performed a simple linear regression analysis between variables that presented a significant regression coefficient ($p > 0.05$) in the multiple linear regression analysis [17]. The data were analyzed with the aid of the statistical program R [29].

3. Results

3.1. Urban Forests: Species, Richness, and Diversity

Were identified 43 scientific articles reporting results about urban forest species richness and diversity, as shown in Appendix B. They were published from 2010 to 2019, covering 29 cities in the nine states of the Brazilian Amazon, as shown in Figure 1.

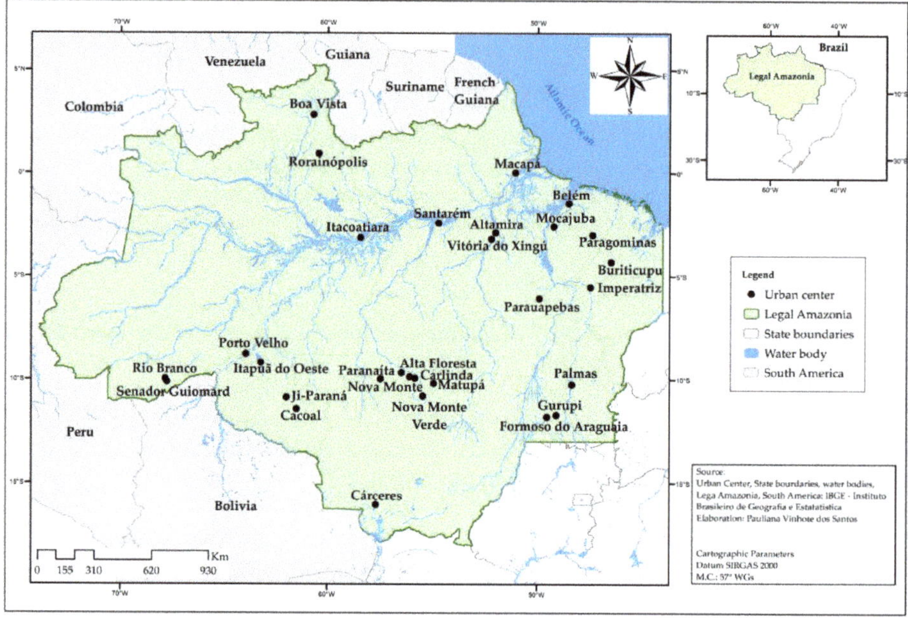

Figure 1. Location of the 29 Brazilian Amazonian cities with scientific articles on urban forestry including data about species diversity.

In total, 530 taxa were cataloged, 479 of which were identified at the species level and 51 only up to genus. Four articles presented species that were not identified at the species level, nor the genus. One study grouped five individuals into "other species". The species belong to 71 botanical families, with Fabaceae having a greater number of species, as shown in Table 1. The Chrysobalanaceae family presented only five species, but together they registered 8281 individuals.

Table 1. Main botanical families, number of species, and individuals reported in studies on urban forestry in 29 cities of Brazilian Amazonia.

Family	Number of Species	Number of Individuals
Fabaceae	151	5997
Arecaceae	46	2478
Bignoniaceae	33	2411
Myrtaceae	26	3202
Malvaceae	27	1363
Moraceae	21	4852
Anacardiaceae	17	4578
Rutaceae	16	444
Apocynaceae	14	178
Rubiaceae	12	488

Of the species reported in studies on urban forestry in the Brazilian Amazon, the most frequent were *Ficus benjamina* and *Mangifera indica* (86.2% of cities) and *Licania tomentosa* (82.8%), as shown in Figure 2. These three species are also the ones with the highest density of individuals reported in the urban green areas studied in the articles, with *L. tomentosa* presenting 8018 individuals, which represents 20.6% of all individuals inventoried in the studies of the 29 cities.

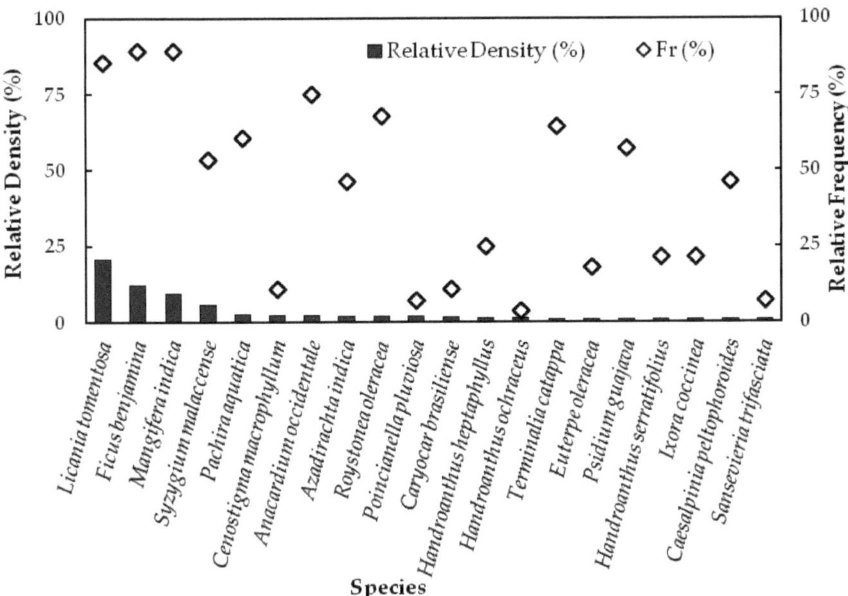

Figure 2. Density and relative frequency of the main plant species grown in urban forests in 29 cities of Brazilian Amazonia.

The average richness was 43.1 plant species per city, inventoried in scientific studies. The average density among the studies was approximately 1341 individuals per city. When analyzing the richness of species and number of individuals per city, it appears that the city of Rio Branco presented the greatest inventoried richness (179 species) and Boa Vista the highest number of individuals cataloged (6913), as shown in Figure 3. Belém and Itacoatiara had only one species studied by the articles.

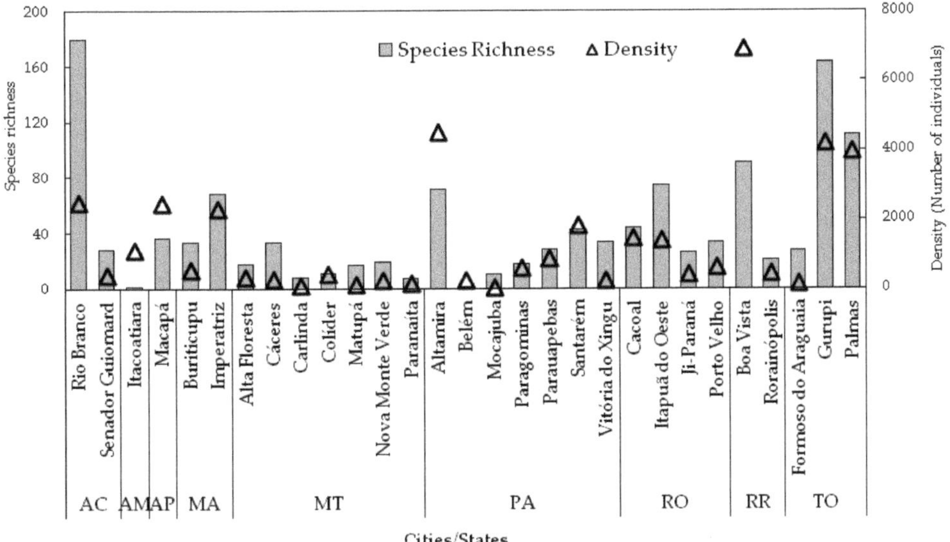

Figure 3. Species richness and number of individuals in urban forestry of 29 cities of Amazonia.

The average diversity among the 29 cities was 2.082 (H′), with the city of Rio Branco presenting a greater diversity of species (H′ = 4.183), followed by Palmas (H′ = 3.524) and Itapuã do Oeste (H′ = 3.167). These three cities also obtained the lowest values for the Simpson Index (0.0271; 0.0525; 0.090, respectively), confirming a high diversity of plants in the urban forests of these cities.

Among the identified species, 34.7% are exotic and 65.3% are native to Brazil. Among the native ones, 68.7% occur in the Amazon and also in other Brazilian biomes, and 14.1% have the Amazon as their only phytogeographic domain. On urban forestry of the cities studied, the average richness of exotic plant species was 9.4 and of native species was 7.7. In these 29 cities, there was a higher relative frequency of exotic species, as shown in Figure 4. Regarding the number of individuals of each species by origin, there was no significant difference between cities.

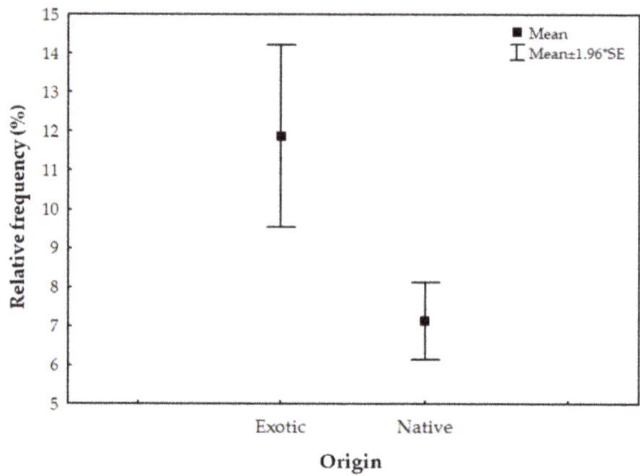

Figure 4. The *t* test for relative frequency of native and exotic plant species in urban forestry in the Brazilian Amazonia.

3.2. Socioeconomic and Climatic Characteristics

The 29 municipalities studied by the articles of urban forestry have a territorial area ranging from 871 km^2 (Mocajuba) to 159,533 km^2 (Altamira), with an average of 14,005 km^2. The population average was 187,066 people, with the city of Nova Monte Verde having the lowest number of people and Belém the largest, as shown in Table 2. The budgets for the creation and/or maintenance of urban green areas were on average 2,414,530.105 €, which represents 5.8 € per inhabitant/year. Information from two cities was missing.

The number of vehicles in the 29 cities is 2,320,963 and together, these cities had 38,882 individuals of plant species inventoried. We understand that the studies carried out sampling, but to arrive at the proportion of a tree or shrub for each car, 59.7 times the number of inventoried plants would be needed. City halls raise funds from tax collection to provide improvements in the infrastructure of cities, including urban green areas. In this sense, it is important to highlight that the Federal Constitution of Brazil provides that 50% of the collection of taxes from the ownership of licensed motor vehicles belongs to the municipalities where the vehicles are licensed [30]. Belém was the city with the largest fleet of vehicles (451,776), with one vehicle per 3.3 inhabitants. However, the city of Alta Floresta has a higher proportion of vehicles per inhabitant, with one vehicle per 1.2 people.

Table 2. Territorial, social, and demographic characteristics of 29 Brazilian Amazonian cities with studies on urban forestry.

Brazil States (acronyms)	Cities	A (km^2)	UPR %	UFor.R %	Population	GDP/person	HDI	Fleet
Acre (AC)	Rio Branco	8834.94	20.4	13.8	407,319	4623.10	0.727	179,395
	Senador Guiomard	2322.03	4.2	13.3	23,024	3468.30	0.640	6501
Amazonas (AM)	Itacoatiara	8891.91	11.9	57.9	101,337	4476.52	0.644	23,368
Amapá (AP)	Macapá	6563.85	8.8	66	503,327	4304.46	0.733	152,475
Maranhão (MA)	Buriticupu	2544.86	0.1	59.4	72,358	1476.72	0.556	16,164
	Imperatriz	1368.99	22.4	69.7	258,682	5897.64	0.731	152,881
	Alta Floresta	8953.19	3	37.7	51,782	6786.22	0.714	44,699
	Cáceres	24,593.12	23.2	79.2	94,376	4058.20	0.708	49,271
	Carlinda	2416.14	0	34.8	10,305	3328.32	0.665	5136
Mato Grosso (MT)	Colíder	3103.96	7.1	83.6	33,438	5821.51	0.713	23,878
	Matupá	5219.03	1.1	16.3	16,566	8372.36	0.716	11,156
	Nova Monte Verde	5150.56	1.7	25.5	9178	4197.79	0.691	4468
	Paranaíta	4796.01	1.8	92.8	11,225	8011.01	0.672	6669
	Altamira	159,533.33	22.7	44.3	114,594	4826.70	0.665	61,868
	Belém	1059.46	36.1	22.3	1,492,745	4366.95	0.746	451,776
	Mocajuba	871.17	0	19.7	31,136	2012.23	0.575	1830
Pará (PA)	Paragominas	19,342.57	5.1	12.9	113,145	5270.45	0.645	41,564
	Parauapebas	6886.21	21.8	30.5	208,273	9,546.46	0.715	91,669
	Santarém	17,898.39	7.8	43.3	304,589	3332.92	0.691	102,101
	Vitória do Xingu	3089.54	2.4	73.4	15,134	4,922.76	0.596	2459
	Cacoal	3792.89	11.7	86.6	85,359	4749.09	0.718	66,914
Rondônia (RO)	Itapuã do Oeste	4081.58	0	-	10,458	3251.61	0.614	3068
	Ji-Paraná	6896.65	6.4	17.3	128,969	4879.70	0.714	92,828
	Porto Velho	34,090.95	21.7	40	494,013	6188.08	0.736	276,601
Roraima (RR)	Boa Vista	5687.04	4.3	47.5	399,213	5333.16	0.752	190,870
	Rorainópolis	33,579.74	5.8	23.7	30,163	3913.10	0.619	6132
	Formoso do Araguaia	13,423.38	1.4	96.9	18,440	4239.82	0.670	7300
Tocantins (TO)	Gurupi	1836.09	0.4	88.7	86,647	5158.13	0.759	60,893
	Palmas	2218.94	31.3	79.9	299,127	6217.63	0.788	187,029

Note: A= area; UPR = Urbanization of Public Roads Rate; UFor.R = Urban Forestry Rate; Population; GDP = Gross Domestic Product in euros; HDI = Human Development Index; Fleet = Fleet of vehicles; 1.00 € = 4.66 R$, in February, 05 2020.

The average minimum and maximum temperatures in cities were 25.9 and 27.1 °C, respectively. The average minimum humidity was 72% and the maximum was 77.4%. The annual rainfalls of cities ranged from 373.40 to 4208 mm, with an average of 1717.03 mm, showing favorable conditions for the establishment of tropical plants. Brazil's climate does not impose many restrictions on the adaptation of plant species in the urban space, especially to native species. Cities were similar in terms of environmental conditions and are grouped by these similarities, as shown in Figure 5.

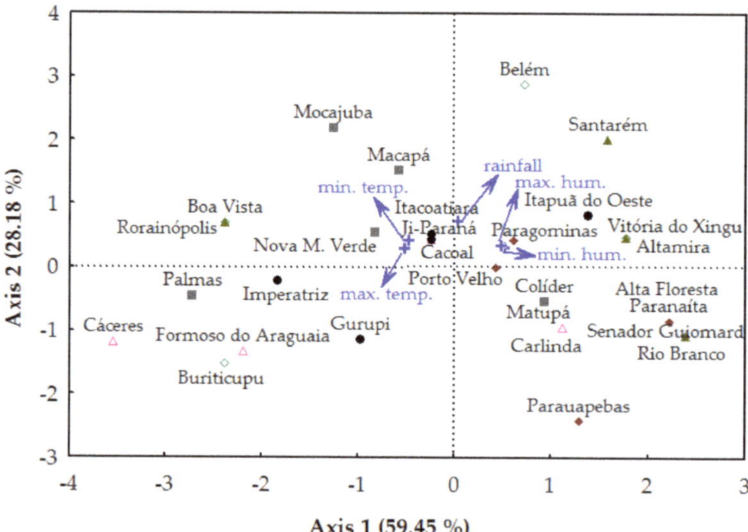

Figure 5. Principal component analysis (PCA) demonstrating climatic similarities between the 29 cities with studies on urban forestry in the Brazilian Amazonia. Note: min. temp. = minimum temperature; max. temp. = maximum temperature; min. hum. = minimum humidity; max. hum. = maximum humidity.

The average rate of tree-lined roads in cities was 49.2% and ranged from 12.9% (Paragominas) to 96.9% in Formoso do Araguaia. The average Urbanization of Public Roads Rate was 10%, and Belém was the city with higher urbanization of its streets (36.1%).

The studies about urban forestry were carried out in 29 cities in the Brazilian Amazonia, which among them still have cities with a zero urbanization rate of public roads and the highest rate is 36.1% (Belém), as shown in Table 2. This parameter is calculated by the Brazilian Institute of Geography and Statistics [22] and indicates the percentage of households located on paved streets. Formoso do Araguaia, a city with a higher rate of tree lined in roads (Urban Forestry Rate—UFor.R), allocates around 33.90 €/person/year for carrying out environmental management in the municipality, almost 6 times more than the average of 29 cities.

Among the socioeconomic and ecological variables, the Human Development Index of the cities had a significant influence on ecological variables of richness and density, as shown in Figure 6A,B.

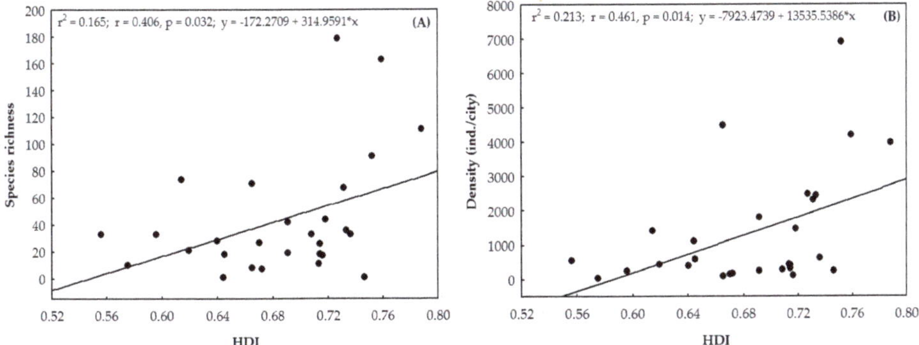

Figure 6. Linear regression analysis between the Human Development Index (HDI), density (**A**), and richness of species (**B**) vegetation inventoried on urban forestry by studies in the Brazilian Amazonia.

4. Discussion

The results revealed that in 29 Amazonian cities, the exotic species were more frequent than the native. Besides, the three most frequent species had almost 42% of the inventoried individuals. The local population act in the design of urban forests, but without necessary monitoring by the public authorities. Results exposed that research on urban forestry in the Amazon is recent (last decade). The number of articles published in scientific journals seems to be low and shows under publication. Although it is known that there is a high number of research works at concluding undergraduate courses, and even at a master's and doctorate level, most have not been published as articles in scientific journals. Research results that are not published in scientific journals can limit access to information by other researchers, making knowledge of scientific evidence unfeasible [31].

Brazilian legislation supports the creation and maintenance of urban green areas. The Brazilian Constitution provides that urban complexes and sites of landscape value constitute Brazilian cultural heritage and that everyone has the right to an ecologically balanced environment, and for this, public authorities and the community have a role in defending and preserving [30]. In addition, for the protection and management of urban green areas, the 2012 Brazilian Forest Code provides that city halls may require green areas in allotments, commercial enterprises, and in the implementation of infrastructure in cities.

Brazilian States have to take actions that guarantee a healthy urban environment. Urban forestry may play a key role in sequestering carbon emissions in cities. A study carried out in a city in China, between 2004 and 2006, showed that the average annual emissions of C by the combustion of fossil fuels was 11.16 million tons, with the C stored by the urban forests of this city corresponding to 3.02% of annual average C emissions [32].

Studies about urban forestry in the Brazilian Amazonia have shown that people have been primarily responsible for planting and cultivating plant species located on the streets [28,33]. In this sense, the municipal government needs to prioritize the planning of actions related to urban forestry. Considering these people's actions on the composition and management of the species, the preference for *F. benjamina* and *L. tomentosa* occur because they form leafy tops and provide shade throughout the year, being species of rapid growth and easy adaptation to the urban environment [34].

Additionally, the three most frequent species are also the ones with the highest relative density, so that together they account for almost 42% of all individuals inventoried by studies on forest ballots in the Brazilian Amazonia. The use of the species *L. tomentosa*, as shown in Figure 7, that produces edible fruits of high nutritional value [35], can increase city resilience in times of crisis. Meanwhile, this was considered disproportionate in a study carried out in three cities of Mato Grosso [27] and for this reason the authors recommended the introduction of new shrub and tree species native to the region in the urban green spaces studied.

Figure 7. *Licania tomentosa* cultivated in Belém streets, Brazilian Amazonia. Photo: Vieira, L.A.

On the other hand, it is interesting to note that 129 species inventoried in urban forestry in 29 cities were represented by only one individual. This shows disproportionality in almost 25% of the species grown in these areas. This was pointed out as a problem for urban forestry by research carried out in cities in the state of Mato Grosso [27,36].

Plants that occupy urban space are more propitious to attack by plagues and diseases as a result of the anthropized environment, and greater diversity can reduce the risks to plant health of these plants [37]. In this sense, in order to increase the diversity of plant species, it is recommended not to plant more than 10% of any species; no more than 20% of any genus; and no more than 30% of any botanic family [38].

In general, there is a lack of specific legislation to address the creation, management, and maintenance of urban green areas in municipalities. However, it should be noted that the city of Belém, capital of the state of Pará, instituted Municipal Law No. 8909, of March 29, 2012, which resulted in a Technical Guidance Manual for Urban Forestry in Belém [39]. This lack of legislation can contribute to irregular distribution among plant species [27], often resulting in inadequate attitudes by residents and even public managers, who have little technical information on urban green areas [40].

It was reported that plants were poorly distributed geographically in the city of Vitória do Xingu, and that the central region is more wooded than the peripheral ones, demonstrating the lack of planning by the government [41]. The urban forest can help environmental balance, such as sequestering carbon, but it is necessary that the technicians and the population understand that the planting must consider the correct species in the right place, in order to reduce early tree mortality, optimize ecosystem services, and maintain biodiversity [12].

The adoption of species by the population also occurs, in part, as a reflection of old landscape trends, because from the aesthetic point of view, species of great beauty were distributed all over the world, and not only in a geographical or restricted vegetal formation, so the choice for many exotic species resulted [42]. This trend dates back to the Brazilian colonial period, when propagules of plant species from all over the world were collected to be cultivated in botanical gardens and therefore, even today, native species seem to be a secondary element in the urban landscape [43].

Exotic species can have a high capacity for adaptation in non-native environments, competing for resources (light, water, and nutrients), inhibiting the growth of native species, being potentially toxic to local fauna [44], and can present invasive species behavior [45,46]; directly affecting biodiversity, the economy, and human health [47]. The priority for exotic species ends up disregarding the native's potential [48]. The cultivation of native species is technically recommended in order to guarantee the co-evolutionary ecological and genetic relationships, of dispersal of propagules (pollen and seeds), involving fauna and flora, within the urban environment and also for the conservation of native genetic material [46]. In addition, aesthetically, a wooded city with species characteristic of its region, would make the urban environment unique, with differentiated aspects from other cities and much more attractive [48].

Considering the importance of urban forestry, we mention that an inventory of the afforestation of the city of Altamira (PA) was carried out, through a partnership between the City Hall, higher education institutions, and the Public Ministry [28]. The results showed a great presence of exotic species, and for that reason, actions of production of seedlings of native species have been carried out for cultivation in the urban forestry of this city, such as *Handroanthus serratifolius*, *Andira parviflora*, and *Clitoria racemosa*.

Bifurcation at low heights is a problem reported by the articles, as it makes it difficult for pedestrians to move on sidewalks [49–51]. For these cases, it is recommended that the crown be lifted, an intervention that aims to suppress the lower branches [49]. Damage to the sidewalks caused by plant roots was mentioned, in general due to the little space available for root growth and the improper choice of the species considering this space [28,52,53], or by soil conditions, which were generally compacted [54].

In many cases, plants considered inadequate in the occupied urban green space must be replaced, especially those that are generating more disservices than benefits [55]. However, this measure must be well grounded and planned. Based on a technical report from the environmental agency, suppression can be authorized in cases of danger of falling or of increasing and irreversible damage to property; due to the irrecoverable phytosanitary state; or when it is dead [39]. The spacing between plants, area free of pavement, need for staking and protection grids, and soil available for each plant should follow adequate arboriculture techniques [56].

These results show that the selection of species to be cultivated in Amazonian urban streets is a priority. This choice must meet the demand for shade (related to the canopy architecture and the deciduous behavior of the species) and contribute to the thermal comfort and well-being of the population [43]. The selection of species made by people in the cities studied shows that there is interest in shade and fruit production, such as *Mangifera indica*, *Syzygium malaccense*, *Anacardium occidentale*, *Euterpe oleracea*, *Psidium guajava*, etc. Although fruit species are of importance to the human population and attractive to avifauna, planting these species on sidewalks should be avoided, as these fruits can cause damage to vehicles and pedestrians circulating on the site, for this, one must choose species that do not produce large and fleshy fruits [39].

Our study showed the plant diversity in urban forests in 29 cities. Meanwhile, most studies in the Brazilian Amazonia focus their results in problems related to urban forests. Among the tree maintenance difficulties mentioned were reports of phytosanitary problems (diseases and insect attacks), mechanical injuries in some part of the plant (usually in the bark of the stem), and presence of hemiparasites. Problems related to infrastructure were conflicts caused by deficient crown and root maintenance, damage on the sidewalks due to lack of adequate space for root growth, crown conflicts with electric cables, bifurcated individuals at short heights, and incorrect pruning. On the other hand, urban trees provide multiple ecosystem services and contribute to the improvement of people's quality of life and to the balance of the urban environment.

Human Development Index Influencing the Plant Density and Richness

As mentioned above, we showed there is a correlation between density and diversity of urban tree species and the Human Development Index (HDI) of the Amazonian cities. The positive correlation indicates that cities with higher human development indexes have higher values of plant diversity and density in urban forests. In our study, 15 cities have an HDI considered as high (0.700–0.799). In these cities, the average plant density in urban forests was 1808 inventoried plants and the average richness was 57 species/city. The other 14 cities had an average HDI (0.550–0.699) and the average density was 840 inventoried plants and a richness of 28 species/city.

The HDI is an important social metric and highlights that sustainability has to be based on the rationale that high human development facilitates sustainable development [57]. A research of sub-Saharan African cities revealed that one of the barriers to the sustainable delivery of ecosystem services is social inequality and urban planning [58]. In Central Asia, the Human Development Index is high in countries with an urbanization level over 70 percent, and probably, a higher socio-economical level leads to larger interest for environmental and sustainable solutions among society [59]. A study about environmental justice in accessibility to green infrastructure in two European cities show that deprived neighborhoods with minorities had less availability to quality green spaces [60].

A study in 100 cities around the world, including four Brazilian cities (Curitiba, São Paulo, Porto Alegre, and Manaus), highlighted that the more affluent cities tend to have a greater biomass of vegetation, involving the maintaining of larger areas of vegetation and larger tree populations in streets [61]. Plant diversity may reflect social, economic, and cultural influences. In Arizona city, a study aimed to investigate the influence of biotic, abiotic, and human-related variables with richness of perennial plants (including both exotic and native) and showed that plant diversity at higher income neighborhoods was on average twice that found in the landscapes of less wealthy areas [62]. Thus, municipalities with a low Human Development Index should adopt policies for viable and diverse urban forests, providing knowledge to the population and aiming at well-being for society.

5. Conclusions and Recommendations

Investments should be made in research on species suitable for urban forestry in Amazonian cities. It is expected that there will be an appreciation of native species, aiming to increase their diversity in the urban environment, seeking, in them, positive characteristics, not only of aesthetic parameters, but functional. In this sense, considering that the capacity of species to transform CO_2 into biomass, through photosynthesis, in urban green areas has been little studied [31], we recommend future research to investigate the potential of this transformation by the species used in urban forestry in the Brazilian Amazonia.

Investments should be made in periodic surveys and systematization of information on plant individuals in urban forestry, through continuous inventories. Thus, the performance of forestry professionals should be valued, and who must compose teams for planning, management, and monitoring of urban forestry. It is important that municipalities periodically monitor indicators for urban forestry, in order to optimize the financial resources invested in these areas, as well as to achieve the objectives of ecosystem services and other benefits of trees in cities.

Professionals must also consider the concept of green infrastructure, involving environmental strategies based on multidisciplinary teams of forestry, architecture, construction, and urbanism, aiming at sustainability in cities. Urban trees provide numerous ecosystem services to both inhabitants and visitors, thus local authorities should consider how to plan, manage, and promote urban green infrastructure as part of the tourism offer [63].

It is necessary that the teams of the management bodies are trained to manage these areas sustainably and improve the social perception of biodiversity and the importance of sustainable development. Different strategies should be performed to promote the sustainable use of trees in the cities, including land stewardship, and involving civil society in conservation based on environmental volunteering. Through environmental education actions, the population must be aware of the role of urban green infrastructure, especially the native species, and how citizens can contribute to the conservation of urban forestry in the Brazilian Amazonia.

Finally, city halls and city councils must make efforts to establish study committees that propose policies for the management of urban green areas in the cities under their jurisdiction. In addition, it is important that annual budgets provide for financial resources for the maintenance and restoration of urban green areas, or the implementation of new ones. The collaboration between the local administration and universities is of great importance for the main goals of sustainable, livable, and carbon neutral cities.

Author Contributions: Conceptualization, T.A.V. and T.P.; methodology, T.A.V. and T.P.; validation, T.A.V. and T.P.; formal analysis, T.A.V.; investigation, T.A.V.; writing—original draft preparation, T.A.V.; writing—review and editing, T.A.V. and T.P.; supervision, T.P.; project administration, T.P.; funding acquisition, T.P. All authors have read and agreed to the published version of the manuscript.

Funding: This research was funded by the Foundation for Science and Technology under grand UIDB/04020/2020.

Acknowledgments: This research has been developed during the sabbatical study at the Research Center for Tourism, Sustainability and Well-being, University of Algarve, supported by the Federal University of Western Pará; and by project "Improving life in a changing urban environment through Biophilic Design" PTDC/GES-URB/31928/2017.

Conflicts of Interest: The authors declare no conflict of interest.

Appendix A

Table A1. Results of multiple regression analyses for species richness, density, and diversity (H′) as dependent variables (Y) and population size, Gross Domestic Product (GDP), and Human Development Index (HDI) as independent variables (X_1, X_2, and X_3) under an urban forestry context in Brazilian Amazonia. SE = standard error.

	Richness				Density				Diversity (H′)			
	β	SE	t	p-level	β	SE	t	p-level	β	SE	t	p-level
Population	−0,000025	0,000032	−0,78367	0,4405	−0,00038	0,001137	−0,33563	0,7399	−0,00045	0,00091	−0,49046	0,6280
GDP/person	−0,00038	0,000504	−0,75383	0,4579	−0,01431	0,017917	−0,79883	0,4319	−0,01484	0,014335	−1,03502	0,3105
HDI	384,4758	166,1848	2,313543	0,0292	14887,52	5911,218	2,51852	0,0185	4070,547	4729,542	0,860664	0,3976
	F = 0,162712405741611				F = 0,0860500508959708				F = 0,680173534037709			
	R^2 = 0,182250658037764				R^2 = 0,227989944982515				R^2 = 0,0574862399773954			

Appendix B

Table A2. List of articles consulted on urban forestry in Brazilian Amazonia.

Article titles	Authors	Journal	Year	Link to Access
Inventário e diagnóstico da arborização urbana viária de Rio Branco, AC	Paiva, A.V., et al.	*Revista da Sociedade Brasileira de Arborização Urbana*	2010	https://revistas.ufpr.br/revsbau/article/view/66256
Análise da arborização urbana de três cidades da região norte do Estado de Mato Grosso	Almeida, D.N.; Rondon Neto, R.M.	*Acta Amazonica*	2010	http://www.scielo.br/pdf/aa/v40n4/v40n4a03
Análise da arborização urbana de duas cidades da região norte do estado de Mato Grosso	Almeida, D.N.; Rondon Neto, R.M.	*Revista Árvore*	2010	https://www.redalyc.org/pdf/488/48815860015.pdf
Diagnóstico da arborização urbana da cidade de Cacoal-RO	Almeida, J.R.; Barbosa, C.G.	*Revista da Sociedade Brasileira de Arborização Urbana*	2010	https://revistas.ufpr.br/revsbau/article/view/66239
Levantamento censitário da arborização urbana viária de Senador Guiomard, Acre	Maranho, A.S., et al.	*Revista da Sociedade Brasileira de Arborização Urbana*	2012	https://revistas.ufpr.br/revsbau/article/view/66532
The floristic composition of urban afforestation of the city Altamira, Pará State, Brazil.	Parry, M.M., et al.	*Revista da Sociedade Brasileira de Arborização Urbana*	2012	https: //revistas.ufpr.br/revsbau/article/download/66550/38357
Diagnóstico quali-quantitativo da arborização das praças do município de Altamira, Pará	Souza, O.P.S., et al.	*Enciclopédia Biosfera*	2013	http://www.conhecer.org.br/enciclop/2013b/ CIENCIAS%20AGRARIAS/diagnostico%20quali.pdf
A arborização pública e a eficiência do sombreamento da superfície urbana em bairros residenciais de Porto Velho, RO	Santos, J.A., et al.	*Revista da Sociedade Brasileira de Arborização Urbana*	2013	https://revistas.ufpr.br/revsbau/article/view/66440
Fitossociologia e diversidade de espécies arbóreas das praças centrais do município de Gurupi-TO	Santos, A.F.; José, A.C.; Sousa, P.A.	*Revista da Sociedade Brasileira de Arborização Urbana*	2013	https://revistas.ufpr.br/revsbau/article/view/66511
Diversidade em uma área verde urbana: avaliação qualitativa da arborização do campus da Universidade Federal do Acre, Brasil	Maranho, A.S.; Paula, S.R.P.	*Revista Agro@mbiente On-Line*	2014	https://revista.ufrr.br/agroambiente/article/view/1868
Quantitative survey of afforestation of squares in the city of Cáceres, Mato Grosso State, Brazil	Assunção, K.C., et al.	*Revista da Sociedade Brasileira de Arborização Urbana*	2014	https://revistas.ufpr.br/revsbau/article/view/66598
Levantamento quali-quantitativo de espécies arbóreas e arbustivas na arborização urbana do município de Paranaíta, Mato Grosso	Mamede, J.S.S., et al.	*Biodiversidade*	2014	http://periodicoscientificos.ufmt.br/ojs/index.php/ biodiversidade/article/view/1956/1452
Espécies empregadas na arborização urbana do bairro Santiago, Ji-Paraná/RO	Santos, J.A.; Costa, L.M.	*Revista da Sociedade Brasileira de Arborização Urbana*	2014	https://revistas.ufpr.br/revsbau/article/view/66595
Inventário quali-quantitativo da arborização viária de um trecho da rodovia PA-275 no município de Parauapebas-PA	Ferro, C.C.S., et al.	*Revista da Sociedade Brasileira de Arborização Urbana*	2015	https://revistas.ufpr.br/revsbau/article/view/63071

Table A2. *Cont.*

Article titles	Authors	Journal	Year	Link to Access
A geotecnologia como ferramenta para o diagnóstico da arborização urbana: o caso de Macapá, Amapá	Castro, H.S., et al.	*RA'EGA*	2016	https://revistas.ufpr.br/raega/article/view/42281
Análise quali-quantitativa da arborização de uma praça urbana do norte do Brasil	Gomes, E.M.C., et al.	*Nativa*	2016	http://periodicoscientificos.ufmt.br/ojs/index.php/nativa/article/view/3180/2642
Diagnóstico florístico da praça Floriano Peixoto NA cidade de Macapá, Amapá	Dantas, A.R., et al.	*Revista da Sociedade Brasileira de Arborização Urbana*	2016	https://revistas.ufpr.br/revsbau/article/view/63494
Análise da composição florística de Boa Vista-RR: subsídio para a gestão da arborização de ruas	Lima Neto, E.M., et al.	*Revista da Sociedade Brasileira de Arborização Urbana*	2016	https://revistas.ufpr.br/revsbau/article/view/63390
Inventário da arborização urbana das principais avenidas do município de Rorainópolis, Roraima	Veloso, J.N.	*Boletim do Museu Integrado de Roraima*	2016	https://www.uerr.edu.br/bolmirr/wp-content/uploads/2016/11/BOLMIRR-v102-Veloso.pdf
Composição de espécies e índices arbóreos nos pátios de três escolas de Gurupi-Tocantins	Batista, E.M.C., et al.	*Revista de estudos ambientais (Online)*	2016	https://proxy.furb.br/ojs/index.php/rea/article/view/5796/3593
Diagnóstico da arborização do parque urbano Tucumã, em Rio Branco-AC	Santos, L.R., et al.	*Revista da Sociedade Brasileira de Arborização Urbana*	2017	https://revistas.ufpr.br/revsbau/article/view/63529
Aspectos dendrométricos e qualitativos de *Licania tomentosa* (Benth.) Fritsch na arborização urbana de Itacoatiara, Amazonas	Gomes, I.B., et al.	*Revista IGAPO*	2017	http://200.129.168.183/ojs/index.php/igapo/article/view/559/473
Observational and experimental evaluation of hemiparasite resistance in trees in the urban afforestation of Santarém, Pará, Brazil	Silva, F.P.; Fadini, R.F.	*Acta Amazonica*	2017	http://www.scielo.br/scielo.php?script=sci_arttext&pid=S0044-59672017000400311&lng=en&tlng=en
Diversidade florística e índices arbóreos de escolas no município de Formoso do Araguaia, Tocantins	Santos, A.F., et al.	*Revista Verde de Agroecologia e Desenvolvimento Sustentável*	2017	https://www.gvaa.com.br/revista/index.php/RVADS/article/view/4353
Diagnóstico da arborização nas calçadas de Gurupi, TO	Oliveira, L.M., et al.	*Revista da Sociedade Brasileira de Arborização Urbana*	2017	https://revistas.ufpr.br/revsbau/article/view/63515/pdf
Estudo quali-quantitativo e percepção ambiental da arborização do Setor Jardim Sevilha, Gurupi - TO	Wanderley, R.J.C., et al.	*Revista da Sociedade Brasileira de Arborização Urbana*	2017	https://revistas.ufpr.br/revsbau/article/view/63579
Inventário e análise da arborização nas calçadas da região central de Gurupi-TO	Rabêlo, D., et al.	*Revista da Sociedade Brasileira de Arborização Urbana*	2017	https://revistas.ufpr.br/revsbau/article/view/63606
Análise dos principais conflitos e espécies inadequadas presentes na arborização viária na região central do município de Imperatriz (MA)	Silva, R.V., et al.	*Revista da Sociedade Brasileira de Arborização Urbana*	2018	https://revistas.ufpr.br/revsbau/article/view/63656

Table A2. *Cont.*

Article titles	Authors	Journal	Year	Link to Access
Avaliação qualitativa da arborização com *Mangifera indica* nas ruas de Belém – PA	Silva, D.A., et al.	*Acta Biológica Catarinense*	2018	http://186.237.248.25/index.php/ABC/article/view/432
Avaliação da organização árborea e a percepção dos usuários das praças do município de Mocajuba, Estado do Pará, Brasil	Barros, V.S., et al.	*Revista da Sociedade Brasileira de Arborização Urbana*	2018	https://revistas.ufpr.br/revsbau/article/view/63664
Diagnóstico visual e fitossociologia na arborização de praças em Paragominas, Pará	Silva, I.R., et al.	*Revista da Sociedade Brasileira de Arborização Urbana*	2018	https://revistas.ufpr.br/revsbau/article/view/63567
Use of morphometry in the arborization of Paragominas city, Pará, Brazil, with *Handroanthus impetiginosus* (Mart. ex DC.) Mattos (Bignoniaceae)	Oliveira, V.P., et al.	*Revista Agro@mbiente On-Line*	2018	https://revista.ufrr.br/agroambiente/article/view/4975
Arborização urbana com nim indiano na cidade de Santarém, Pará, Brasil	Dantas, R.C.O., et al.	*Revista da Sociedade Brasileira de Arborização Urbana*	2018	https://revistas.ufpr.br/revsbau/article/view/63643
Diagnóstico da arborização urbana da cidade de Vitória do Xingu, Pará, Brasil	Silva, L.A., et al.	*Revista da Sociedade Brasileira de Arborização Urbana*	2018	https://revistas.ufpr.br/revsbau/article/view/63622/pdf
Caracterização quantitativa da arborização urbana no município de Itapuã Do Oeste/RO-Brasil	Rocha, C.L.D., et al.	*Revista Saber Científico*	2018	http://revista.saolucas.edu.br/index.php/resc/article/view/764
Colonização por cupins (Isoptera) da arborização urbana e implicações socioambientais em Porto Velho, Rondônia	Santos Junior, A.; Sampaio, F.T.S.	*Boletim do Observatório Ambiental Alberto Ribeiro Lamego*	2018	http://www.essentiaeditora.iff.edu.br/index.php/boletim/article/view/10024/9140
Impacto da implantação do BRT na arborização da região central de Palmas, Tocantins	Pinheiro, R.T., et al.	*Desenvolvimento e Meio ambiente*	2018	https://revistas.ufpr.br/made/article/view/55981
Levantamento quali-quantitativo da arborização urbana no município de Buriticupu, MA	Sousa, L.A., et al.	*Revista da Sociedade Brasileira de Arborização Urbana*	2019	https://revistas.ufpr.br/revsbau/article/view/65372
Silvicultura urbana: levantamento e caracterização da arborização em uma área central na cidade de Cáceres-MT	Santos, C.T.F., et al.	*Scientia Tec*	2019	https://periodicos.ifrs.edu.br/index.php/ScientiaTec/article/view/3194
Composição florística e fitossanidade das praças Barão e Liberdade, Santarém, Pará	Pires, O.V., et al.	*Revista Ibero-Americana de Ciências Ambientais*	2019	http://www.sustenere.co/index.php/rica/article/view/3121
Espécies arbóreas tóxicas presentes na arborização urbana do município de Santarém, Pará	Baumann, S.S.R.T., et al.	*Revista Ibero-Americana de Ciências Ambientais*	2019	http://sustenere.co/index.php/rica/issue/view/155
Espécies frutíferas na arborização urbana do município de Santarém, Pará	Rabelo, L.K.L., et al.	*Revista Ibero-Americana de Ciências Ambientais*	2019	http://sustenere.co/index.php/rica/article/view/3097
Arborização das praças de Gurupi – TO – Brasil: composição e diversidade de espécies	Silva, A.D.P., et al.	*Revista da Sociedade Brasileira de Arborização Urbana*	2019	https://revistas.ufpr.br/revsbau/article/view/67547

References

1. Trindade Júnior, S.C.C. Cidades na floresta: Os "grandes objetos" como expressões do meio técnico-científico informacional no espaço amazônico. *Rev. Inst. Estud. Bras.* **2010**, *50*, 113–138. [CrossRef]
2. Salbitano, F.; Borelli, S.; Conigliaro, M.; Chen, Y. *Guidelines on Urban and Peri-Urban Forestry*; FAO: Rome, Italy, 2016.
3. Padoch, C.; Brondizio, E.; Costa, S.; Pinedo-Vasquez, M.; Sears, R.R.; Siqueira, A. Urban forest and rural cities: Multi-sited households, consumption patterns, and forest resources in Amazonia. *Ecol. Soc.* **2008**, *13*, 2. [CrossRef]
4. Narducci, J.; Quintas-Soriano, C.; Castro, A.; Som-Castellano, R.; Brandt, J.S. Implications of urban growth and farmland loss for ecosystem services in the western United States. *Land Use Policy* **2019**, *86*, 1–11. [CrossRef]
5. Figueiredo, S.L.; Ravena, N. Cidades, fronteiras e diversidades na Amazônia. *R. B. Estud. Urbanos E Reg.* **2014**, *16*, 241–246. [CrossRef]
6. O'Brien, L.; De Vreese, R.; Atmis, E.; Olafsson, A.S.; Sievänen, T.; Brennan, M.; Sánchez, S.; Panagopoulos, T.; de Vries, S.; Kern, M.; et al. Social and Environmental Justice: Diversity in Access to and Benefits from Urban Green Infrastructure—Examples from Europe. In *The Urban Forest*; Springer: Cham, Switzerland, 2017; pp. 153–190. [CrossRef]
7. Morgenroth, J.; Östberg, J.; van den Bosch, C.K.; Nielsen, A.B.; Hauer, R.; Sjöman, H.; Chen, W.; Jansson, M. Urban tree diversity—Taking stock and looking ahead. *Urban For. Urban Green.* **2016**, *15*, 1–5. [CrossRef]
8. Nero, B.F.; Kwapong, N.A.; Jatta, R.; Fatunbi, O. Tree species diversity and socioeconomic perspectives of the urban (food) forest of Accra, Ghana. *Sustainability* **2018**, *10*, 3417. [CrossRef]
9. Nilsson, K.; Randrup, T.B.; Wandall, B.M. Trees in the Urban Environment. In *The Forests Handbook*; Evans, J., Ed.; Wiley-Blackwell: Hoboken, NJ, USA, 2008. [CrossRef]
10. Terkenli, S.T.; Zivojinovic, I.; Tomićević-Dubljević, J.; Panagopoulos, T.; Straupe, I.; Toskovic, O.; Kristianova, K.; Straigyte, L.; O'Brien, L.; Bell, S. Recreational Use of Urban Green Infrastructure: The Tourist's Perspective. In *The Urban Forest*; Pearlmutter, D., Calfapietra, C., Samson, R., O'Brien, L., Krajter Ostoić, S., Sanesi, G., Alonso del Amo, R., Eds.; Springer: Cham, Switzerland, 2017; pp. 191–216. [CrossRef]
11. Cardoso, S.L.C.; Figueiredo, S.L. Arquitetura Ecológica: Modelos paisagísticos, requalificação e refuncionalização de espaços públicos verdes urbanos. *Paisagens Híbridas* **2018**, *2*, 34–53.
12. Rowntree, R.A.; Nowak, D.J. Quantifying the role of urban forests in removing atmospheric carbon dioxide. *J. Arboric.* **1991**, *17*, 269–275.
13. FAO—Food and Agriculture Organization of the United Nations. Benefits of Urban Trees. Available online: http://www.fao.org/3/a-c0024e.pdf (accessed on 10 February 2020).
14. Yan, P.; Yang, Y. Species diversity of urban forests in China. *Urban For. Urban Green.* **2017**, *28*, 160–166. [CrossRef]
15. Thomsen, P.; Bühler, O.; Kristoffersen, P. Diversity of street tree populations in larger Danish municipalities. *Urban For. Urban Green.* **2016**, *15*, 200–210. [CrossRef]
16. Staudhammer, C.L.; Escobedo, F.J.; Blood, A. Assessing methods for comparing species diversity from disparate data sources: The case of urban and peri-urban forests. *Ecosphere* **2018**, *9*, e02450. [CrossRef]
17. Pereira, H.S.; Kudo, S.A.; Silva, S.C.P. Topophilia and environmental valuation of urban forest fragments in an Amazonian city. *Ambiente Soc.* **2018**, *21*, e01590. [CrossRef]
18. IBGE—Instituto Brasileiro de Geografia e Estatísticas. *População Estimada*; Ibge/DPE/Copis: Rio de Janeiro, Brazil, 2019.
19. IBGE—Instituto Brasileiro de Geografia e Estatísticas. *Área da Unidade Territorial: Área terriTorial Brasileira*; Ibge/DPE/Copis: Rio de Janeiro, Brazil, 2019.
20. IBGE—Instituto Brasileiro de Geografia e Estatísticas. *PIB per Capita—2017*; Ibge: Rio de Janeiro, Brazil, 2017.
21. IBGE—Instituto Brasileiro de Geografia e Estatísticas. *Arborização de Vias Públicas*; Ibge: Rio de Janeiro, Brazil, 2010.
22. IBGE—Instituto Brasileiro de Geografia e Estatísticas. *Urbanização de Vias Públicas*; Ibge: Rio de Janeiro, Brazil, 2010.

23. PNUD—Programa das Nações Unidas para o Desenvolvimento. Ranking IDHM Municípios 2010—Brasil. Available online: https://www.br.undp.org/content/brazil/pt/home/idh0/rankings/idhm-municipios-2010.html (accessed on 20 December 2019).

24. Denatran—Departamento Nacional de Trânsito. *Frota de veículos—2018*; Departamento Nacional de Trânsito: Denatran, Brasília, 2018.

25. Inmet—Instituto Nacional de Meteorologia. *Estação Meteorológica de Observação de Superfície Automática*; Inmet: Madrid, Brasília, 2019.

26. Magurran, A.E. *Ecological Diversity and its Measurement*; British Library: Cambridge, UK, 1988; 179p.

27. Almeida, D.N.; Rondon Neto, R.M. Análise da arborização urbana de três cidades da região norte do Estado de Mato Grosso. *Acta Amaz.* **2010**, *40*, 647–656. [CrossRef]

28. Parry, M.M.; Silva, M.M.; Sena, I.S.; Oliveira, F.P.M. The floristic composition of urban afforestation of the city Altamira, Pará State, Brazil. *Revsbau* **2012**, *7*, 143–158. [CrossRef]

29. R Core Team. *R. A language and environment for statistical computing*; R Foundation for Statistical Computing: Vienna, Austria, 2018; Available online: http://www.R-project.org/ (accessed on 7 June 2019).

30. Congresso Nacional. *Constituição da República Federativa do Brasil*; Congresso Nacional: Brasília, Brazil, 1988.

31. Asnake, M. The importance of scientific publication in the development of public health. *Ciênc. saúde coletiva.* **2015**, *20*, 1972–1973. [CrossRef]

32. Liu, C.; Li, X. Carbon storage and sequestration by urban forests in Shenyang, China. *Urban For. Urban Green.* **2012**, *11*, 121–128. [CrossRef]

33. Dantas, R.C.O.; Bezerra, T.G.; Vieira, T.A. Arborização urbana com nim indiano na cidade de Santarém, Pará, Brasil. *Revsbau* **2018**, *13*, 37–46. [CrossRef]

34. Almeida, D.N.; Rondon Neto, R.M. Análise da arborização urbana de duas cidades da região norte do estado de Mato Grosso. *Revista Árvore* **2010**, *34*, 899–906. [CrossRef]

35. Lima de Medeiros, J.; Almeida, T.S.; Neto, J.J.L.L.; Almeida Filho, L.C.P.; Ribeiro, P.R.V.; Brito, E.S.; Morgano, M.A.; Gomes da Silva, M.; Farias, D.F.; Carvalho, A.F.U. Chemical composition, nutritional properties, and antioxidant activity of Licania tomentosa (Benth.) fruit. *Food Chem.* **2020**, *313*, 126117. [CrossRef]

36. Assunção, K.C.; Luz, P.B.; Neves, L.G.; Sobrinho, S.P. Quantitative survey of afforestation of squares in the city of Cáceres, Mato Grosso State, Brazil. *Revsbau* **2014**, *9*, 115–124. [CrossRef]

37. Araújo, A.C.B.; Gracioli, C.R.; Grimm, E.L.; Longhi, S.J. The assessment of floristic composition, size, health and current tree conservation of the International Park in Sant'ana do Livramento Brazil and Rivera Uruguay. *Revsbau* **2012**, *7*, 66–74. [CrossRef]

38. Santamour, F.S., Jr. Trees for Urban Planting: Diversity, Uniformity, and Common Sense. In Proceedings of the 7th Conf. Metropolitan Tree Improvement Alliance (METRIA), Lisle, IL, USA, 11–12 June 1990; pp. 57–65.

39. Porto, L.P.M.; Brasil, H.M.S. *Manual de Orientação Técnica da Arborização Urbana de Belém: Guia para Planejamento, Implantação e Manutenção da Arborização em Logradouros Públicos*; UFRA: Belém, Brasil, 2013.

40. Gomes, E.M.C.; Rodrigues, D.M.S.; Santos, J.T.; Barbosa, E.J. Análise quali-quantitativa da arborização de uma praça urbana do Norte do Brasil. *Nativa* **2016**, *4*, 179–186. [CrossRef]

41. Silva, L.A.; Sousa, C.S.; Parry, M.M.; Herrera, R.C.; Oliveira, F.P.M.; Parry, S.M. Diagnóstico da arborização urbana da cidade de Vitória do Xingu, Pará, Brasil. *Revsbau* **2018**, *13*, 57–72. [CrossRef]

42. Dantas, A.R.; Gomes, E.M.C.; Pinheiro, A.P. Diagnóstico florístico da Praça Floriano Peixoto na cidade de Macapá, Amapá. *Revsbau* **2016**, *11*, 32–46. [CrossRef]

43. Moro, M.F.; Westerkamp, C.; Araújo, F.S. How much importance is given to native plants in cities' treescape? A case study in Fortaleza, Brazil. *Urban For. Urban Green.* **2014**, *13*, 365–374. [CrossRef]

44. Sousa, L.A.; Cajaiba, R.L.; Martins, J.S.C.; Colácio, D.S.; Sousa, E.S.; Pereira, K.S. Levantamento quali-quantitativo da arborização urbana no município de Buriticupu, MA. *Revsbau* **2019**, *14*, 42–52. [CrossRef]

45. Moro, M.F.; Westerkamp, C. The alien street trees of Fortaleza (NE Brazil): Qualitative observations and the inventory of two districts. *Ciência Florest.* **2011**, *21*, 789–798. [CrossRef]

46. de Paiva, A.V.; Lima, A.B.M.; Carvalho, A.; Junior, A.M.; Gomes, A.; Melo, C.S.; Farias, C.O.; Reis, C.; Bezerra, C.; Junior, E.A.; et al. Inventário e diagnóstico da arborização urbana viária de Rio Branco, AC. *Revsbau* **2010**, *5*, 144–159. [CrossRef]

47. Blum, C.T.; Borgo, M.; Sampaio, A.C.F. Espécies exóticas invasoras na arborização de vias públicas de Maringá-PR. *Revsbau* **2008**, *3*, 78–97. [CrossRef]

48. Maranho, A.S.; de Paula, S.R.P.; Lima, E.; Paiva, A.V.; Alves, A.P.; Nascimento, D.O. Levantamento censitário da arborização urbana viária de Senador Guiomard, Acre. *Revsbau* **2012**, *7*, 44–56. [CrossRef]
49. Gomes, I.B.; Pinto, L.A.A. Aspectos dendrométricos e qualitativos de Licania tomentosa (Benth.) Fritsch na arborização urbana de Itacoatiara, Amazonas. *Igapó* **2017**, *11*, 35–46.
50. Castro, H.S.; Dias, T.C.A.C.; Amanajás, V.V.V. A geotecnologia como ferramenta para o diagnóstico da arborização urbana: O caso de Macapá, Amapá. *Raega—O Espaço Geográfico Em Análise* **2016**, *38*, 146–168. [CrossRef]
51. Ferro, C.C.S.; Oliveira, R.S.; Andrade, F.W.C.; Souza, S.M.A.R. Inventário quali-quantitativo da arborização viária de um trecho da rodovia PA-275 no município de Parauapebas-PA. *Revsbau* **2015**, *10*, 73–84. [CrossRef]
52. Santos, L.C.; Santos, E.A.; Pinheiro, R.M.; Ferreira, E.J.L. Diagnóstico da arborização do parque urbano tucumã, em Rio Branco-AC. *Revsbau* **2017**, *12*, 103–117. [CrossRef]
53. Pires, O.V.; Araújo, N.M.; Silva, J.R.P.; Bonfim, M.C.S.; Sousa, S.F.; Maestri, M.P. Composição florística e fitossanidade das praças Barão e Liberdade, Santarém, Pará. *Rev. Ibero Am. De Ciências Ambient.* **2019**, *10*, 228–237. [CrossRef]
54. Silva, D.A.; Batista, D.B.; Batista, A.C. Avaliação qualitativa da arborização com Mangifera indica nas ruas de Belém—PA. *Acta Biológica Catarin.* **2018**, *5*, 34–45. [CrossRef]
55. Castillo Rodríguez, L.C.; Pastrana Falcón, J.C. Diagnóstico del arbolado viario de El Vedado: Composición, distribución y conflictos. *Arquit. Y Urban.* **2015**, *36*, 92–118.
56. Silva, O.H.; Locastro, J.K.; Sanches, S.P.; Angelis Neto, G.; Angelis, B.L.D.; Caxambú, M.G. Avaliação da arborização viária da cidade de São Tomé, Paraná. *Ciência Florest.* **2019**, *29*, 371–384. [CrossRef]
57. Wilson, J.; Tyedmers, P.; Pelot, R. Contrasting and comparing sustainable development indicator metrics. *Ecol. Indic.* **2007**, *7*, 299–314. [CrossRef]
58. Du Toit, M.J.; Cilliers, S.S.; Dallimer, M.; Goddard, M.; Guenat, S.; Cornelius, S.F. Urban green infrastructure and ecosystem services in sub-Saharan Africa. *Landsc. Urban Plan.* **2018**, *18*, 249–261. [CrossRef]
59. Akerlund, U.; Knuth, L.; Randrup, T.B.; Schipperijn, J. Urban and Peri-Urban Forestry and Greening in West and Central Asia—Experiences, Constraints and Prospects. LSP Working Paper 36 Access to Natural Resources Sub-Programme. 2006. Available online: http://www.fao.org/3/a-ah238e.pdf (accessed on 20 December 2019).
60. De Sousa Silva, C.; Viegas, I.; Panagopoulos, T.; Bell, S. Environmental justice in accessibility to green infrastructure in two European cities. *Land* **2018**, *7*, 134. [CrossRef]
61. Dobbs, C.; Nitschke, C.R.; Kendal, D. Global drivers and tradeoffs of three urban vegetation ecosystem services. *PLoS ONE* **2014**, *9*, e113000. [CrossRef] [PubMed]
62. Hope, D.; Gries, C.; Zhu, W.; Fagan, W.F.; Redman, C.L.; Grimm, N.B.; Nelson, A.L.; Martin, C.; Kinzig, A. Socioeconomics drive urban plant diversity. *Proc. Natl. Acad. Sci. USA* **2003**, *22*, 8788–8792. [CrossRef] [PubMed]
63. Terkenli, T.; Bell, S.; Toskovic, O.; Tomicevic, J.D.; Panagopoulos, T.; Straupe, I.; Kristianova, K.; Staigyte, L.; 'O'Brien, L.; Zivozjinovic, I. Tourist perceptions and uses of urban green infrastructure: An exploratory cross-cultural investigation. *Urban For. Urban Green.* **2020**, *49*, 126624. [CrossRef]

Article

Model-Based Selection of Cost-Effective Low Impact Development Strategies to Control Water Balance

Johannes Leimgruber *, Gerald Krebs, David Camhy and Dirk Muschalla

Graz University of Technology, Institute of Urban Water Management and Landscape Water Engineering, Stremayrgasse 10/I, 8010 Graz, Austria; gerald.krebs@tugraz.at (G.K.); camhy@tugraz.at (D.C.); d.muschalla@tugraz.at (D.M.)
* Correspondence: leimgruber@tugraz.at

Received: 4 April 2019; Accepted: 18 April 2019; Published: 25 April 2019

Abstract: Urbanization induces an increase of runoff volume and decrease of evapotranspiration and groundwater recharge. Low impact development (LID) strategies aim to mitigate these adverse impacts. Hydrologic simulation is a reasonable option to assess the LID performance with respect to the water balance and is applicable to planning purposes. Current LID design approaches are based on design storm events and focus on the runoff volume and peak, neglecting evapotranspiration and groundwater recharge. This contribution presents a model-based design approach for the selection of cost-effective LID strategies. The method is based on monitored precipitation time series and considers the complete water balance and life-cycle-costs, as well as the demand for land. The efficiency of LID strategies (E_{LID}) is introduced as an evaluation measure which also accounts for emphasizing different goals. The results show that there exist several pareto-optimal LID strategies providing a reasonable basis for decision-making. Additionally, the application of LID treatment trains emerges as an option of high potential.

Keywords: life cycle costs; stormwater management; storm water management model

---

1. Introduction

The increase of impervious land cover caused by urbanization considerably affects the water balance [1]. While the runoff increases, the evapotranspiration and groundwater recharge decreases. This results in several negative impacts, like higher runoff peak rates, larger runoff volumes, higher potential of flooding events, urban heat islands, etc. [2–4].

Low impact development (LID) strategies are a widely known and implemented concept in stormwater management. They aim to replicate hydrologic characteristics of natural catchments, thus mitigating the adverse impacts of urbanization [5,6]. LID strategies are applied to maintain or restore the pre-development hydrologic regime [5,7]. In order to evaluate the LID performance with respect to this purpose, the pre- and post-development hydrologic conditions of a catchment are analyzed. Hydrologic simulations are a reasonable and common option for such assessments. Several modeling tools allow for the simulation of hydrologic processes of LIDs (compare overview of Jayasooriya and Ng [8]). The US EPA Storm Water Management Model (SWMM) [9] was selected for this study as it is currently one of the most sophisticated tools for the hydrologic simulation of LIDs [8].

The design of LIDs (particularly infiltration systems), e.g., infiltration swales or infiltration trenches, implies the calculation of the required retention volume. Basically, it is the difference between the stormwater volume collected by the LID and the stormwater volume that infiltrates through the LID into the soil underneath. Planning guidelines or design manuals often propose to use design storm events of a certain duration and return period in order to determine the required LID retention volume (e.g., [10–12], for an overview of international approaches compare Ballard et al. [13]). Such approaches do not consider the actual storm characteristics, e.g., the time-variant intensity, affecting the performance of LIDs. In addition, conditions at the start of a storm event, e.g., soil moisture or storage capacity due to antecedent storm events and dry periods, are not taken into account. This can result in a divergent assessment of LID performance compared to monitoring in the field. Thus, long-term and continuous simulations have to be used, even if dealing with single storm event evaluations (compare [14]).

Planning guidelines (e.g., [10,11]), evaluation approaches (e.g., [15]), as well as previous studies dealing with LID performance (compare reviews by Ahiablame et al. [6] and Eckart et al. [7]) or LID effectiveness (e.g., [16]) focus on the runoff and neglect the groundwater recharge and evapotranspiration, although they control groundwater levels and the micro-climate by means of cooling and prevent urban heat island effects [4,17]. Therefore, in terms of an environmentally sustainable and reasonable application of LIDs, all components of the water balance have to be considered. Consequently, holistic approaches (e.g., [18,19]) are a suitable basis for planning purposes. Furthermore, the assessment of LID performance is conducted on site scale, as suggested by Burns et al. [20], in order to restore/protect natural hydrologic processes at small scales. That is reasonable considering micro-climate issues and the restoration of natural flow regimes at larger scales downstream.

Various LID design approaches aim to design a certain LID strategy but do not provide recommendations for the selection of the proper LID strategy. Furthermore, little attention is paid to the possibility of combining LIDs to LID treatment trains, which can be well-performing LID strategies as well (e.g., [21–24]). Of course, the selection of LID strategies is also influenced by the cost-effectiveness (e.g., [25–28]), considering the life cycle costs. Several cost-estimating tools for LIDs have been developed (e.g., [29,30]).

Although individual approaches considering the water balance, cost-effectiveness (life cycle costs), or LID treatment trains exist, recommendations for a combined and holistic assessment are not available. This paper presents an approach for selecting suitable LID strategies considering a combined evaluation of the complete water balance (runoff volume, evapotranspiration, groundwater recharge) and the cost-effectiveness for both stand-alone LIDs and treatment trains.

2. Materials and Methods

2.1. Case Studies and Data

The study was conducted using three case studies that represent characteristic urban areas: Two residential areas and one commercial area (Figure 2). The commercial area is 100% impervious and covers 16000 m^2, including a roofage of 6000 m^2. Both residential areas cover 1100 m^2 each. They differ in the degree of development (dod), which is the proportion between built-up area and building site area. The first residential area (low-developed) has a roofage of 200 m^2 (dod = 0.18) whereas the second residential area (high-developed) has a roofage of 600 m^2 (dod = 0.55). Both residential areas comprise a driveway of 40 m^2, while the remaining plot is covered by lawn. All roofs are tiled in the initial state.

The precipitation series used for the long-term simulations was obtained in Graz/Austria, has a length of 10 years (1996–2006), and an average annual precipitation depth of 783 mm. It was provided by the Austrian Water and Waste Management Association (OEWAV) [31]. Daily minimum–maximum temperatures for the computation of evaporation rates, using the Hargreaves method [32], were provided by the Central Institute for Meteorology and Geodynamics (ZAMG) [33].

2.2. Investigated LID strategies and Model Development

The US EPA Storm Water Management Model (SWMM) [9], which was used in this study, is a dynamic rainfall-runoff model. It can be used for a single event or a continuous long-term simulation and simulates hydrologic processes on the surface as well as routing of runoff in the sewer system. SWMM accounts for a variety of hydrologic processes, like time-varying precipitation, interception of depression storage, evaporation of surface water, evapotranspiration out of the soil and/or LIDs, infiltration of stormwater into the soil, and percolation of infiltrated water into groundwater.

The following LIDs, which are frequently implemented in stormwater management projects, were selected for this study: Green roof, infiltration trench, bio-retention cell, infiltration swale, and blind drain.

Green roofs consist of an engineered and (partially) vegetated soil mixture above a drainage mat that serves as stormwater conveying layer. Infiltration swales are depressions that retain and infiltrate stormwater, whereas infiltration trenches are ditches filled with gravel, providing retention volume for stormwater to infiltrate into the native soil below. Bio-retention cells are a combination of infiltration swale and infiltration trench. They provide retention volume through a surface depression as well as an engineered and (partially) vegetated soil mixture and an underlying gravel storage bed. Blind drains are underground infiltration bodies filled with gravel or other filling material.

The mentioned LIDs, except for the blind drain, were simulated with a soil moisture model comprising different layers, e.g., surface, soil, and storage, which is implemented in SWMM (see Figure 1a–d). The layers simulate the different hydrologic functions of the LID. The surface layer accounts for the runoff generation and allows for infiltration into the soil or storage layer. Optionally, a retention volume on the surface can be defined. The soil and storage layer provide retention volume as well and permit infiltration into the native soil. The drainage mat conveys percolated stormwater off the roof. The LID parameters were chosen in agreement with literature parameter ranges (e.g., [34–36]). In order to facilitate a comparison, all LIDs were similarly parameterized to provide comparable retention capacities and hydrologic behavior (Table 1). LIDs collect direct rainfall as well as runoff from other catchments. The runoff from LIDs was directed to the sewer system or to another LID catchment (LID treatment train). For additional information about the LID simulation in SWMM, the reader is referred to Rossman et al. [34]. The blind drain was simulated with a storage node that allows for infiltration to the native soil and prevents evapotranspiration while simulating the surface above the blind drain as well (Figure 1e). The blind drain storage depth was defined to 30cm.

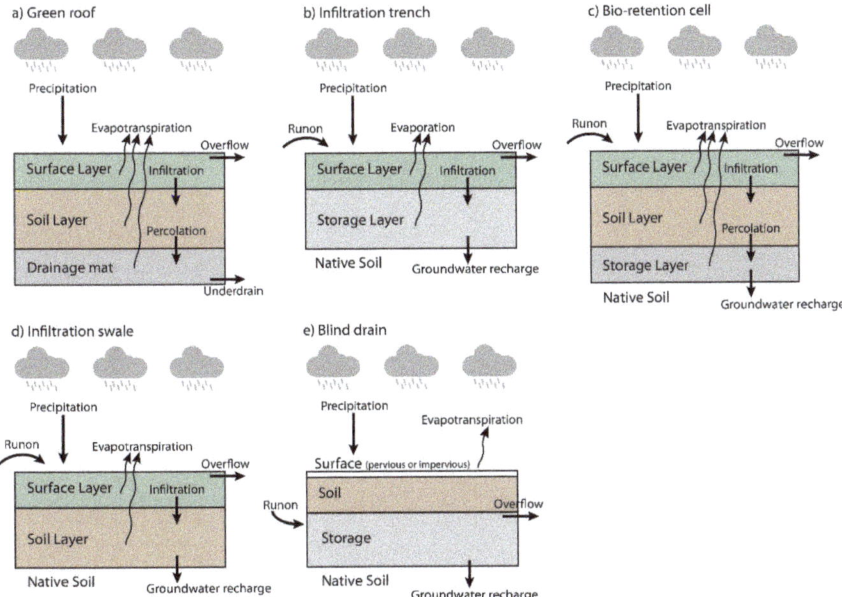

Figure 1. Scheme and simulated processes of the investigated low impact development (LID) strategies: (**a**) Green roof; (**b**) infiltration trench; (**c**) bio-retention cell; (**d**) infiltration swale; (**e**) blind drain.

Table 1. Parameters of investigated LIDs.

Green roof			Infiltration trench		
Parameter		Unit	Parameter		Unit
Berm height	10	mm	Berm height	300	mm
Vegetation volume	0.2	%	Vegetation volume	0.0	
Surface roughness	0.1	$s/m^{1/3}$	Surface roughness	0.02	$s/m^{1/3}$
Surface slope	1.0	%	Surface slope	1.0	%
Soil thickness	100	mm	Storage thickness	1000	mm
Porosity	0.55	-	Storage void ratio	0.3	-
Field capacity	0.4	-	Storage seepage rate	10	mm/h
Wilting point	0.1	-			
Conductivity	50	mm/h			
Conductivity slope	30	-			
Suction head	65	mm			
Drainage mat thickness	30	mm			
Drainage mat void fraction	0.4	-			
Drainage mat roughness	0.02	$s/m^{1/3}$			
Bio-retention cell			**Infiltration swale**		
Parameter		Unit	Parameter		Unit
Berm height	300	mm	Berm height	300	mm
Vegetation volume	0.1	fraction	Vegetation volume	0.1	fraction
Surface roughness	0.16	$s/m^{1/3}$	Surface roughness	0.16	$s/m^{1/3}$
Surface slope	1	%	Surface slope	1.0	%
Soil thickness	300	mm	Soil thickness	300	mm
Porosity	0.5	-	Porosity	0.5	-
Field capacity	0.2	-	Field capacity	0.2	-
Wilting point	0.1	-	Wilting point	0.1	
Conductivity	120	mm/h	Conductivity	120	mm/h
Conductivity slope	40	-	Conductivity slope	40	-
Suction head	50	mm	Suction head	50	mm
Storage thickness	100	mm			
Storage void fraction	0.3	-			
Storage seepage rate	10	mm/h			

A potential total LID area (A_{pot}) was assigned to the three areas according to the space available (Figure 2): 2500 m^2 for the commercial area, 60 m^2 for the low-developed residential area, and 120 m^2 for the high-developed residential area. The maximal extent of the underground blind drain was selected accordingly. Each A_{pot} was divided into 50 sections that consequently had a dimension of 50 m^2, 1.2 m^2, and 2.4 m^2 per section, respectively. Each section could be occupied by an LID type or left in the initial state. The sections were incrementally used for the application of a LID type (e.g., infiltration swale applied to 1, 2, 3 … 50 sections) and a simulation was conducted for every state. In addition, two different LID types were applied to the sections, directing the runoff from the first LID to the second LID. Thus, different LID treatment trains were simulated (Figure 3). Again, the application of LIDs to the sections was executed incrementally.

$$A_{LID1} = \frac{A_{pot}}{50} \cdot n_{LID1}$$
$$A_{LID2} = \frac{A_{pot}}{50} \cdot n_{LID2}$$
$$\max(n_{LID1} + n_{LID2}) = 50 \tag{1}$$
$$\text{for single LID strategies}: n_{LID2} = 0,$$

where A_{LID1} is the area of LID1 in m^2, A_{LID2} is the area of LID2 in m^2, A_{pot} is the potential LID area for the respective case study, n_{LID1} is the number of sections occupied by LID1, and n_{LID2} is the number of sections occupied by LID2.

With respect to the roof, the green roof system was not applied incrementally. Only the two options "tiled roof" and "green roof" covering the complete roofage were simulated (Figure 3).

The potential total LID area could theoretically be divided into an infinite number of sections in order to get continuous results, but this would result in high computational effort. Therefore, the discrete results for the water balance using the grid of 50 sections were used as supporting points for a linear interpolation.

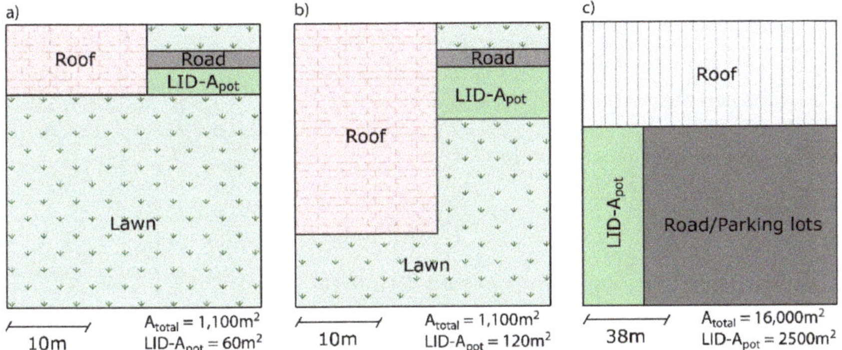

Figure 2. Schematic setting of the investigated case studies: (**a**) Low-developed residential area; (**b**) high-developed residential area; (**c**) commercial area. A$_{total}$ is the total area of the case study and LID-A$_{pot}$ is the total potential LID area.

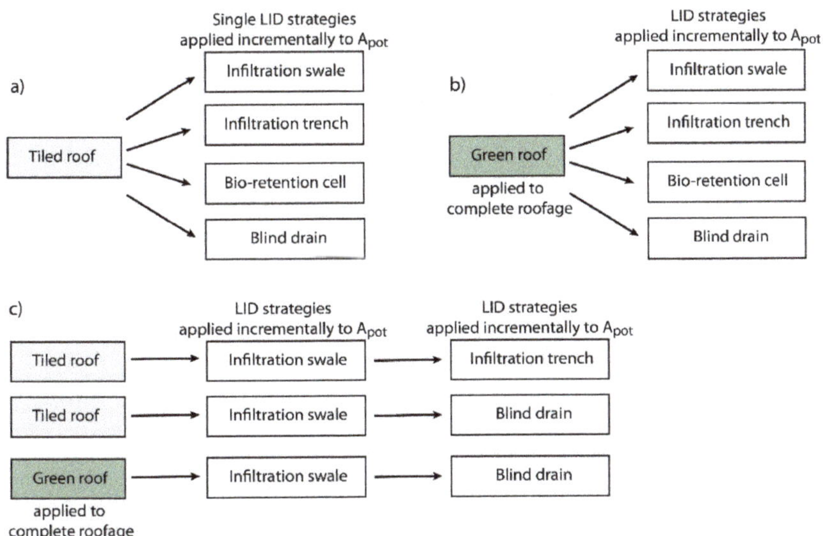

Figure 3. Investigated LID strategies. (**a**) Single LID strategies; (**b**) LID treatment trains with green roof; (**c**) two-part and three-part LID treatment trains.

2.3. Relations between Water Balance, Life Cycle Costs, and Demand for Land

All three areas (Figure 2) were simulated for the investigated LID strategies. Based on the SWMM simulation results, the water balance can be computed:

$$\Delta S = P - ET - R - GR, \tag{2}$$

where P is the precipitation (mm), R is the runoff volume (mm), ET is the evapotranspiration (mm), GR is the groundwater recharge (mm), and ΔS is the change in system storage (mm).

The water balance components can also be expressed as fraction of the precipitation:

$$1 = \frac{\Delta S + ET + R + GR}{P} \tag{3}$$

The life cycle costs, including construction and maintenance costs, were calculated for every LID strategy based on the size (number of sections) of each LID and following a dynamic cost comparison calculation [37]. The interest rate was assumed to be 3% and the intended life of LID practice based on routine maintenance was assumed to be 30 years [38]. According to the investigated references (see Table 2), 5% of the construction costs were used as annual maintenance costs. The reference date was defined at the start of the LID life span. The singular construction costs were distributed uniformly and added to the annual maintenance costs:

$$TC_a = C_0 \cdot \frac{i \cdot (1+i)^n}{(1+i)^n - 1} + C_0 \cdot p \tag{4}$$

where TC_a is the total annual cost per unit (€/year), C_0 is the construction cost per unit (€), n is the life span (years), i is the interest rate (%) to discount future costs, and p is the proportion of maintenance to construction costs (%).

Table 2. Construction costs and maintenance costs for the investigated LIDs.

LID	Construction Costs (C_0)		Maintenance Costs ($C_0 \cdot p$)		Total Costs (TC_a)		Reference (Values Adapted)
Green roof	35	€/m^2	1.75	€/(m^2·year)	3.54	€/(m^2·year)	[39–41]
Infiltration swale	30	€/m^2	1.5	€/(m^2·year)	3.03	€/(m^2·year)	[38–41]
Infiltration trench	105	€/m^3	5.25	€/(m^3·year)	10.61	€/(m^3·year)	[38–41]
Bio-retention cell	135	€/m^3	6.75	€/(m^3·year)	13.64	€/(m^3·year)	[38–41]
Blind drain	105	€/m^3	5.25	€/(m^3·year)	10.61	€/(m^3·year)	[38–41]

Besides the LID performance with respect to the water balance and the economic aspect regarding the construction and maintenance costs, the demand for land is an additional important factor that has to be evaluated. Especially in highly urbanized areas, available land is rare and/or expensive. Consequently, the demand for land (d_{land}) is used as a further indicator of LID performance:

$$\text{For bio-retention cell, infiltration swale, and infiltration trench}:$$
$$d_{land} = A_{LID}$$
$$\text{For blind drain}: \tag{5}$$
$$d_{land} = 0$$

where d_{land} is the demand for land and A_{LID} is the area of the LID (see also Equation (1)).

2.4. Assessment and Efficiency of LID Strategies

LID strategies can be used in order to achieve, or at least approximate, a certain targeted water balance with a limited budget regarding the costs and/or demand for land. The challenge is to identify an LID strategy that meets the desired water balance while resulting in minimum costs and demand for land. Usually, there is not one optimal solution that equally satisfies the mentioned requirements. Thus, the relation between the water balance, costs, and demand for land has to be identified in order to find a reasonable LID strategy as a kind of trade-off.

The obtained simulation results are used to calculate the deviation from a targeted water balance. This deviation is defined as the sum of the absolute deviations of the particular water components:

$$D_{WB} = \overline{R_{sim} - R_t} + \overline{ET_{sim} - ET_t} + \overline{GR_{sim} - GR_t}, \tag{6}$$

where D_{WB} is the deviation from a targeted water balance (in percentage points), R is the runoff volume (in % of precipitation depth), ET is the evapotranspiration (in % of precipitation depth), GR is the groundwater recharge (in % of precipitation depth), sim denotes the simulated value, and t denotes the value of target state.

The targeted water balance can be either defined by stakeholders based on case-specific boundary conditions like the capacity of the present sewer system or based on hydrologic simulations, aiming for natural conditions (e.g., [19]). For demonstration purposes, an arbitrary defined targeted water balance with a runoff volume of 5%, an evapotranspiration of 45%, and a groundwater recharge of 50% is used.

The deviation from the targeted water balance (Equation (6)), the costs (Equation (4)), and the demand for land (Equation (5)) have to be minimized. This requirement is used to identify all nondominated (pareto-optimal) results. The approach of gridding methods (e.g., compare [42]) was used for this purpose, as the mentioned objectives were evaluated for a defined number of points (grid of 50 sections).

The deviation from the targeted water balance and the demand for land are used to evaluate the effect of invested money. The efficiency of LID strategies, as a function of costs, is computed as the sum of the normalized deviation from the targeted water balance and the normalized demand for land. Additionally, weighting factors are introduced to emphasize a certain goal:

$$E_{LID}(C) = 1 - \left(w_{land} * \frac{d_{land}}{\max(d_{land})} + w_{WB} * \frac{D_{WB}}{\max(D_{WB})} \right) \tag{7}$$
$$\text{with}: \ w_{land} + w_{WB} = 1,$$

where E_{LID} is the efficiency of LID strategies, C is the cost, d_{land} is the demand for land, D_{WB} is the deviation from the targeted water balance, w_{land} is the weighting factor for the demand for land, and w_{WB} is the weighting factor for the deviation from targeted water balance.

3. Results and Discussion

The change in system storage is almost zero or at least negligibly small compared to the other water balance components for the long-term assessment. Consequently, it is not further taken into consideration. Concerning the investigated LID treatment trains, only results for selected strategies, that show high potential, are illustrated.

3.1. Relations between Water Balance, Life Cycle Costs, and Demand for Land

3.1.1. Single LID Strategies

The qualitative results regarding the runoff volume are the same for all three investigated areas while the absolute values reveal some differences. A decrease in runoff volume is identified with an increasing number of LID sections and an associated increase of costs (Figures 4a, 5a and 6a). The larger the LID area, the more stormwater can be retained, resulting in smaller runoff volumes. The decrease curves start steep and flatten, converging to a runoff volume of zero. Thus, the effect of invested money on the runoff volume decreases with an increasing LID area. The results for the infiltration swale show the steepest costs-runoff-curve due to the smallest costs per implemented section. However, regarding the demand for land, the bio-retention cell and infiltration trench show a better and similar performance (Figures 4d, 5d and 6d) because they provide a larger and similar retention volume per LID section. The infiltration trench has smaller costs per section compared to the bio-retention-cell and a larger retention volume due to the surface storage (berm height) compared to the blind drain. Consequently, the performance is better regarding the costs (Figures 4a, 5a and 6a). The underground blind drain does not require land and is a suitable option when land is rare and/or expensive (Figure 4d,e, Figure 5d,e and Figure 6d,e).

The evapotranspiration is expectedly constant for the underground blind drain for all three areas (Figure 4b,e, Figure 5b,e and Figure 6b,e). The value of this constant evapotranspiration depends on the investigated area, thus on the kind of surface above the blind drain (impervious road/parking lot for commercial area, pervious lawn for residential areas) and the ratio of impervious to pervious surface in the area (different dod for residential areas). The kind of surface of sections that are not used for applying LIDs is also the reason for different results between the residential and commercial area with respect to the infiltration trench and evapotranspiration (Figure 4b,e, Figure 5b,e and Figure 6b,e). Applying an infiltration trench to residential areas shows an almost constant evapotranspiration as the evapotranspiration performance of an infiltration trench is similar to those of the appropriate lawn area; stormwater infiltrates into the native soil and is not available for evapotranspiration for a longer period. In contrast, applying an infiltration trench to the commercial area results in an increasing evapotranspiration as the reference evaporation from the road/parking lot is very small. The infiltration swale and bio-retention cell show an equal increase of evapotranspiration with an increasing LID area for all investigated areas as stormwater is retained in the soil layer and available for evapotranspiration (Figure 4b,e, Figure 5b,e and Figure 6b,e). The increase is linear, as SWMM does not account for the response of evapotranspiration to the soil moisture variation [43].

The consequence of a constant evapotranspiration for the blind drain and infiltration trench applied to residential areas is that the groundwater recharge shows the complete opposite of the runoff volume (Figure 4c,f, Figure 5c,f and Figure 6c,f). An increasing LID size results in an increase of the groundwater recharge. The increase curve mirrors the runoff volume decrease curve. The result for the infiltration trench applied to the commercial area, the infiltration swale, and the bio-retention cell is similar, except a small decrease of groundwater recharge for larger LIDs (Figure 4c,f, Figure 5c,f and Figure 6c,f). This is caused by the increasing retention volume, resulting in a runoff volume that converges to zero and an evapotranspiration that increases linearly.

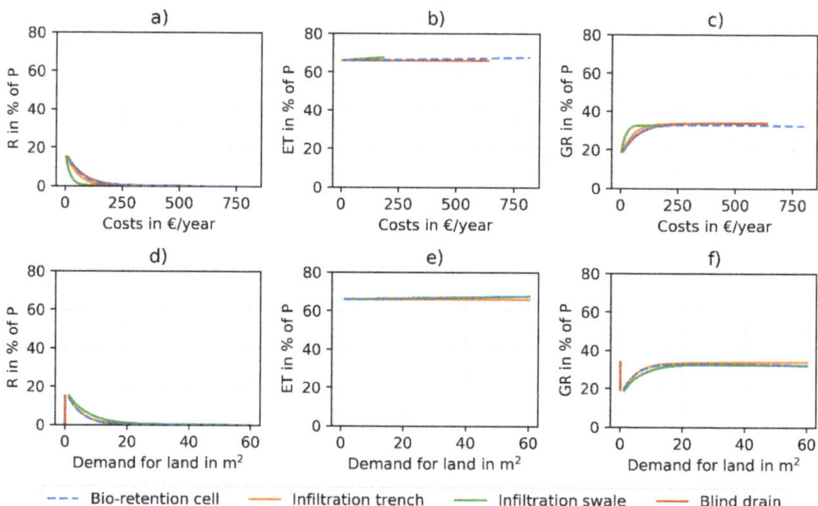

Figure 4. Simulated long-term water balance for the low-developed residential area applying single LID strategies of increasing size. Relation between costs and (**a**) runoff volume, (**b**) evapotranspiration, (**c**) groundwater recharge. Relation between demand for land and (**d**) runoff volume, (**e**) evapotranspiration, (**f**) groundwater recharge.

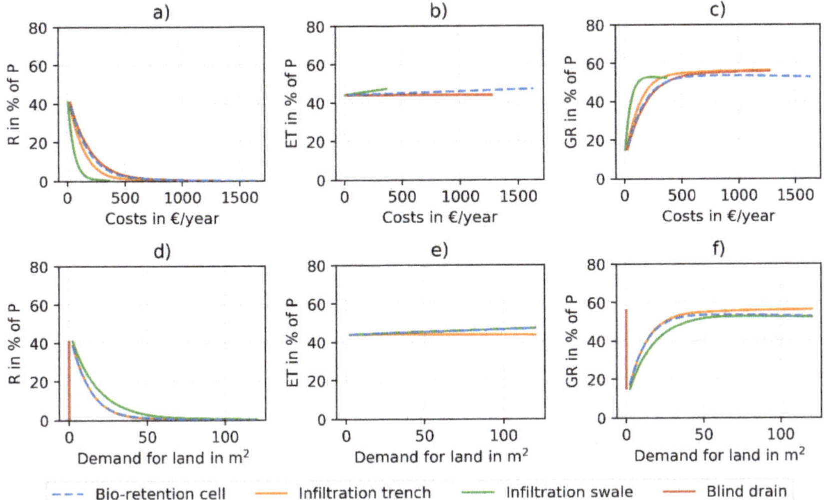

Figure 5. Simulated long-term water balance for the high-developed residential area applying single LID strategies of increasing size. Relation between costs and (**a**) runoff volume, (**b**) evapotranspiration, (**c**) groundwater recharge. Relation between demand for land and (**d**) runoff volume, (**e**) evapotranspiration, (**f**) groundwater recharge.

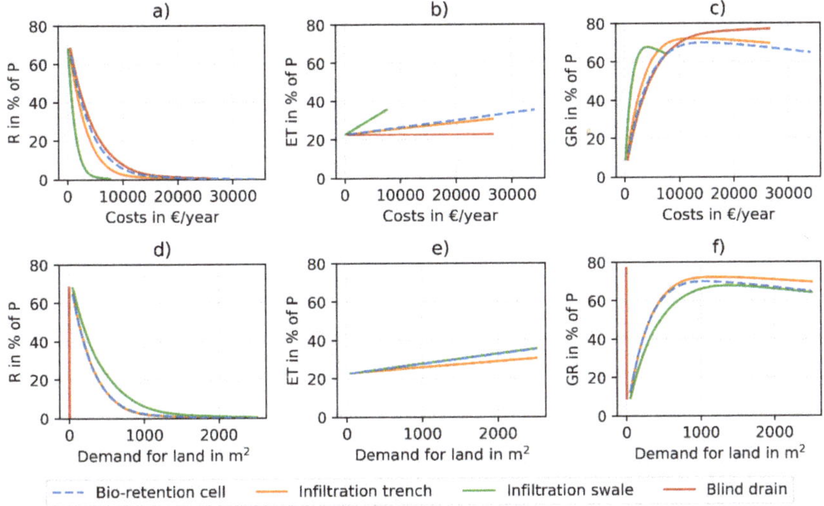

Figure 6. Simulated long-term water balance for the commercial area applying single LID strategies of increasing size. Relation between costs and (**a**) runoff volume, (**b**) evapotranspiration, (**c**) groundwater recharge. Relation between demand for land and (**d**) runoff volume, (**e**) evapotranspiration, (**f**) groundwater recharge.

It is obvious that the extent of the mentioned effects of applying single LIDs on the water balance differs between the investigated types of area. It increases with an increasing degree of imperviousness. The low-developed residential area already has a large lawn area resulting in a small runoff volume and high evapotranspiration in the initial state, whereas the commercial area shows the highest potential of applying LIDs.

3.1.2. Two-Part LID Treatment Train with Green Roof

The application of a green roof within the scope of a LID treatment train shows two general effects on the water balance. The first is related to the different hydrologic performance of the green roof itself compared to a tiled roof. The second is related to the consequently changed runoff volume to the downstream LID.

The green roof retains stormwater, which is consequently available for evapotranspiration. Thus, the runoff volume from the roof decreases, whereas evapotranspiration increases compared to the scenarios with a tiled roof (compare subplots a and b of Figures 4–6 with tiled roof and Figures 7–9 with green roof).

Consequently, the runoff to the downstream LID is reduced compared to scenarios with a tiled roof, resulting in an overall reduced runoff volume, whereas the groundwater recharge is decreased. The overall evapotranspiration increases due to the substantially increase of roof evapotranspiration. The effect of increasing the downstream LID area of infiltration swale, infiltration trench, bio-retention cell, and blind drain is basically the same as for the single LID investigations; the runoff volume decreases whereas the groundwater recharge increases. The evapotranspiration increases for the downstream bio-retention cell and infiltration swale and is constant for the blind drain. The application of the infiltration trench shows the already mentioned difference between residential and commercial areas, namely a constant evapotranspiration for the residential areas and an increasing evapotranspiration for the commercial area.

The magnitude of effects applying a green roof differs again between the investigated areas. The results for the low-developed residential area show that downstream LIDs have very little impact on the water balance (Figure 7). The green roof and lawn area generate small runoff volumes and a high evapotranspiration. Implementing an LID treatment train with green roof on a high-developed residential area shows larger but still small effects on the water balance (Figure 8). In contrast, as a large part of the commercial area consists of an impervious road/parking lot, the application of downstream LIDs shows the largest effect (Figure 9).

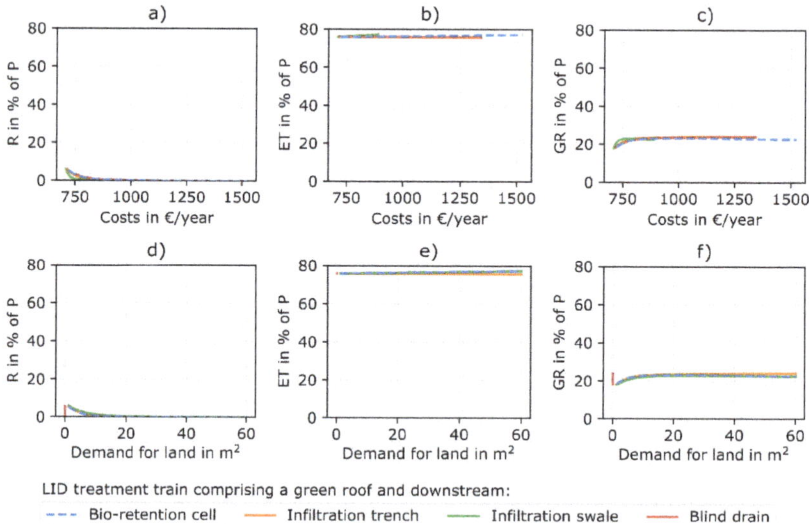

Figure 7. Simulated long-term water balance for the low-developed residential area applying a green roof and different downstream LIDs of increasing size (LID treatment train). Relation between costs and (**a**) runoff volume, (**b**) evapotranspiration, (**c**) groundwater recharge. Relation between demand for land and (**d**) runoff volume, (**e**) evapotranspiration, (**f**) groundwater recharge.

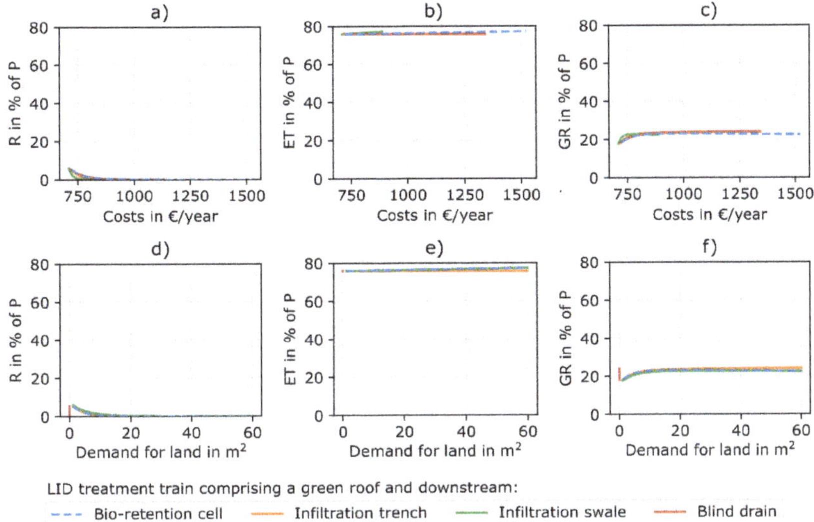

Figure 8. Simulated long-term water balance for the high-developed residential area applying a green roof and different downstream LIDs of increasing size (LID treatment train). Relation between costs and (**a**) runoff volume, (**b**) evapotranspiration, (**c**) groundwater recharge. Relation between demand for land and (**d**) runoff volume, (**e**) evapotranspiration, (**f**) groundwater recharge.

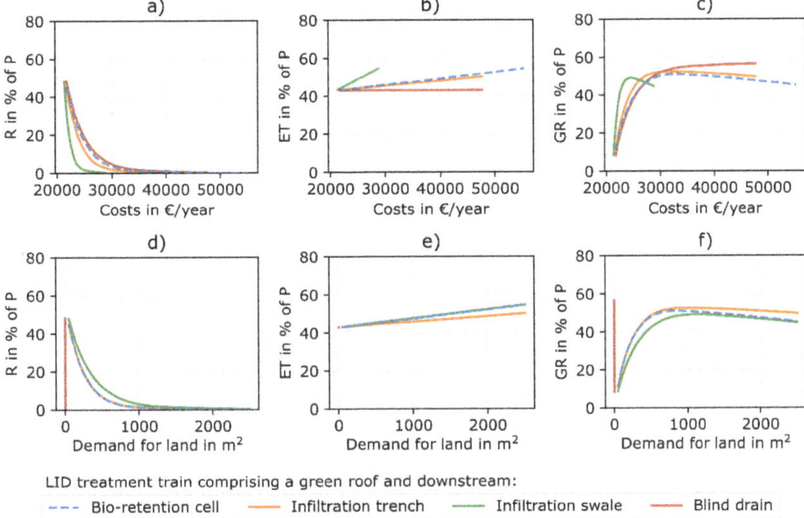

Figure 9. Simulated long-term water balance for the commercial area applying a green roof and different downstream LIDs of increasing size (LID treatment train). Relation between costs and (**a**) runoff volume, (**b**) evapotranspiration, (**c**) groundwater recharge. Relation between demand for land and (**d**) runoff volume, (**e**) evapotranspiration, (**f**) groundwater recharge.

3.1.3. Two-Part LID Treatment Train: Infiltration Swale—Infiltration Trench

The results for the single LID strategies and the two-part LID treatment trains with green roof showed that the largest effect on the water balance is obtained for the commercial area, whereas the impact is small for the residential areas, especially for the low-developed residential area.

As the qualitative performance is similar, the commercial area is used for illustrating effects of other LID strategies.

The investigations on applying single LIDs showed that the infiltration swale performs well regarding the costs, but has some shortcomings regarding the demand for land, e.g., compared to an infiltration trench. Consequently, an LID treatment train comprising an infiltration swale and a downstream infiltration trench is promising. This assumption is verified by the conducted simulations (Figure 10).

LID treatment trains provide the possibility of selecting LID strategies as a kind of trade-off between water balance, costs, and demand for land. An example illustrates this conclusion: Assuming a targeted runoff volume of 10%, applying only an infiltration swale results in costs of €2500 per year and a demand for land of 815 m^2, whereas applying an infiltration trench results in costs of €6100 per year and a demand of land of 573 m^2 (Figure 10). The mentioned strategies with a single LID would result in an evapotranspiration of 27% (infiltration swale) or 24.7% (infiltration trench). In contrast, an LID treatment train with a demand for land of 694 m^2 comprising equal fractions of infiltration swale and infiltration trench results in costs of €4733 per year. The mentioned LID treatment train results in the targeted 10% runoff volume and an evapotranspiration of 25.8%.

Selecting different proportions for the infiltration trench and infiltration swale on the total LID area moves the results in a certain direction. Assuming a certain limit for costs, increasing the proportion of the infiltration swale results in smaller runoff volumes and larger evapotranspiration, but is associated with a larger demand for land. On the other hand, assuming a certain limit for the demand of land, increasing the proportion of the infiltration swale results in larger runoff volumes and larger evapotranspiration, associated with lower costs. Thus, certain goals (e.g., desired runoff volume, evapotranspiration, groundwater recharge, maximal costs, or demand for land) can be achieved by selecting the proportion of the infiltration trench and infiltration swale within the scope of an LID treatment train.

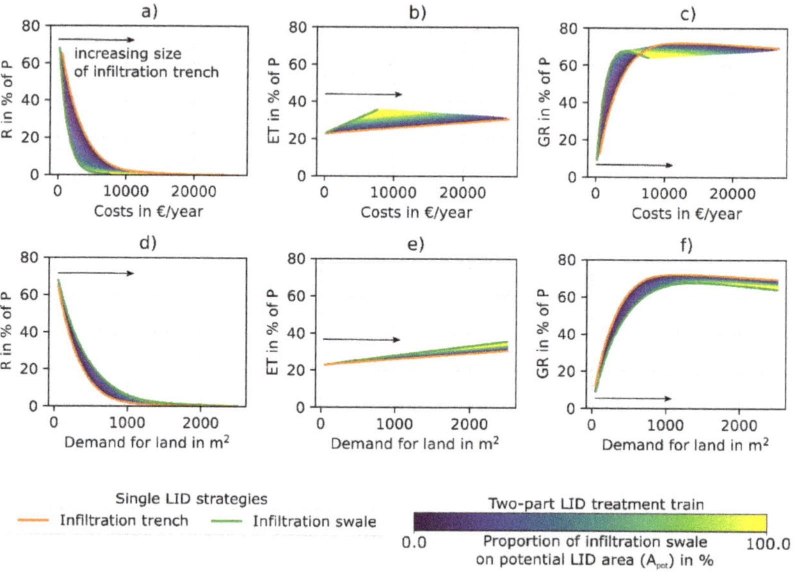

Figure 10. Simulated long-term water balance for the commercial area applying an LID treatment train comprising an altering proportion of infiltration swale on total LID area (A$_{pot}$) and a downstream infiltration trench of increasing size. Each colored line of the LID treatment train illustrates the simulation results for a constant proportion of infiltration swale on A$_{pot}$ and an increasing size of the infiltration trench (indicated by the arrow). Relation between costs and (**a**) runoff volume, (**b**) evapotranspiration, (**c**) groundwater recharge. Relation between demand for land and (**d**) runoff volume, (**e**) evapotranspiration, (**f**) groundwater recharge.

3.1.4. Two-Part LID Treatment Train: Infiltration swale—Blind Drain

The investigations on single LIDs revealed a good performance in runoff volume reduction and an increase of groundwater recharge with an outstanding demand for land of zero for the application of a blind drain. However, the evapotranspiration performance is basically null. Combining the blind drain with an infiltration swale in an LID treatment train can mitigate this fundamental shortcoming (Figure 11).

The infiltration swale accounts for an increase of evapotranspiration (Figure 11b,e), while the downstream blind drain decreases the runoff volume and increases the groundwater recharge without causing an additional demand for land (Figure 11d,f). Thus, this LID treatment train is suitable to control/improve the complete water balance, especially when land is rare and/or expensive. The size of the infiltration swale can be chosen due to the maximal land available and/or due to economic aspects. The size of the blind drain is either limited by a defined limit of costs or can be determined to control the runoff volume of the LID treatment train.

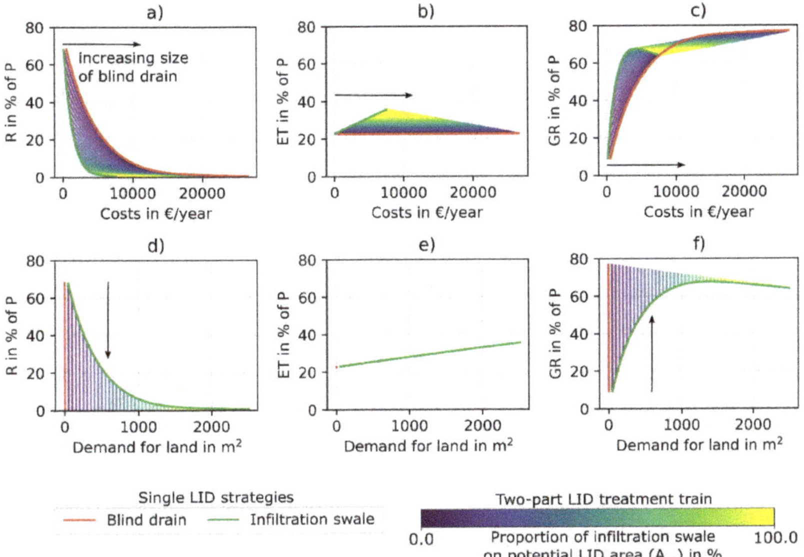

Figure 11. Simulated long-term water balance for the commercial area applying an LID treatment train comprising an infiltration swale with altering proportion on potential LID area (A_{pot}) and a downstream blind drain of increasing size. Each colored line of the LID treatment train illustrates the simulation results for a constant proportion of infiltration swale on A_{pot} and an increasing size of the blind drain (indicated by the arrow). Relation between costs and (**a**) runoff volume, (**b**) evapotranspiration, (**c**) groundwater recharge. Relation between demand for land and (**d**) runoff volume, (**e**) evapotranspiration, (**f**) groundwater recharge.

3.1.5. Three-Part LID Treatment Train: Green Roof—Infiltration Swale—Blind Drain

The application of a green roof within a three-part LID treatment train with a downstream infiltration swale and a blind drain shows the same effects as identified for the two-part LID treatment trains with a green roof (see Section 3.1.3): The overall runoff volume and groundwater recharge decrease, whereas the evapotranspiration increases (Figure 11a,b and Figure 12a,b) as stormwater is retained and evaporated on the green roof.

The green roof is especially valuable for the evapotranspiration (Figures 11b and 12b, increase of ca. 21 percentage points) while causing substantially higher costs (additional €21,240 per year).

The demand for land in order to achieve a certain runoff volume decreases when implementing an upstream green roof as the runoff to the infiltration swale is reduced.

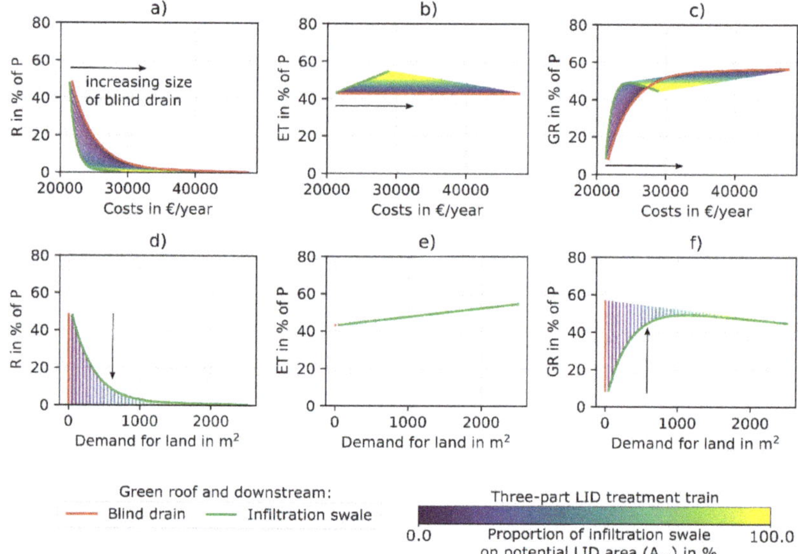

Figure 12. Simulated long-term water balance for the commercial area applying an LID treatment train comprising a green roof, an infiltration swale with altering proportion on total LID area (A_{pot}), and a downstream blind drain of increasing size. Each colored line of the LID treatment train illustrates the simulation results for a constant proportion of infiltration swale on A_{pot} and an increasing size of the blind drain (indicated by the arrow). Relation between costs and (**a**) runoff volume, (**b**) evapotranspiration, (**c**) groundwater recharge. Relation between demand for land and (**d**) runoff volume, (**e**) evapotranspiration, (**f**) groundwater recharge.

The effects of applying LIDs are in agreement with many field and laboratory studies, as well as evaluations based on hydrologic simulations (for an overview compare [6,7]). All LID strategies decrease the runoff volume due to the provided retention volume. The decrease curve starts steep and flattens, converging to zero. The green roof, bio-retention cell, and infiltration swale provide an increase of evapotranspiration. The increase is linear, as SWMM does not account for the response of evapotranspiration to the soil moisture variation [43]. In contrast, the infiltration trench applied to residential areas and the underground blind drain do not affect the evapotranspiration, but substantially increase the groundwater recharge.

The results indicate that the potential of applying LIDs is increasing, with an increasing imperviousness of the investigated area as slightly impervious areas already show a relatively small runoff volume and high evapotranspiration. Nevertheless, LIDs are applicable for both residential and commercial areas (in agreement with Dietz et al. [44]).

The green roof as part of an LID treatment train shifts the water balance components compared to the LID applications without a green roof (Figures 7–9 and 12) as stormwater is retained in the soil layer and available for evapotranspiration. This is in agreement with several field, laboratory, and modeling studies (for overview compare Ahiablame et al. [6] or Eckart et al. [7]). Consequently, the overall runoff volume and groundwater recharge are decreased.

3.2. Assessment and Efficiency of LID Strategies

The results for the commercial area (Figure 13) show a minimum deviation from the targeted water balance of 28 percentage points for the application of an infiltration swale, but at the same time result in a maximal demand for land. Assuming the same cost limit, the infiltration swale generally shows best results regarding the deviation from targeted water balance compared to other LID strategies. On the other hand, with respect to the demand for land, the blind drain shows expectable good results. However, at a certain point (ca. €4800 per year), the application of additional blind drain volume only results in higher costs without further reducing the deviation from targeted water balance.

It is obvious that strategies with a very small runoff volume going below the targeted runoff volume may increase the deviation from targeted water balance. The same can occur for strategies resulting in a groundwater recharge larger than the targeted one. However, following a holistic approach considering the complete water balance, the challenge is to find a solution that addresses the deviation from the complete targeted water balance and not a solution that only considers the deviation from target state of a particular water balance component. However, investigations on a larger scale can shift the point of view. LID strategies applied to a site, resulting in an exceedance of a certain component of the targeted water balance, can also be reasonable. They are applicable to counterbalance the respective component of the targeted water balance component of another site where it cannot be achieved or only associated with very high costs or demand for land. Nevertheless, the assessment on a site scale should be preferred, as suggested by Burns et al. [20].

All single LID strategies show a range of nondominated options. Thus, all single LID strategies provide pareto-optimal options. However, LID strategies resulting in small costs but a large deviation from targeted water balance will not be suitable in practice. Nevertheless, the results can be used to select a reasonable LID strategy. Stakeholders have the opportunity to emphasize a certain goal (deviation from targeted water balance, costs, demand for land) in the decision process.

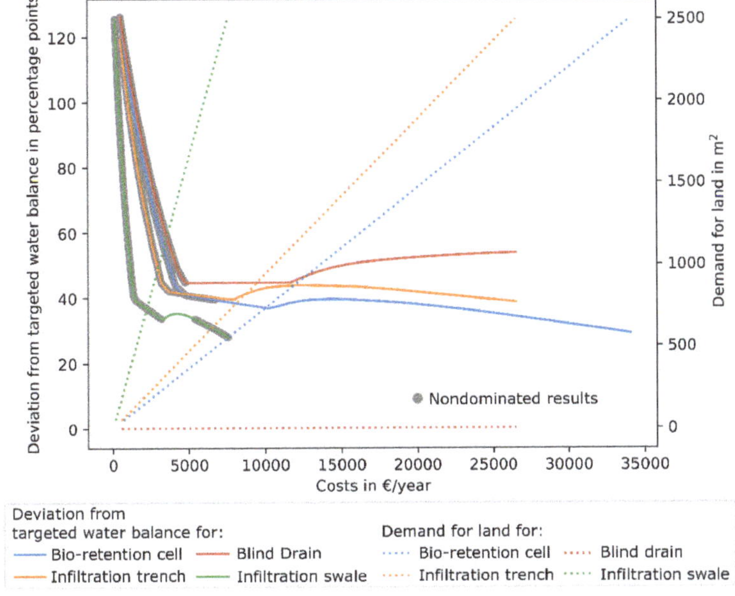

Figure 13. Assessment of applying single LID strategies to the commercial area with respect to a targeted water balance and demand for land. The nondominated results (grey-bold) are only illustrated for the relationship between costs and deviation from targeted water balance.

The trend in the results for the high-developed area is similar to those of the commercial area (Figure 14). In contrast to the commercial area, a deviation from targeted water balance of almost zero is achieved, applying an infiltration swale or a bio-retention cell. The costs to obtain this condition are higher for the bio-retention cell (€360 per year) than for the infiltration swale (€116 per year), but the demand for land is smaller for the bio-retention cell (26.4 m^2) than for the infiltration swale (38.4 m^2).

Once again, all single LID strategies show a range of nondominated options. As already mentioned, the decision process can be seen as a trade-off between the deviation from targeted water balance, costs, and demand for land.

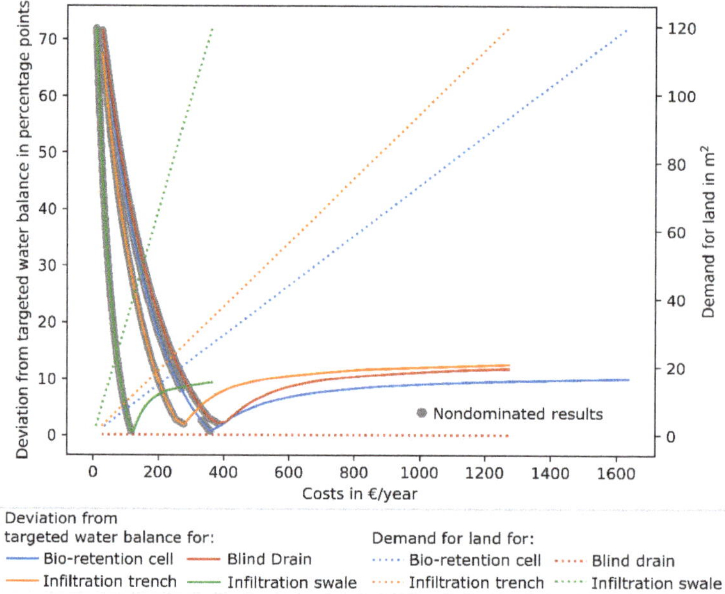

Figure 14. Assessment of applying single LID strategies to the high-developed residential area with respect to a targeted water balance and demand for land. The nondominated results (grey-bold) are only illustrated for the relationship between costs and deviation from targeted water balance.

The efficiency of LID strategies shows that the infiltration swale provides the best results when only the deviation from the targeted water balance is considered ($w_{land} = 0.0$, $w_{WB} = 1.0$, Figure 15a, compare also Figure 13). An increasing weighting factor for the demand for land results in an increasing E_{LID} for LID strategies comprising a blind drain (Figure 15). This is valid for a single blind drain as well as for an LID treatment train comprising an infiltration swale and a downstream blind drain, providing pareto-optimal results. Thus, when land is rare, the application of a blind drain can be a reasonable option. Implementing it as part of an LID treatment train with an infiltration swale is especially valuable. The infiltration swale is cost-saving and accounts for evapotranspiration, while the blind drain collects and infiltrates possibly occurring runoff from the infiltration swale while causing no further demand for land and.

If only the deviation from the targeted water balance is considered for E_{LID} ($w_{land} = 0.0$, $w_{WB} = 1.0$, Figure 15a), the improvement of E_{LID} is small at a certain point (ca. €4000 per year for the infiltration trench and the bio-retention cell) as the deviation from the targeted water balance can only be reduced slightly while the demand for land and costs increase. Concerning the blind drain, E_{LID} even decreases as the deviation from the targeted water balance increases, caused by an overly high groundwater recharge (compare also Figure 6). Emphasizing the demand for land, E_{LID} also decreases more and more for the other single LIDs (infiltration trench, bio-retention cell, infiltration swale) as the increase of demand for land exceeds the reduction of deviation from targeted water balance.

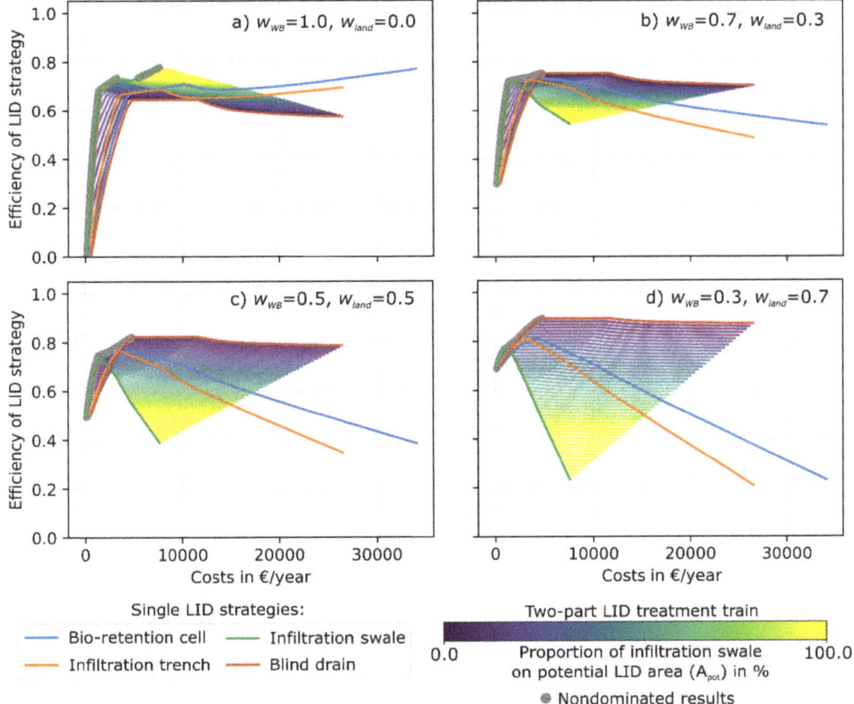

Figure 15. Assessment of applying different single LID strategies of increasing size and an LID treatment train consisting of an infiltration swale with altering proportion on total LID area (A_{pot}) and a downstream blind drain of increasing size to the commercial area. Calculation of LID efficiency (E_{LID}) with varying weights deviation from targeted water balance (w_{WB}) and demand for land (w_{land}). Each colored line of the LID treatment train illustrates E_{LID} for a constant proportion of infiltration swale on A_{pot} and an increasing size of the blind drain. The nondominated results are illustrated in grey-bold.

The assessment of LID strategies with respect to a targeted water balance shows that the decision is dependent on the main goal of the stormwater management project. Besides the deviation from the targeted water balance, the demand for land and costs have to be taken into consideration. Depending on the emphasis given on the individual goals, different LID strategies can appear to be "most effective".

4. Conclusions

This paper introduces a method for a model-based selection of cost-effective LID strategies to control water balance. The method is based on a holistic approach considering the complete water balance. The objectives within the design and selection process are the deviation from the targeted water balance, the demand for land and the costs. The efficiency of LID strategies (E_{LID}) is defined as a measure to evaluate the investigated LID strategies, providing also the possibility of weighting the individual objectives.

The conducted simulations illustrate how LID strategies affect the water balance depending on the applied size of LID: Reduction of runoff volume, increase of evapotranspiration, and groundwater recharge. The results are valuable for the planning process in order to estimate the respective effect on the water balance components of different LID strategies.

The investigations revealed that there is not one specific optimal LID strategy when the water balance, as well as costs and demand for land, are taken into consideration. Nevertheless, the method's results provide a well-founded and holistic basis for the selection of a reasonable LID strategy. Stakeholders can choose from several nondominated results, emphasizing a certain objective.

The application of an LID treatment train shows high potential. It is especially valuable combining a cost-saving LID that accounts for evapotranspiration (e.g., infiltration swale) and a downstream LID that accounts for infiltration and results in no further demand for land (e.g., blind drain).

The quantitative results are restricted to the investigated areas and their hydrologic boundary conditions, the precipitation time series, the assumed costs, and the LID strategies used. However, the developed method is applicable to other areas, other precipitation time series, and other LID strategies. Further research is related to this assumption, as well as using the method's findings on a larger scale.

Author Contributions: Conceptualization and methodology, J.L., G.K., and D.M.; investigation, J.L.; resources, D.C.; data curation, J.L. and D.C.; validation, J.L. and G.K.; writing—original draft preparation, J.L.; writing—review and editing, J.L., G.K., D.C., and D.M.; visualization, J.L. and D.C.; supervision, D.M.

Funding: The project was funded by the Austrian Federal Ministry for Sustainability and Tourism and the State of Styria.

Acknowledgments: The research is part of the project "FlexAdapt – Entwicklung flexibler Adaptierungskonzepte für die Siedlungsentwässerung der Zukunft". The project was funded by the Austrian Federal Ministry for Sustainability and Tourism and the State of Styria.

Conflicts of Interest: The authors declare no conflict of interest.

References

1. Shuster, W.D.; Bonta, J.; Thurston, H.; Warnemuende, E.; Smith, D.R. Impacts of impervious surface on watershed hydrology: A. review. *Urban Water J.* **2005**, *2*, 263–275. [CrossRef]

2. Cheng, S.J.; Wang, R.Y. An approach for evaluating the hydrological effects of urbanization and its application. *Hydrol. Process.* **2002**, *16*, 1403–1418. [CrossRef]

3. Haase, D. Effects of urbanisation on the water balance—A long-term trajectory. *Environ. Impact Assess. Rev.* **2009**, *29*, 211–219. [CrossRef]

4. Fletcher, T.D.; Andrieu, H.; Hamel, P. Understanding, management and modelling of urban hydrology and its consequences for receiving waters: A state of the art. *Adv. Water Resour.* **2013**, *51*, 261–279. [CrossRef]

5. US EPA. *Low Impact Development (LID)—A Literature Review*; US EPA Office of Water (4203): Washington, DC, USA, 2000.

6. Ahiablame, L.M.; Engel, B.A.; Chaubey, I. Effectiveness of low impact development practices: Literature review and suggestions for future research. *Water Air Soil Pollut.* **2012**, *223*, 4253–4273. [CrossRef]

7. Eckart, K.; McPhee, Z.; Bolisetti, T. Performance and implementation of low impact development—A review. *Sci. Total Environ.* **2017**, *607*, 413–432. [CrossRef]

8. Jayasooriya, V.M.; Ng, A.W.M. Tools for Modeling of Stormwater Management and Economics of Green Infrastructure Practices: A Review. *Water Air Soil Pollut.* **2014**, *225*, 2055. [CrossRef]

9. Rossman, L. *Storm Water Management Model—User's Manual Version 5.1*; US EPA Office of Research and Development: Washington, DC, USA, 2015.

10. DWA. *Arbeitsblatt DWA-A 138—Planung, Bau und Betrieb von Anlagen zur Versickerung von Niederschlagswasser*; Deutsche Vereinigung für Wasserwirtschaft, Abwasser und Abfall e.V.: Hennef, Germany, 2005.

11. ON. *ÖNORM B 2506-1: Regenwasser-Sickeranlagen für Abläufe von Dachflächen und befestigten Flächen—Teil 1: Anwendung, hydraulische Bemessung, Bau und Betrieb*; Austrian Standards: Vienna, Austria, 2013.

12. MDE. *Maryland Stormwater Design Manual*; Maryland Department of the Environment (MDE): Baltimore, MD, USA, 2000.

13. Ballard, B.W.; Udale-Clarke, H.; Kellagher, R.; Dou, V.; Powers, D.; Gerolin, A.; McCloy, A.; Schmitt, T. International approaches to the hydraulic control of surface water runoff in mitigating flood and environmental risks. *E3S Web Conf.* **2016**, *7*, 12004. [CrossRef]
14. Leimgruber, J.; Krebs, G.; Camhy, D.; Muschalla, D. Sensitivity of Model-Based Water Balance to Low Impact Development Parameters. *Water* **2018**, *10*, 1838. [CrossRef]
15. Coffman, L. *Low-Impact Development Design Strategies, An Integrated Design Approach*; Department of Environmental Resources, Programs and Planning Division: Prince George's County, MD, USA, 2000.
16. Seo, M.; Jaber, F.; Srinivasan, R. Evaluating various low-impact development scenarios for optimal design criteria development. *Water Switz.* **2017**, *9*, 270. [CrossRef]
17. Goebel, P.; Coldewey, W.G.; Dierkes, C.; Kories, H.; Meßer, J.; Meißner, E. Impacts of green roofs and rain water use on the water balance and groundwater levels in urban areas. *Grundwasser* **2007**, *12*, 189–200. [CrossRef]
18. Henrichs, M.; Langner, J.; Uhl, M. Development of a simplified urban water balance model (WABILA). *Water Sci. Technol.* **2016**, *73*, 1785–1795. [CrossRef]
19. Henrichs, M.; Steinbrich, A.; Leistert, H.; Scherer, I.; Schuetz, T.; Uhl, M.; Weiler, M. Model Based Estimation of a Natural Water Balance as Reference for Planning in Urban Areas. In *New Trends in Urban Drainage Modelling—UDM 2018, Proceedings of International Conference on Urban Drainage Modelling, Palermo, Italy, 23–26 September 2018*; Springer International Publishing: Cham, Switzerland, 2019; pp. 953–957.
20. Burns, M.J.; Fletcher, T.D.; Walsh, C.J.; Ladson, A.R.; Hatt, B.E. Hydrologic shortcomings of conventional urban stormwater management and opportunities for reform. *Landsc. Urban Plan.* **2012**, *105*, 230–240. [CrossRef]
21. Kim, R.; Lee, J.; Lee, D. Development of Treatment-Train-Package Novel System for the Road Runoff Controlling in Urban Area. *Int. J. Control. Autom.* **2015**, *8*, 155–162. [CrossRef]
22. Xu, C.; Hong, J.; Jia, H.; Liang, S.; Xu, T. Life cycle environmental and economic assessment of a LID-BMP treatment train system: A case study in China. *J. Clean. Prod.* **2017**, *149*, 227–237. [CrossRef]
23. She, N.; Liu, J.; Lucas, W.; Li, T.; Wu, L. Performance of LID Treatment Trains in Shenzhen University during Extreme Storm Events. In Proceedings of the International Low Impact Development Conference 2015, Houston, TX, USA, 19–21 January 2015.
24. Auger, S.; Van Seters, T.; Singh, A.; Antoszek, J. Water Quality Target Assessment Using LID TTT for Better SWM Designs in Ontario. In Proceedings of the International Low Impact Development Conference 2018, Nashville, TN, USA, 12–15 August 2018.
25. MacMullan, E.; Reich, S. *The Economics of Low-Impact Development: A Literature Review*; ECONorthwest: Eugene, OR, USA, 2007.
26. Liao, Z.; Chen, H.; Huang, F.; Li, H. Cost–effectiveness analysis on LID measures of a highly urbanized area. *Desalin. Water Treat.* **2015**, *56*, 2817–2823. [CrossRef]
27. Chui, T.F.M.; Liu, X.; Zhan, W. Assessing cost-effectiveness of specific LID practice designs in response to large storm events. *J. Hydrol.* **2016**, *533*, 353–364. [CrossRef]
28. Montalto, F.; Behr, C.; Alfredo, K.; Wolf, M.; Arye, M.; Walsh, M. Rapid assessment of the cost-effectiveness of low impact development for CSO control. *Landsc. Urban Plan.* **2007**, *82*, 117–131. [CrossRef]
29. Houdeshel, C.; Pomeroy, C.; Hair, L.; Moeller, J. Cost-Estimating Tools for Low-Impact Development Best Management Practices: Challenges, Limitations, and Implications. *J. Irrig. Drain. Eng.* **2010**, *137*, 183–189. [CrossRef]
30. Yu, Z.; Aguayo, M.; Montalto, F.; Piasecki, M.; Behr, C. Developments in LIDRA 2.0: A Planning Level Assessment of the Cost-Effectiveness of Low Impact Development. In Proceedings of the World Environmental and Water Resources Congress 2010, Providence, RI, USA, 16–20 May 2010; pp. 3261–3270, ISBN 978-0-7844-1114-8.
31. OEWAV. *ÖWAV—Leitfaden—Niederschlagsdaten zur Anwendung der ÖWAV-Regelblätter 11 und 19*; Österreichischer Wasser-und Abfallwirtschaftsverband: Vienna, Austria, 2007.
32. Hargreaves, G.H.; Samani, Z.A. Reference Crop Evapotranspiration from Temperature. *Appl. Eng. Agric.* **1985**, *1*, 96–99. [CrossRef]
33. ZAMG. Jahrbuch—ZAMG. Available online: https://www.zamg.ac.at/cms/de/klima/klimauebersichten/jahrbuch (accessed on 28 November 2017).

34. Rossman, L.A.; Huber, W.C. *Storm Water Management Model Reference Manual Volume III—Water Quality*; US EPA National Risk Management Research Laboratory: Cincinnati, OI, USA, 2016.

35. FLL. *Dachbegrünungsrichtlinie—Richtlinie für die Planung Ausführung und Pflege von Dachbegrünungen*; Forschungsgesellschaft Landschaftsentwicklung und Landschaftsbau e.V.: Bonn, Germany, 2008.

36. Rossman, L.A.; Huber, W.C. *Storm Water Management Model Reference Manual Volume I—Hydrology*; US EPA National Risk Management Research Laboratory: Cincinnati, OI, USA, 2016.

37. *Leitlinien zur Durchführung dynamischer Kostenvergleichsrechnungen (KVR-Leitlinien)*; 8. überarb. Aufl.; Deutsche Vereinigung für Wasserwirtschaft, Abwasser und Abfall (Ed.) DWA: Hennef, Germany, 2012; ISBN 978-3-941897-55-7.

38. Leimbach, S.; Brendt, T.; Ebert, G.; Jackisch, N.; Zieger, F.; Kramer, S. *Regenwasserbewirtschaftungsanlagen in der Praxis: Betriebssicherheit, Kosten und Unterhaltung*; Universität.: Freiburg, Germany, 2018.

39. Matzinger, A.; Riechel, M.; Remy, C.; Schwarzmüller, H.; Rouault, P.; Schmidt, M.; Offermann, M.; Strehl, C.; Nickel, D.; Sieker, H.; et al. *Zielorientierte Planung von Maßnahmen der Regenwasserbewirtschaftung—Ergebnisse des Projektes KURAS*; Konzepte für urbane Regenwasserbewirtschaftung und Abwassersysteme: Berlin, Germany, 2017.

40. Sieker Sieker—Die Regenwasserexperten. Available online: https://www.sieker.dede/home.html (accessed on 5 December 2018).

41. Muschalla, D.; Gruber, G.; Scheucher, R. *ECOSTORMA—Handbuch—Ökologische und ökonomische Maßnahmen der Niederschlagswasserbewirtschaftung*; Ministerium für ein lebenswertes Österreich: Wien, Germany, 2014.

42. Schuetze, M.; Butler, D.; Beck, M.B. *Modelling, Simulation and Control of Urban Wastewater Systems*; Springer: London, UK, 2002; ISBN 1-85233-553-X.

43. Youcan, F.; Steven, B. Improving Evapotranspiration Mechanisms in the U.S. Environmental Protection Agency's Storm Water Management Model. *J. Hydrol. Eng.* **2016**, *21*, 06016007.

44. Dietz, M.E. Low Impact Development Practices: A Review of Current Research and Recommendations for Future Directions. *Water. Air. Soil Pollut.* **2007**, *186*, 351–363. [CrossRef]

 *sustainability*

Article

The Edge Effect on Plant Diversity and Soil Properties in Abandoned Fields Targeted for Ecological Restoration

Sheunesu Ruwanza

Department of Environmental Science, Rhodes University, Grahamstown 6140, South Africa;
ruwanza@yahoo.com; Tel.: +27-46-603-7009

Received: 11 September 2018; Accepted: 30 November 2018; Published: 28 December 2018

Abstract: Changes in biotic and abiotic factors may create opportunities for biodiversity recovery in abandoned agricultural fields. This study examined the natural/old field edge effect on plant diversity and soil properties at Lapalala Wilderness in Limpopo Province, South Africa. Detailed vegetation surveys and soil measurements were conducted in three old fields that share a natural/old field road edge boundary. On each site, three transects, each with four plots (10 × 10 m), located 10 m into the natural area and 10, 30 and 50 m into the old field from the edge, were setup. Plant diversity and composition measurements were conducted on each plot. Soil moisture and total N, C and P were measured at the center of each plot. Results indicate that abundance of some woody species was significantly ($P < 0.001$) higher close to the edge than far into the old fields. However, this was not the case for herbs and grasses which did not increase with edge proximity. All measured soil properties were significantly ($P < 0.001$) higher close to the edge than far into the old fields. The study concludes that both vegetation and soil properties are influenced by proximity to the edge.

Keywords: old field succession; tree establishment; microclimate; plant-soil interactions; soil nutrients; Lapalala Wilderness

1. Introduction

Land abandonment has been on the rise in South Africa due to several factors, such as rural depopulation, poor land management, shifts in global markets for agricultural products and decline in soil fertility [1,2]. A recent survey showed that abandonment of cropland in South Africa was widespread between 1950 and 2010, with peak abandonment rates occurring between 1970 and 1990 [2]. In view of the high land abandonment rates in the country, ecological restoration of abandoned agricultural fields (here after referred to as old fields) is very important due to direct and indirect ecological benefits associated with restored ecosystems [3–5]. Restored old fields improve the flow of ecosystem services by conserving and improving natural capital, soil stability, water quality, biodiversity, and global climate stability [6]. From an economic stand point, restored old fields create employment through ecotourism [6,7]. However, the successful recovery of vegetation and soil in old fields is determined by the interaction of several biotic and abiotic factors e.g., soil seed bank, seed dispersal mechanisms, propagule source, species interactions, soil nutrient availability and temperature regimes [8,9]. To further complicate the restoration process in old fields, the extent to which both vegetation and soil properties were damaged during cultivation, has a bearing on the recovery process [4]. For example, cultivation has been shown to negatively affect soil nutrients and soil stored seed banks [4,8,10], and the extent to which the above-mentioned soil properties are damaged may slow down the natural succession process after the land is abandoned.

Recent studies have shown that vegetation and soil recovery in old fields can be affected by proximity to forest edge [5,11,12]. The edge effect is defined as the difference in biotic and abiotic factors that exist at the border of a fragmented habitat relative to the interior environment [12,13]. Increased seed dispersal rates and the creation of microclimates at the natural/old field edge have been reported to favor establishment of plant communities that are different from those found in the old field interior [5,10,14]. Some studies have shown that proximity to the natural/old field edge facilitates the establishment of woody species over other growth forms [10,15,16]. In contrast, some studies have concluded that the edge effect does not influence vegetation recovery in degraded ecosystems [17–19]. Given the mixed results on how edge proximity affects vegetation, there exist a need to assess the edge effect across a variety of environments to develop predictive hypotheses across varying contexts.

Few studies have looked at the edge effect on soil properties in old fields targeted for ecological restoration [5]. Bueno et al. [5] and Ramırez et al. [20] reported that creation of conducive soil microclimates underneath the canopy of recruiting plants can facilitate germination and growth of more plants at the edge. Recruiting plants can in turn exert a facilitative effect through litter accumulation and creation of suitable conditions for litter decomposition, thus subsequently increase soil nutrients [5,21]. In old fields where the edge has a positive effect on vegetation abundance and diversity, there exists a hypothesis that soil properties will also be positively affected by edge proximity, which is likely due to vegetation changes [5]. This has never been tested in old fields of savanna bushveld were ecological restoration initiatives are taking place at Lapalala Wilderness. To make recommendations for ecological restorations in old fields targeted for ecological restoration at Lapalala Wilderness, this study examined how natural/old field edges affect both plant diversity and soil properties. The main research question was: does proximity to natural/old field edges affect plant (species diversity and composition) and soil properties (soil moisture and macronutrients of N, C, and P). The study hypothesis is that changes in vegetation diversity and cover due to edge proximity will influence changes in soil properties. The results are intended to guide future soil and vegetation recovery studies in old fields targeted for ecological restoration in South African ecosystems.

2. Materials and Methods

2.1. Study Area

The study was conducted at Lapalala Wilderness reserve, which is in Limpopo Province, South Africa (Figure 1). The privately-owned reserve is approximately 45,000 ha and is home to several wild animals e.g., white rhinoceros, zebra, impala, wildebeest, giraffe and kudu. Mucina and Rutherford [22] classified vegetation in the reserve as Waterberg Mountain Bushveld. Although soils in the reserve are predominantly sandy from the Waterberg groups, the old fields where the study was conducted are dominated by clay and loam soils from basic norite/epidiorite substrates [22,23]. A recent study at the same study site showed that soils are generally acidic with an average pH of 4.73 [23]. Annual average rainfall ranges from 400 to 600 mm and most rain falls in summer between September and April. Temperatures are hot in summer (maximum 30 °C in January) and cold in winter (minimum 4 °C winter). Winter months commonly have frost, averaging 61 to 90 days annually. Mean annual evaporation ranges between 2200 to 2400 mm [22,24].

Figure 1. Location of the study area at Lapalala Wilderness, Limpopo province of South Africa.

2.2. Sampling Design

Three old field sites that share a farm road border with natural sites were identified towards the end of the rainy season (April 2017) to optimize plant identification. The three sites were approximately 300 m apart to allow independence of measurements [25]. The 35-year old fields were previously used for tobacco farming but are currently used for wild animal grazing. At each of the three sites, three line transects that were 20 m apart were setup (Figure 2). The line transects which run perpendicular to the farm road edge extended 10 m into the natural area and 50 m into the old field. At 10 m (from the road edge into the natural side), and at 10, 30, and 50 m (from the road edge into the old fields) plots measuring 10 m x 10 m were setup for plant and soil measurements.

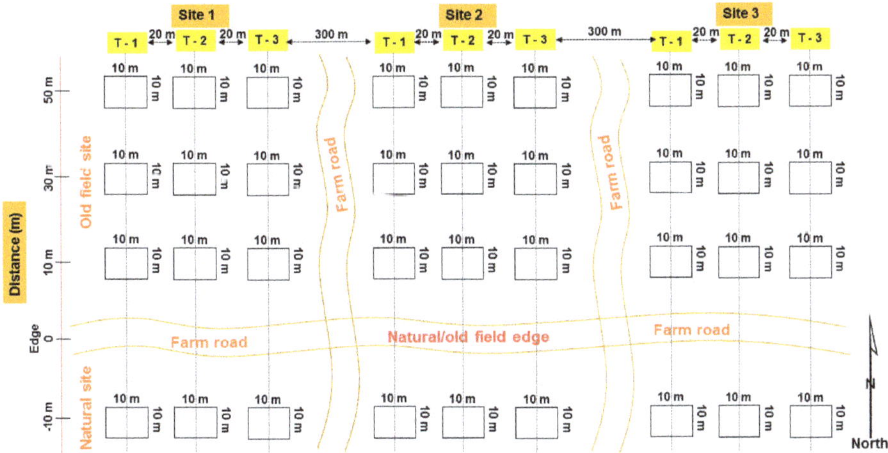

Figure 2. Schematic diagram showing the three sites, transects (T-1, T-2 and T-3) and sampled plots at various distances (-10 m, 10 m, 30 m, and 50 m) from the natural/old field edge.

2.3. Vegetation Surveys

Within each plot, a detailed vegetation survey was conducted in April (end of summer). Conducting vegetation surveys during April allowed easily identification of plant species. Plant species richness and abundance of all trees and shrubs was determined by counting the species present in the 10 m $\times$ 10 m plot. Species richness and abundance for herbs and grasses was determined by counting all the species present in a 1 m $\times$ 1 m quadrate placed at the center of the plot. Given that some grasses and herbs are small and numerous, the above-mentioned quadrate size was used to allow accurate count and cover estimates for these two growth forms. Cover of all the plant species present in the plot were visually estimated to the nearest 5% and 1% when species occupied less than 5% of the plot. A sample of all the plant species present in the plot and quadrate was collected for identification at University of Venda in the Department of Botany herbarium. Plant species were assigned to four growth forms, namely trees, shrubs, herbs and grasses based on phytomorphology.

2.4. Soil Sampling

At the center of each plot, two soil cores (30 cm apart) were collected for total N, C and P measurements, using a soil core measuring 10 cm in diameter, at 8 cm depth. Soil cores were collected once due to limited finances and the expectation that there would not be any marked differences that would warrant repeated measures within a season. After collection, soils were sieved using a 2 mm sieve to remove stones and debris. Soils from the first collecting point were measured for gravimetric soil moisture, whilst soils from the second collecting point were measured for macronutrients of total N, C, and P. Gravimetric soil moisture was measured by first weighing the soils wet, oven drying them at 105 °C for 72 h, and then re-weighing them to obtain the water content, which was expressed as a percentage [26]. Total soil N was determined by complete combustion using an Elemental Analyser (EuroVector Euro EA 3000, Milan, Italy), whilst total soil C was determined using the Walkley-Black method [27]. Soil P was analysed using a Bray-2 extract [28].

2.5. Statistical Analysis

For each plot and quadrate, species richness, Shannon-Wiener (H'), Simpson's index of diversity (1-D) [29,30] and Evenness index (J) using Pielou's 'J' [31] were calculated. To avoid pseudo-replication, plant and soil results from the three transects per site and from the same distance from the edge were averaged [32]. Data were tested for normality and proof of homogeneity of variance using the Shapiro-Wilk's and Levene's test, respectively, and data were normally distributed. To examine the effects of edge on measured plant and soil variables, one-way ANOVA was performed in Statistica version 13.1 [33]. Correspondence analysis (CA) was performed using Canoco for Windows 5 [34] to investigate how distance from the edge affected the plant species assemblage. Plant species abundance data were log-transformed before CA was performed to eliminate the influence of extreme values on ordination scores.

3. Results

3.1. The Edge Effect on Vegetation

Species richness and Shannon-Wiener showed significant ($P < 0.01$) differences between the measured distances from the natural/old field edge (Figure 3a,b). Species richness was significantly ($P < 0.001$) higher at 10 m from the edge into the natural area compared to all other measured distances into the old fields. However, comparisons between distances from the edge into the old fields on species richness showed no significant ($P > 0.05$) differences (Figure 3a). Shannon-Wiener was significantly ($P < 0.01$) higher at 10 m from the edge of both the natural and old field, compared to 30 m and 50 m into the old fields (Figure 3b). Comparisons on Shannon-Wiener between distances from the edge into the old field showed no significant ($P > 0.05$) differences (Figure 3b). Both evenness and Simpsons index of diversity showed no significant ($P > 0.05$) differences at measured distances from the edge (Figure 3c,d).

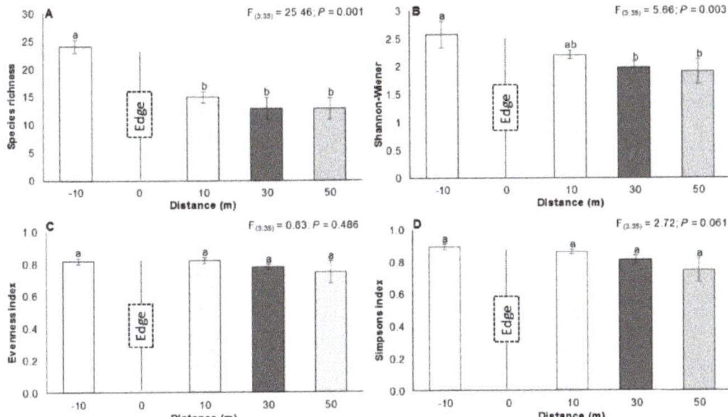

Figure 3. Comparison of indices of diversity (**a**—species richness, **b**—Shannon-Wiener, **c**—evenness index and **d**—Simpsons index) at various sampling distance from the natural/old field edge. Data are means ± SE and significant ANOVA results are shown.

Species richness for trees and shrubs significantly ($P < 0.001$) decreased with distance from the edge into the old field (Figure 4a). Species richness for herbs significantly ($P < 0.05$) increased with distance from the edge, but there were no significant ($P > 0.05$) differences between 10 m into the natural areas compared to 30 and 50 m into the old field (Figure 4b). Species richness for grasses significantly ($P < 0.001$) increased with distance from edge into the old field (Figure 4c). Cover of trees and shrubs significantly ($P < 0.001$) decreased with increase in distance from the edge (Figure 4d). Cover for herbs was significantly ($P < 0.05$) higher at 50 m compared to 10 m from the edge into the old field. Comparisons of the cover of herbs between 10 m into natural areas and 30 m into the old fields showed no significant ($P > 0.05$) differences (Figure 4e). Cover for grasses significantly ($P < 0.001$) increased with an increase in distance from the edge (Figure 4f).

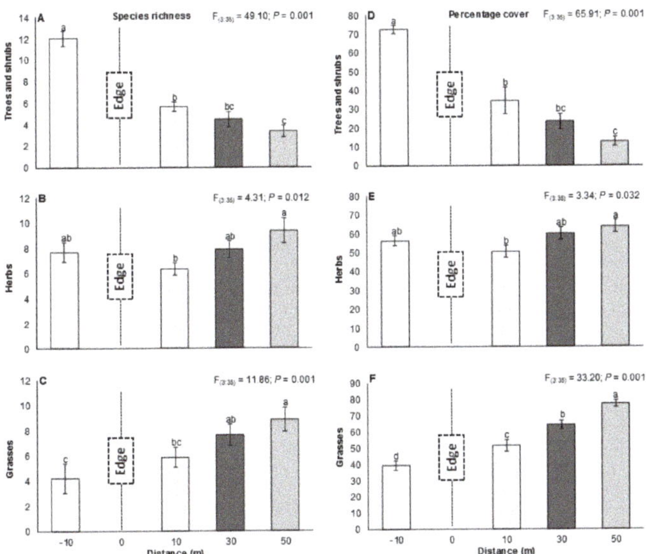

Figure 4. Comparison of species richness (**A–C**) and vegetation cover (**D–F**) for trees and shrubs, herbs, and grasses at various sampling distance from the natural/old field edge. Data are means ± SE and significant ANOVA results are shown.

A list of the 36 plant species that were identified at various distances from the natural/old field edge are presented in the Appendix A (Table A1), with their occupancy frequencies shown. Correspondence analysis for trees and shrubs showed that most species assembled in the natural areas, 10 m from the edge (Figure 5a). However, the woody species of *Peltophorum africanum, Senegalia nigrescens, S. senegal, Vachellia tortilis, V. nilotica, V. karroo,* and *Lippia javanica* appeared at all measured distances from the edge, but their occupancy frequencies in the old fields decreased with an increase in distance from the edge (Figure 5a). Correspondence analysis for herbs showed that the species *Felicia sp* dominated 10 m into the natural area from the edge, whereas *Solanum panduriforme* and *S. incanum* dominated 10 m into the old field. The herb species of *Xerochrysum bracteatum, Gomphrena celosioides* and *Datura stramonium* assembled 30 m and 50 m into the old field from the edge (Figure 5b). Correspondence analysis for grasses showed only two distinct groups, where *Aristida congesta, Digitaria eriantha* and *Eragrostis inamoena* assembled 10 m into the old field from the edge and *Panicum maximum, Cynodon dactylon* and *Eragrostis lehmanniana* assembled 30 m and 50 m into the old field as well as 10 m into the natural area from the edge (Figure 5c).

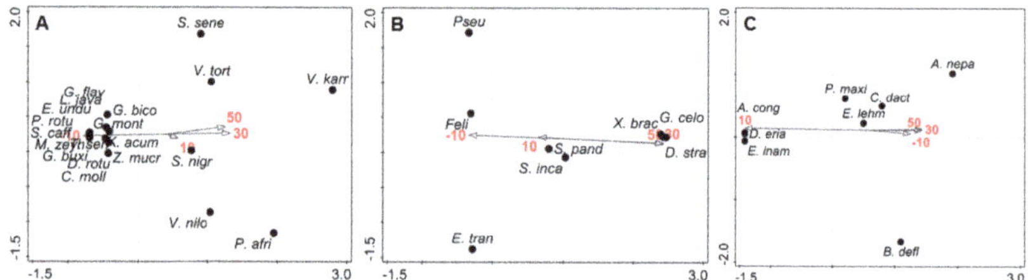

Figure 5. Correspondence analysis (CA) ordination of species (•) along sampled distance from the natural/old field edge for (**a**) trees and shrubs, (**b**) herbs, and (**c**) grasses. The first four letters of the species names are presented with full names in the Appendix A (Table A1).

3.2. The Edge Effect on Soils

Generally, measured soil parameters decreased with an increase in distance from the edge into the old fields. Gravimetric soil moisture was significantly ($P < 0.001$) higher 10 m into both the natural and old fields from the edge compared to 30 and 50 m into the old field from the edge (Figure 6a). All measured soil macro nutrients of total N, C and P significantly ($P < 0.001$) decreased with an increase in distance from the edge into the old fields (Figure 6a–d). However, for soil C and P, there were no significant ($P > 0.05$) differences at 30 and 50 m distances from the edge into the old fields (Figure 6c,d).

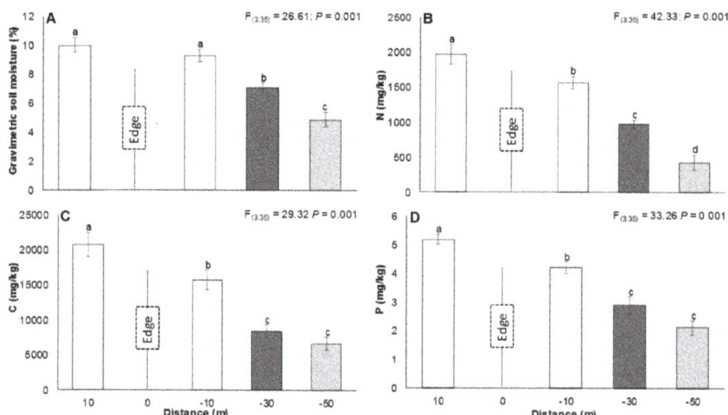

Figure 6. Comparison of soil moisture, soil N, soil C and soil P at various sampling distance from the natural/old field edge. Data are means ± SE and significant ANOVA results are shown.

4. Discussion

4.1. The Edge Effect on Vegetation

Results of this study indicate that the edge effect can influence plants in degraded old fields at Lapalala Wilderness. This is because species richness and cover of some woody species e.g., *P. africanum, S. nigrescens, S. Senegal, V. tortilis, V. nilotica,* and *V. karroo* decreased with increasing distance from the edge into the old field. The above results concur with previous studies that have shown that the establishment of woody species in degraded old fields is strongly influenced by proximity to natural edge areas [5,10,15,35,36]. However, some studies have shown the opposite, where species richness and composition do not respond to the edge effect [17–19]. One of the factors that could explain the reported high abundance of some woody species closer to the old field edge, is increased seed dispersal in areas close to the edge (this is regarded as the concept of proximity to seed source). Previous studies have shown that proximity to seed sources is linked with increased seed dispersal and plant establishment near the edge as compared to far from the edge [37,38]. A study by Cubina and Aide [39] showed that both seed rain and soil seed bank are higher near the seed source and tend to decrease with increasing distance from the source. For example, Krug and Krug [40] reported that seed density of native species was closely related to distance from the natural vegetation boundary. In another study linked to seed dispersal, Donaldson et al. [41] showed that pollination rates are higher close to the natural edge and decrease with distance from the edge boundary. This indicates that a higher number of pollinators are found near the source (natural forested area) as compared to far from the source, this is likely to contribute to variations in seed availability between areas close to the edge and those that are far.

The role of animal grazing on seed dispersal along the natural/old field edges was not tested in this study, though it could explain this study's observation on plant diversity. The dominance of animal grazing on edge boundaries is viewed as an animal safety measure because animals can retreat into the closed canopy natural vegetation for safety from predators [42–44]. This grazing behavior (commonly referred to as herbivore-predator avoidance strategy) will leave areas close to the edge being heavily grazed of grasses and herbs, compared to those far from the edge which become under grazed. Besides grazing, competitive interactions between plant species as well as edge size could also explain the observed effects of edge proximity on species composition [45]. Competition for resources (e.g., water) by woody species has the potential to negatively affect germination and establishment of grasses and herbs as compared to trees, this following results by [45] who showed that trees do suppress grass species growth through competition, though this is more common in wet compared to dry ecosystems. The size of the edge is known to affect seed dispersal [46,47], particularly animal movement between the natural and old field areas. For example, if the edge size is wide and open, some seed dispersers (e.g., ants) might avoid edge crossing due to predation fears [46].

4.2. The Edge Effect on Soil

Results of the study showed that soil moisture and nutrients of N, C, and P were high close to the edge as compared to far from the edge. Piessens et al. [48] reported varied effects of the edge on soil nutrients with such effects being limited to approximately 8 m from the edge. Riedel and Epstein [10] showed that soil carbon did not increase with proximity to the forest edge, suggesting that soil carbon is not influenced by the edge. In this study, the reported changes in soil properties in relation to distance from the edge could be a result of the creation of microclimatic conditions by recruiting woody species at the edge as compared to far from the edge. Previous studies have shown that soil temperature, which subsequently affects soil moisture and bacterial activities, is lower near the forest edge compared to far into the old field [49]. This points to the creation of favorable microclimates near the forest edge compared to far from the edge [50]. For example, increased vegetation abundance and diversity (in this case woody species) near the edge, has the potential to increase soil moisture content through plant canopy shading [51,52]. Similarly, increased vegetation abundance and diversity near the edge has the potential to increase litter fall and composition. Under favorable soil moist conditions created by the woody plants, increased bacterial and fungal activity will result in increased litter decomposition, thus increased soil nutrient content in moist areas near the edge compared to dry areas far from the edge.

The observed high species diversity for trees and shrubs and measured soil properties at the edge compared to far in the old field points to a positive plant-soil interaction [53,54]. A study by Normann et al. [47] concluded that changes in environmental and soil factors (e.g., soil pH and light penetration) at the edge might drive changes in species composition. Similarly, Wirth et al. [55] reported high soil chemical inputs at the edge, due to high litter deposition and decomposition, thus confirming the suspected positive plant-soil interaction. However, positive plant-soil interactions and feedbacks might lead to monodominance of some single species, thus leading to bush encroachment and plant invasion at the edge.

5. Conclusions

The hypothesis that changes in vegetation diversity and cover due to edge proximity will influence changes in soil properties is accepted. This is because results show that both vegetation diversity and measured soil properties decreased with increasing distance from the edge into the old field. Given the limited resources that ecological restoration initiatives receive, results of this study suggest that restoration initiatives should prioritize natural/old field edges. Prioritization can be in the form of fencing the edge (e.g., creating a 10 m buffer zone around the natural/old field edge) to facilitate vegetation and soil recovery. Fencing the edge has the potential to reduce animal grazing which could be causing substantial loss of grasses and herbs close to the edge. Secondly, if fencing the whole edge is expensive, protecting (through fencing) woody species that commonly occur at the edge e.g., *P. africanum*, *S. nigrescens*, *S. Senegal*, *V. tortilis*, *V. nilotica*, and *V. karroo* should be considered. The woody species recruiting at the edge have the potential to act as restoration foci, were vegetation and soil recovery is likely to start [56]. These plants can act as nurse plants, which have been shown to facilitate ecological restoration in degraded ecosystems [51]. Lastly, proximity to the natural/old field edge seems to be positively influencing vegetation (some woody plant species) and soil (moisture and macronutrients of N, C and P) recovery. The abundance and diversity of some woody species, as well as soil nutrient content, were higher near the natural/old field edge compared to far into the old field. However, to fully understand the edge effect on both plant and soil, it is important to conduct research on all the processes occurring at natural/old field edges e.g., seed dispersal, creation of microclimates, seed bank and plant establishment and growth.

Funding: This research received no external funding.

Acknowledgments: I am grateful to the owners, the managing director (Anton Walker), the conservation manager (Hermann Muller) and the research coordinator (Annemieke van der Goot) at Lapalala Wilderness for permission to conduct this study and for logistical help during field work.

Conflicts of Interest: The author declares no conflict of interest.

Appendix A

Table A1. Thirty-six frequently occurring species in relation to distance from the natural/old field edge. (*) Indicates that the species was present and is based on calculated species occupancy frequencies categorized as * (1–20%), ** (21–40%), *** (41–60%), **** (61–80%) and ***** (81–100) with (–) indicating that the species was not present.

Species Name	Distance from Edge (in m)			
	−10	10	30	50
Trees and Shrubs				
Dombeya rotundifolia	***	–	–	–
Euclea undulata	***	–	–	–
Grewia monticola	***	–	–	–
Gymnosporia buxifolia	***	–	–	–
Kirkia acuminata	***	–	–	–
Mimusops zeyheri	***	–	–	–
Pterocarpus rotundifolius	***	–	–	–
Senegalia caffra	***	–	–	–
Peltophorum africanum	***	**	**	*
Senegalia nigrescens	****	***	*	*
Senegalia senegal	***	**	*	*
Vachellia tortilis	***	***	**	**
Vachellia nilotica	***	***	**	**
Vachellia karroo	***	**	**	*
Combretum molle	****	**	–	–
Ziziphus mucronata	***	**	–	–
Grewia bicolor	***	**	–	–
Grewia flava	***	*	–	–
Lippia javanica	****	*	*	*
Terminalia sericea	****	*	–	–
Herbs				
Gomphrena celosioides	–	***	***	***
Xerochrysum bracteatum	–	***	***	***
Solanum incanum	**	****	***	***
Datura stramonium	–	***	****	***
Solanum panduriforme	–	***	–	–
Emilia transvaalensis	**	***	–	–
Felicia sp	**	***	–	–
Pseudognaphalium sp	**	***	–	–
Grasses				
Arundinella nepalensis	*	***	****	****
Brachiaria deflexa	*	****	***	****
Cynodon dactylon	*	****	***	****
Panicum maximum	*	****	***	****
Eragrostis lehmanniana	*	****	***	****
Aristida congesta	–	***	–	–
Digitaria eriantha	–	****	–	–
Eragrostis inamoena	–	****	–	–

References

1. De la Hey, M.; Beinart, W. Why have South African smallholders largely abandoned arable production in fields? A case study. *J. S. Afr. Stud.* **2017**, *43*, 753–770. [CrossRef]
2. Blair, D.; Shackleton, C.M.; Mograbi, P.J. Cropland abandonment in South African smallholder communal lands: Land cover change (1950–2010) and farmer perceptions of contributing factors. *Land* **2018**, *7*, 121. [CrossRef]

3. Myster, R.W. Post-agricultural invasion, establishment, and growth of Neotropical trees. *Bot. Rev.* **2004**, *70*, 381–402. [CrossRef]

4. Cramer, V.A.; Hobbs, R.J.; Standish, R.J. What's new about old fields? Land abandonment and ecosystem assembly. *Trends Ecol. Evol.* **2008**, *23*, 104–112. [CrossRef] [PubMed]

5. Bueno, A.; Llamb, L.D. Facilitation and edge effects influence vegetation regeneration in old-fields at the tropical Andean forest line. *Appl. Veg. Sci.* **2015**, *18*, 613–623. [CrossRef]

6. Blignaut, J.; Aronson, J.; de Wit, M. The economics of restoration: Looking back and leaping forward. *Ann. N. Y. Acad. Sci.* **2014**, *1322*, 35–47. [CrossRef] [PubMed]

7. Barral, M.P.; Rey Benayas, J.M.; Meli, P.; Maceira, N.O. Quantifying the impacts of ecological restoration on biodiversity and ecosystem services in agroecosystems: A global meta-analysis. *Agric. Ecosyst. Environ.* **2015**, *202*, 223–231. [CrossRef]

8. Guariguata, M.R.; Ostertag, R. Neotropical secondary forest succession: Changes in structural and functional characteristics. *For. Ecol. Manag.* **2001**, *148*, 185–206. [CrossRef]

9. Ruwanza, S. Furrows as centers of restoration in old fields of Renosterveld, South Africa. *Ecol. Restor.* **2017**, *35*, 289–291. [CrossRef]

10. Riedel, S.M.; Epstein, H.E. Edge effects on vegetation and soils in a Virginia old-field. *Plant Soil* **2005**, *270*, 13–22. [CrossRef]

11. Ries, L.; Fletcher, R.J.J.; Battin, J.; Sisk, T.D. Ecological responses to habitat edges: Mechanisms, models, and variability explained. *Annu. Rev. Ecol. Evol. Syst.* **2004**, *35*, 491–522. [CrossRef]

12. Lin, L.; Cao, M. Edge effects on soil seed banks and understory vegetation in subtropical and tropical forests in Yunnan, SW China. *For. Ecol. Manag.* **2009**, *257*, 1344–1352. [CrossRef]

13. Murcia, C. Edge effects in fragmented forests: Implications for conservation. *Trends Ecol. Evol.* **1995**, *10*, 58–62. [CrossRef]

14. Gehlhausen, S.M.; Schwartz, M.W.; Augspurger, C.K. Vegetation and microclimatic edge effects in two mixed-mesophytic forest fragments. *Plant Ecol.* **2000**, *147*, 21–35. [CrossRef]

15. Smit, R.; Olff, H. Woody species colonization in relation to habitat productivity. *Plant Ecol.* **1998**, *139*, 203–209. [CrossRef]

16. Meiners, S.J.; Pickett, S.T.A.; Handel, S.N. Probability of tree seedling establishment changes across a forest–old field edge gradient. *Am. J. Bot.* **2002**, *89*, 466–471. [CrossRef]

17. Hester, A.J.; Hobbs, R.J. Influence of fire and soil nutrients on native and non-native annuals at remnant vegetation edges in the Western Australian wheatbelt. *J. Veg. Sci.* **1992**, *3*, 101–108. [CrossRef]

18. Goldblum, D.; Beatty, S.W. Influence of an old field/forest edge on a Northeastern United States deciduous forest understory community. *J. Torrey Bot. Soc.* **1999**, *126*, 335–343. [CrossRef]

19. Phillips, O.L.; Rose, S.; Mendoza, A.M.; Vargas, P.N. Resilience of Southwestern Amazon forests to anthropogenic edge effects. *Conserv. Biol.* **2006**, *20*, 1698–1710. [CrossRef]

20. Ramırez, L.; Rada., F.; Llambı, L.D. Linking patterns and processes through ecosystem engineering: Effects of shrubs on microhabitat and water status of associated plants in the high tropical Andes. *Plant Ecol.* **2015**, *216*, 213–225.

21. Duncan, R.S.; Chapman, C.A. Seed dispersal and potential forest succession in abandoned agriculture in tropical Africa. *Ecol. Appl.* **1999**, *9*, 998–1008. [CrossRef]

22. Mucina, L.; Rutherford, M.C. *The Vegetation of South Africa, Lesotho and Swaziland*; Strelitzia 19; South African National Biodiversity Institute: Pretoria, South Africa, 2006.

23. Ruwanza, S.; Mulaudzi, D. Soil physico-chemical properties in Lapalala Wilderness old agricultural fields, Limpopo Province of South Africa. *Appl. Ecol. Environ. Res.* **2018**, *16*, 2475–2486. [CrossRef]

24. Midgley, D.C.; Pitman, W.V.; Middleton, B.J. *A Guide to Surface Water Resources of South Africa*; Water Research Commission: Johannesburg, South Africa, 1983.

25. Galatowitsch, S.; Richardson, D.M. Riparian scrub recovery after clearing of invasive alien trees in headwater streams of the Western Cape, South Africa. *Biol. Conserv.* **2005**, *122*, 509–521. [CrossRef]

26. Black, C.A. *Methods of Soil Analysis: Part I Physical and Mineralogical Properties*; American Society of Agronomy: Madison, WA, USA, 1965.

27. Chan, K.Y.; Bowman, A.; Oates, A. Oxidizible organic carbon fractions and soil quality changes in an Oxic Paleustalf under different pasture leys. *Soil Sci. Soc. Am. J.* **2001**, *166*, 61–67. [CrossRef]

28. Bray, R.H.; Krutz, L.T. Determination of total, organic and available forms of phosphorus in soils. *Soil Sci.* **1945**, *59*, 39–45. [CrossRef]

29. Shannon, C.E.; Weaver, W. *The Mathematical Theory of Communication*; University of Illinois Press: Urbana, IL, USA, 1949.

30. Spellerberg, I.F.; Fedor, P.J. A tribute to Claude Shannon (1916–2001) and a plea for more rigorous use of species richness, species diversity and the 'Shannon–Wiener' Index. *Glob. Ecol. Biogeogr.* **2003**, *12*, 177–179. [CrossRef]

31. Zar, J.H. *Biostatistical Analysis*; Prentice Hall: London, UK, 1996.

32. Hurlbert, S. Pseudoreplication and the design of ecological field experiments. *Ecol. Monogr.* **1984**, *54*, 187–211. [CrossRef]

33. StatSoft, Inc. Statistica (Data Analysis Software System). Version 13.1. 2017. Available online: http://www.statsoft.com (accessed on 10 July 2018).

34. Šmilauer, P.; Lepš, J. *Multivariate Analysis of Ecological Data Using Canoco 5*, 2nd ed.; Cambridge University Press: Cambridge, UK, 2014.

35. Yao, J.; Holt, R.D.; Rich, P.M.; Marshall, W.S. Woody plant colonization in an experimentally fragmented landscape. *Ecography* **1999**, *22*, 715–728. [CrossRef]

36. Horn, A.; Krug, C.B.; Newton, I.P.; Esler, K.J. Specific edge effects in highly endangered Swartland Shale Renosterveld in the Cape Region. *Ecol. Mediterr.* **2011**, *37*, 63–74.

37. Cramer, V.A.; Hobbs, R.J. *Old Fields: Dynamics and Restoration of Abandoned Farmland*; Island Press: Washington, DC, USA; London, UK, 2007.

38. Mucina, L.; Bustamante-Sánchez, M.A.; Duguy, B.; Holmes, P.; Keeler-Wolf, T.; Armesto, J.J.; Dobrowolski, M.; Gaertner, M.; Smith-Ramírez, C.; Vilagrosa, A. Ecological restoration in mediterranean-type shrublands and woodlands. In *Routledge Handbook of Ecological and Environmental Restoration*; Allison, S., Murphy, S., Eds.; Taylor & Francis: Abingdon, UK, 2017; pp. 173–196.

39. Cubina, A.; Aide, T.M. The effect of distance from forest edge on seed rain and soil seed bank in a tropical pasture. *Biotropica* **2001**, *33*, 260–267. [CrossRef]

40. Krug, C.B.; Krug, R.M. Restoration of old fields in Renosterveld: A case study in a Mediterranean type shrubland of South Africa. In *Old Fields: Dynamics and Restoration of Abandoned Farmland*; Cramer, V.A., Hobbs, R.J., Eds.; Island Press: Washington, DC, USA, 2007; pp. 265–285.

41. Donaldson, J.; Nänni, I.; Zachariades, C.; Kemper, J. Effects of fragmentation on pollinator diversity and plant reproductive success in renosterveld shrublands of South Africa. *Conserv. Biol.* **2002**, *16*, 1267–1276. [CrossRef]

42. Statton, J.; Gustin-Craig, S.; Dixon, K.W.; Kendrick, G.A. Edge effects along a seagrass margin result in an increased grazing risk on *Posidonia australis* transplants. *PLoS ONE* **2015**, *10*, e0137778. [CrossRef] [PubMed]

43. Hovel, K.A.; Regan, H.M. Using an individual-based model to examine the roles of habitat fragmentation and behavior on predator-prey relationships in seagrass landscapes. *Landsc. Ecol.* **2008**, *23*, 75–89. [CrossRef]

44. Sousa, W.P. The role of disturbance in natural communities. *Annu. Rev. Ecol. Syst.* **1984**, *15*, 353–391. [CrossRef]

45. Dohn, J.; Dembélé, F.; Karembé, M.; Moustakas, A.; Amévor, K.A.; Hanan, N.P. Tree effects on grass growth in savannas: Competition, facilitation and the stress-gradient hypothesis. *J. Ecol.* **2013**, *101*, 202–209. [CrossRef]

46. Penido, G.; Ribeiro, V.; Fortunato, D.S. Edge effect on post-dispersal artificial seed predation in the southeastern Amazonia. *Braz. J. Biol.* **2015**, *75*, 347–351. [CrossRef] [PubMed]

47. Normann, C.; Tscharntke, T.; Scherber, C. How forest edge–center transitions in the herb layer interact with beech dominance versus tree diversity. *J. Plant Ecol.* **2016**, *9*, 498–507. [CrossRef]

48. Piessens, K.; Honnay, O.; Devlaeminck, R.; Hermy, M. Biotic and abiotic edge effects in highly fragmented heathlands adjacent to cropland and forest. *Agric. Ecosyst. Environ.* **2006**, *114*, 335–342. [CrossRef]

49. Duncan, R.S.; Duncan, V.E. Forest succession and distance from forest edge in an afro-tropical grassland. *Biotropica* **2000**, *32*, 33–41. [CrossRef]

50. Wright, T.E.; Kasel, S.; Tausz, M.; Bennett, L.T. Edge microclimate of temperate woodlands as affected by adjoining land use. *Agric. For. Meteorol.* **2010**, *150*, 1138–1146. [CrossRef]

51. Ren, H.; Yang, L.; Liu, N. Nurse plant theory and its application in ecological restoration in lower subtropics of China. *Prog. Nat. Sci.* **2008**, *18*, 137–142. [CrossRef]

52. Badano, E.I.; Samour-Nieva, O.R.; Flores, J.; Flores-Flores, J.L.; Flores-Cano, J.A.; Rodas-Ortíz, J.P. Facilitation by nurse plants contributes to vegetation recovery in human-disturbed desert ecosystems. *J. Plant Ecol.* **2016**, *9*, 485–497. [CrossRef]

53. Bever, J.D. Feedback between plants and their soil communities in an old field community. *Ecology* **1994**, *75*, 1965–1977. [CrossRef]

54. Dybzinski, R.; Fargione, J.E.; Zak, D.R.; Fornara, D.; Tilman, D. Soil fertility increases with plant species diversity in a long-term biodiversity experiment. *Oecologia* **2008**, *158*, 85–93. [CrossRef] [PubMed]

55. Wirth, R.; Meyer, S.T.; Leal, I.R.; Tabarelli, M. Plant herbivore interactions at the forest edge. *Prog. Bot.* **2008**, *69*, 423–448.

56. Zahawi, R.A.; Augspurger, C.K. Tropical forest restoration: Tree islands as recruitment foci in degraded lands of Honduras. *Ecol. Appl.* **2006**, *16*, 464–478. [CrossRef]

MDPI

St. Alban-Anlage 66

4052 Basel

Switzerland

Tel. +41 61 683 77 34

Fax +41 61 302 89 18

www.mdpi.com

Sustainability Editorial Office

E-mail: sustainability@mdpi.com

www.mdpi.com/journal/sustainability